Fundamentals of Computational Neuroscience

Fundamentals of
Computational Neuroscience

Thomas P. Trappenberg

Dalhousie University

UNIVERSITY PRESS

OXFORD

UNIVERSITY PRESS

Great Clarendon Street, Oxford OX2 6DP

Oxford University Press is a department of the University of Oxford.
It furthers the University's objective of excellence in research, scholarship,
and education by publishing worldwide in

Oxford New York

Auckland Cape Town Dar es Salaam Hong Kong Karachi
Kuala Lumpur Madrid Melbourne Mexico City Nairobi
New Delhi Shanghai Taipei Toronto

With offices in

Argentina Austria Brazil Chile Czech Republic France Greece
Guatemala Hungary Italy Japan South Korea Poland Portugal
Singapore Switzerland Thailand Turkey Ukraine Vietnam

Oxford is a registered trade mark of Oxford University Press
in the UK and in certain other countries

Published in the United States
by Oxford University Press Inc., New York

First published 2002

Reprinted 2004, 2005, 2006, 2007

A catalogue record for this book is available from the British Library

Library of Congress Cataloging in Publication Data
(Data available)
ISBN 978-0-19-851582-1 (Hbk)
ISBN 978-0-19-851583-8 (Pbk)

10 9 8 7 6 5

Typeset by the author
Printed in Great Britain
on acid-free paper by
Antony Rowe Ltd., Chippenham, Wiltshire

In memory of my father
Rüdiger
dedicated to my children
Kai and **Nami**

Foreword

J.G. Taylor, King's College London

One of the main scientific adventures of this newly started century is that of understanding the brain, both of ourselves and of the animals we care for. New experimental results are pouring in on all sides, with most exciting data coming from the new brain imaging machines, as well as from advances in single cell recordings and from new measurement techniques only now being developed. All these new results have, however, to be understood in terms of some framework strong enough to bear their increasing weight. Without such an apparatus at our disposal we are in danger of not being able to see the wood for the trees, and even not seeing any wood at all. What should such a framework be constructed from in order to bear this increasing weight of experimental data?

Experiment can only progress effectively hand in hand with a developing theory. The two-way dialogue between them is strongly supported by Popper's 'falsification' approach to science: scientific understanding progresses by proposing an explanation of a range of phenomena which itself can be tested and is then modified so as to take account of the difficulties it meets when confronted by new experimental data. It is thus necessary to include a suitably strong theoretical component to this developing framework in order to understand the brain. It is also necessary to allow this theoretical component to develop by strong interaction with experiment.

It is this crucial need of providing a theoretical framework into which the burgeoning study of the brain can be fitted that is provided by the subject of Neural Networks. This discipline, now properly being recognized as such, allows the requisite theoretical ideas to be formulated and developed in order to make predictions testable against experiment in a falsifiable manner.

Neural networks have been developed for over 50 years, since their early description by McCulloch and Pitts to give a remarkable result as to the logical powers of the simplest form of neural network, composed of elegantly simple neurons. This result helped spark off a revolution in artificial intelligence, both in leading to the development of the digital computer and in the creation of an array of artificial neural networks for industrial applications. The subject of neural networks has now amply justified its continued existence in these areas. However, it now comes back to its roots when we turn back to use it, not in the important but still limited industrial applications I just mentioned. It is now necessary to attempt to create really intelligent machines, even up to seemingly conscious ones (if not actually so) in order to be able to build a theoretical framework which can talk in a useful dialogue to the data of neuroscience.

Here is the nub: how do we move neural networks towards the greater complexity of the brain? At the same time how can we develop them so they include both a non-conscious as well as a conscious component. The latter has so far been in the domain of artificial intelligence, working on a symbol system. The problem faced by neural networks is to create through them a powerful enough architecture to encompass both the non-conscious and conscious processing levels. This is not yet achievable, but progress is being made.

This book lays the foundations for such an approach. Its author has had considerable experience in applying neural networks to neuro-scientific modelling tasks. Yet he writes to emphasize the basic concepts involved in applying neural networks to the brain. As such this book is an essential component in the learning process for any who wish to join those trying to go from models of unconscious brain activity to those of a higher order, and who wish to complete their education on how lower level models can be constructed and understood.

Preface

Computational neuroscience is still a very young and dynamically developing discipline. Many different facets are explored by a growing number of scientists. It is beyond the scope of this book to review the research literature extensively, and I was not able to avoid some personal bias in the choice of topics included in this book. Some colleagues might disagree on the emphasis I have given specific topics, such as the detailed discussion on the information processing in networks of relatively simple processing devices. However, my aim in writing this book is to offer a trail through some of the exciting discoveries in this area, and to introduce some of the basic thoughts that guide the current thinking of many researchers about information processing in the brain. I believe that a basic understanding of many of the concepts outlined in this book is becoming mandatory for all neuroscientists, and I hope that I can help to build some foundations in theoretical techniques that are becoming important in neuroscience. I believe that these attempts will be useful for students and researchers who are interested in finding out 'how the brain computes'.

In writing this book I challenged myself in particular with the task of highlighting some reasons for the assumptions we make commonly in recent models; a discussion that is often omitted in the research literature. Furthermore, I intended to be brief in my description so that a reading of the entire book is possible in reasonable time (note that most of the formulas can be skipped in a first reading). At the same time I did not want to give the impression that the content of this book is the entire working knowledge of computational neuroscience. I did, therefore, not hesitate to mention some very specific terms (usually highlighted in italic), even without sufficient discussions, in order to provide the reader with starting points to search the common literature if further studies of such topics are desired. In line with the intent to provide a teaching book, I concentrated in the main text on describing ideas (rather than outlining a research trail). I have therefore avoided literature references in the text and have instead provided a short and commented 'Further reading' list at the end of each chapter. This list consists mainly of books that can provide further overviews and references for further detailed studies. A few references to some original research literature have been included when we followed such work closely.

I am aware of the fact that many readers interested in this subject may not be mathematically oriented, but I included mathematical formulas for several reasons. The most important is that formulas allow compact yet detailed specifications of models that can be translated directly into computer programs. It is, however, important to see beyond the formulas and to understand their meaning. There is no reason to be afraid of formulas; they only represent a shorthand notation that allows precision while avoiding lengthy verbal descriptions. Not all formulas have to be derived personally and can even be skipped in a first reading. Some formulas are included mainly for completeness and for reference if they should become important in your research. Please ask your instructor or mathematically inclined colleague to discuss with you any formulas that are not directly transparent to you. You don't have to be or become a mathematician to be able to understand the formulas in this book.

All chapters on different network architectures begin with an outline of some neuroscientific issues or some examples of information-processing principles that are aimed to motivate the study of the specific models in the corresponding chapter. The models in the book are, however, relatively general and are intended to illustrate basic information processing mechanisms in the brain. More specific models in the literature are often composed of such basic elements, and a study of the basic models in this book will enable the reader to follow some of the recent literature in this area. I have included only a few sample discussions of original research papers that highlight some of the points I wanted to make. I apologize for not mentioning many other research papers that have had major impacts in the field as well as papers that have guided much of my thinking.

This book is about neuroscience, and several issues mentioned in this book can be found in basic textbooks of neuroscience, in particular some basic descriptions of neurons and brain organizations. I want to encourage every reader who is not familiar with these subjects to consult these books, some examples of which I have included in the 'Further reading' list. In contrast to widely spread neuroscience textbooks, we concentrate in this book on the computational aspects, which are often not the major focus of other books. This book focuses on the modelling aspects, in particular the teaching of modelling, how to make useful abstractions, and how to use tools to analyse models. In contrast to some books on the technological aspects of neural networks I wanted to outline the relations of theoretical concepts or particular technological solutions to brain functions.

In Chapters 2 and 3 we focus our attention on single neurons. These chapters are very brief in comparison to the large amount of knowledge we have gathered in this area over the last decades. We start with a very short review of basic neuronal components and functionalities, mainly to remind us of some terms and basic facts that are well known and treated in the literature. We mention only briefly compartmental models of neurons as these are the subject of several excellent dedicated publications. We concentrate instead on simpler neuron models that are frequently used in network models. The reader familiar with the basic anatomy and functionality of neurons can skip most of this outline and instead concentrate on the models of single neurons in Chapter 3 that are more specific to studies in computational neuroscience.

In Chapter 4 we begin to explore networks of neurons. It includes the formal introduction of node models, which are often viewed as abstractions of single neurons, because I strongly believe that these models are much better motivated on a network level as outlined in the text. These node models will be essential to much of the discussion in the rest of the book. In Chapter 5 we discuss how information is represented and communicated within the nervous system, which motivates many aspects of the subsequent models.

Chapters 6–9 introduce in turn each one of the fundamental network models that we think are essential for information processing in the brain. These include mapping networks such as perceptrons, associators, auto-associators, and continuous attractor and competitive networks. We commonly discuss these models as examples in conjunction with specific brain-processing issues, although such models are found at the heart of many models in computational neuroscience. In Chapter 10 we follow up on some more specific issues on motor control, supervised learning, and reward learning, which are fundamental issues in computational neuroscience. Chapter 11 then discusses combinations of the fundamental network models more formally, which we have to consider when we want to understand information processing in the brain on a system level. Such modular networks include the mixture of experts, product of experts, and coupled auto-associator networks. In addition, we include in this chapter some examples of more specific system level issues on working memory and translation-invariant object recognition.

Most of the models discussed in this book can be implemented and simulated with small computer programs of only a few lines. We have included an introduction to programming such models within the MATLAB ®[1] programming environment in Chapter 12. MATLAB is very convenient for neural network simulations and scientific visualization, and this tool can be mastered within a very short time. The hands-on guide to some of the models in this book is also aimed to further the understanding of such models and the formulas provided in the main text. Many of the simulation details are encapsulated in this chapter to keep the main text focused on the general issue. MATLAB provides a Neural Network Toolbox with ready-to-use routines of basic and advanced techniques in neural computation. We are not using this toolbox in our treatment as most of the basic algorithms, and even advanced research projects, can be programmed easily within the basic MATLAB programming language itself. This gives the greatest flexibility in creating our own solutions and in implementing novel ideas. Employing only the basic MATLAB programming environment will also make the working of neural networks most transparent and will teach us, at the same time, a useful tool.

---

[1] MATLAB is a registered trademark of The MathWorks, Inc.

Acknowledgements

Many colleagues and friends have contributed to this book through personal conversations, their enlightening research, and general guidance, too many indeed to be able to mention them all. Keeping with my theme to be brief I will thus only mention some colleagues who particularly have encouraged me to study computational neuroscience. On the theory side these were Shun-ichi Amari, Edmund Rolls, and John Taylor, and I have benefited much from conversations with Andrew Back, Gustavo Deco, Shiro Ikeda, Simon Stringer, and Masami Tatsuno. On the physiology front I have to thank in particular Doug Munoz and Okahide Hikosaka who gave me the opportunity to learn at first hand from brilliant researchers, as well as the members (and past members) of their active teams including Mike Dorris, Stefan Everling, Johan Lauwereyns, and Masamichi Sakagami. My interest in the brain would not have occurred without the cognitive scientists Raymond Klein and Patricia McMullen; Ray in particular has guided me for most of my way through neuroscience, and I want to thank him for his continuous support, guidance, and collaboration. I am very grateful to Justus Verhagen, Sue Becker, Masami Tatsuno, and Theodore Chiasson for reading through very rough drafts and pointing out many mistakes and unclear sections. Needless to say that I am responsible for all remaining mistakes. Last, but certainly not least, I want to thank my wife Kanayo for her support, without which the writing of this book would not have been possible.

Contents

1 Introduction

We will outline in this introductory chapter the major focus, specific tools, and the strategies of computational neuroscience. The complexity of the brain makes it necessary to clarify how we attempt to describe it and what we expect from explanatory models. We will argue that theoretical and computational studies are important in understanding experimental measurements. We will discuss the specific role of models in computational neuroscience and argue that they have to be chosen carefully if they are to be useful for advancing our knowledge of the information-processing principles in the brain. Models must not only be able to summarize known experimental data, but must also be able to make predictions that can be verified experimentally. We also include a short guide to the book to show the path we are going to take to explore this fascinating scientific subject.

1.1 What is computational neuroscience?

In the scientific area now commonly called computational neuroscience we utilize distinct techniques and ask specific questions aimed at advancing our understanding of the nervous system. This specific scientific area might be defined as:

> *Computational neuroscience is the theoretical study of the brain to uncover the principles and mechanisms that guide the development, organization, information processing, and mental abilities of the nervous system.*

Thus, computational neuroscience is a specialization within neuroscience. Neuroscience itself is a scientific area with many different aspects. Its aim is to understand the nervous system, in particular the central nervous system that we call the brain. The brain is studied by researchers who belong to diverse disciplines such as physiology, psychology, medicine, computer science, and mathematics, to name but a few. Neuroscience emerged from the realization that interdisciplinary studies are vital to further our understanding of the brain. The brain is one of the most complex systems ever encountered in nature, and there are still many questions that we can only attack through combined efforts. How does the brain work? What are the biological mechanisms involved? How is it organized? What are the information-processing principles used to solve complex tasks such as perception? How did it evolve? How does it change during the lifetime of the organisms? What is the effect of damage to particular areas and the possibilities of rehabilitation? What are the origins of degenerative diseases and possible treatments? All these basic questions are asked by neuroscientists.

1.1.1 The tools and specializations in neuroscience

Many techniques are employed within neuroscience to answer these questions. These include genetic manipulation, *in vivo* and *in vitro* recording of cell activities, optical imaging, functional magnetoresonance scanning, psychophysical measurement, and computer simulation. Each of these techniques is complicated and laborious enough to justify a specialization within neuroscience. We therefore speak of neurophysiologists, cognitive scientists, and anatomists. It is, however, vital for any neuroscientist to develop a basic knowledge of all major techniques in order to understand and use the contributions made by them. The significance of any technique has to be evaluated with a view to its specific problems and limitations as well as its specific aim of the technique. Computational neuroscience is an increasingly important area of neuroscience and a basic understanding of this field has become essential for all neuroscientists.

1.1.2 The focus of computational neuroscience

Computational neuroscience attempts to develop and test hypotheses about the functional mechanisms of the brain. A major focus is therefore the development and evaluation of *models*. The scientific area that is the subject of this book is also known as 'theoretical neuroscience'. Computational neuroscience can be viewed as a specialization within theoretical neuroscience, which employs computers to simulate models. The major reason for using computers is the complexity of many models, which are often analytically intractable. For such models we have to employ carefully designed numerical experiments in order to be able to compare these models to experimental data. However, we do not want to restrict our studies to this tool because analytical studies can often give us a deeper and more controlled insight into the features of models and the reasons behind numerical findings. Whenever possible, we try to include analytical techniques, and we will use the term 'computational neuroscience' synonymously with theoretical neuroscience. The word 'computational' also emphasizes that we are interested in particular in the computational aspects of brain functions.

We have included some examples of analytical techniques in order to give you a taste of some of those powerful techniques. Not every neuroscientist has to perform such calculations, but it is necessary to comprehend the general ideas if you are to get support from specialists in these techniques when required in your own research. However, it is very instructive to perform some numerical experiments yourself. We therefore included an introduction to a modern programming environment that is very well suited to many models in neuroscience. Writing programs and creating advanced graphics can be easily learned within a short time even without extensive computer knowledge.

Although computational neuroscience is theoretical by nature, it is important to bear in mind that the models have to be measured against experimental data; they are otherwise useless for understanding the brain. Only experimental measurements on the real brain can verify 'what' the brain does. In contrast to the experimental domain, computational neuroscience tries to speculate 'how' the brain does it. The speculations are developed into hypotheses, realized into models, evaluated analytically or numerically, and tested against experimental data. We will discuss specific examples of models on several levels (and related evaluation techniques) throughout this book.

1.2 Domains in computational neuroscience

The nervous system has many levels of organization on spatial scales ranging from the molecular level of a few ångstroms ($1 \overset{\circ}{A} = 10^{-10}$m) to that of the whole nervous system on the scale of 1 metre (see Fig. 1.1), and biological mechanisms on all these levels are important for the functioning of the brain. Where do we start when trying to explain brain functions? Do we have to rebuild the brain in its entirety, for example, in a computer, in order to 'understand' it? There are arguments suggesting that we can never achieve this. However, even if it were possible to simulate a whole brain on a computer with all the details from its biochemistry to its large-scale organization, this would not necessarily mean that we had a better explanation of brain functions. What we are really looking for is a better comprehension of brain mechanisms on explanatory levels that are, for example, suitable for applications such as the development of advanced medical treatments. It is therefore important to learn about the art of *abstraction*, making a suitable simplification of a system without abolishing the important features we want to understand.

1.2.1 Levels of abstraction

Which level is most appropriate for the investigation and abstraction depends on the scientific question asked. For example, Parkinson disease is 'caused' by the death of dopaminergic neurons in the substantia nigra, and there are signs that a genetic predisposition might act together with biochemical processes in single neurons that can cause the death of dopaminergic neurons. A detailed investigation on a neuronal level with detailed neuron models is thus obvious. However, we also know about the important role of dopamine in the initiation of motor actions, and a more global system level study (often with less detailed neuron models) is necessary to comprehend the full scale of impairment and to develop better methods of coping with the condition. Therefore, the condition must be studied at various levels and connections must be made between the different levels in order to elucidate how small-scale factors, such as genetic mechanisms or biochemical processes in single neurons, can influence the characteristics of large-scale systems, such as the behaviour of an organism.

1.2.2 Levels of organization in the nervous system

Different levels of organization in the nervous system are illustrated in Fig. 1.1. An easily recognizable structure in the nervous system is the neuron, which is a cell that is specialized in signal processing. Depending on environmental conditions it is able to generate an electric potential that is used to transmit information to other cells to which it is connected. Mechanisms on a subcellular level are certainly important for such information-processing capabilities. Some processes in the neuron utilize cascades of biochemical reactions that must be understood on a molecular level. These include, for example, the transcription of genetic information that influences information processing in the nervous system. Many structures within the neuron can be identified with specific functions, for example, mitochondria and synapses, of which the latter are particularly important for our understanding of signal processing in the nervous system. The complexity of a single neuron, and even that of isolated subcellular mechanisms, often makes computational studies essential for the development and verification of hypotheses. The theory of action potential generation in neurons developed by *Hodgkin* and *Huxley* is similar in elegance and importance to *Maxwell*'s theory of electromagnetism.

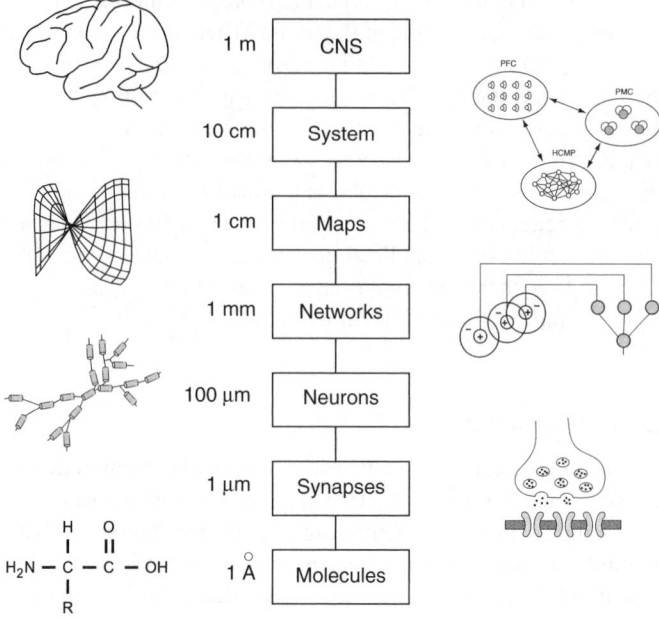

Fig. 1.1 Some levels of organization in the central nervous system on different scales. The illustrations include, from top to bottom, an outline of the brain, a system-level model of working memory (discussed in Chapter 11), a self-organized (Kohonen) map, speculation about the circuit behind orientation-sensitive neurons in V1 by Hubel and Wiesel, a compartment model of a neuron, a chemical synapse, and an amino acid molecule [adapted from Churchland and Sejnowski, *The computational brain*, MIT Press, 1992].

Single neurons are certainly not the whole story. Neurons contact each other and thereby build networks. A small number of interconnected neurons can exhibit complex behaviour and information-processing capabilities not present in a single neuron. The understanding of such networks of interacting neurons is a major domain in computational neuroscience, perhaps because there is so little understanding about such nonlinear interacting systems. Networks have additional information-processing capabilities beyond that of single neurons, such as representing information in a distributed way. Examples of this are topographic maps of sensory stimuli, frequently found in the brain. We concentrate in particular on this level in this book.

And it doesn't stop there. Networks with a specific architecture and specialized information-processing capabilities are incorporated into larger structures that are able to perform even more complex information-processing tasks. A study of the brain on this level is certainly essential to understand higher-order brain functions, and it is at this level where we have probably the least understanding. The central nervous system depends strongly on the interaction of many specialized subsystems, and understanding such interactions seems central for understanding brain functions. It is probable that only the dynamic interaction of several brain areas enables high-level brain functions. We are still unable to reproduce many basic human skills with machines, and a deeper understanding of the function of the brain is desirable for many reasons, including the development of advanced technical applications and medical treatments. The understanding of the brain on the system level is particularly challenging as we have few tools and methods capable of characterizing such large systems. A further difficulty is that the nervous system is not an isolated system but interacts strongly with the environment.

It is important for all neuroscientists to develop a basic understanding of the functionalities on different scales in the brain, although the individual researcher might specialize in mechanisms on a certain scale. Computational neuroscience can help the investigations at all levels of description, and it is not surprising that computational neuroscientists investigate different types of models at different levels of description. The contribution of computational neuroscience is particularly important in understanding the nonlinear interactions among many subprocesses, a characteristic of the brain that is thought to be essential in enabling many brain processes. It is also important to comprehend the interaction between the different levels of description, and computational methods are often suited to investigating these relationships, for example, to bridge the gap between a physiological measurement and the understanding of it in order to explain behavioural correlates of an organism. We will discuss examples of models at different levels of abstraction throughout this book.

1.2.3 The integrated approach

Ideally, we want to integrate experimental facts from different investigational levels into a coherent model of how the brain works as illustrated in Fig. 1.2. Important experimental input to computational neuroscience comes from neurobiology, for example, the mechanisms determining synaptic efficiencies, from neurophysiology in which the behaviour of single neurons is investigated, and from psychology, in which behavioural effects are studied via psychophysical experiments. Computational neuroscientists can then utilize mathematical models for describing the experimental facts, borrowing methods from a wide variety of disciplines such as mathematics, physics, computer science, and statistics. Such formulations and studies of hypotheses with the aid of models should lead to specific experimental predictions that have to be verified experimentally as mentioned above. The comparison of model predictions with experimental data can then be used to refine the hypothesis and to develop more accurate models or models that can shed light on different phenomena. Studies within computational neuroscience can also help to develop applications such as the advanced analysis of brain imaging data, technical applications that utilize brain-like computations, and, ultimately, the development of advanced treatments for patients with brain damage.

Some models in the literature have illustrated that they are able to relate simultaneously to experimental findings on different levels. Those models are often comprised of single elements that rely on neurobiological mechanisms such as biologically plausible synaptic plasticity. The behaviour of the elements in the whole system can, on the other hand, reflect experimental findings from electrophysiological studies, and the activity of specific modules can be related to brain imaging studies. The computational power

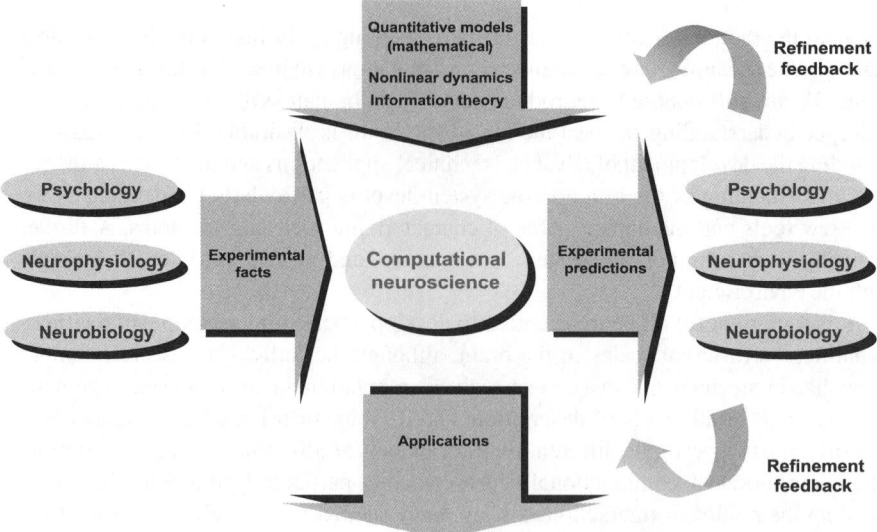

Fig. 1.2 Illustration of the role of computational neuroscience in the integration of experimental facts from different levels of investigation. The models developed in computational neuroscience have to be able to make predictions that can be verified experimentally. The close comparison of experiments with model predictions can then be used to make refinements in the models (or may lead to the development of new approaches) that can further our understanding of brain systems and could also lead to new predictions that have to be verified experimentally. This may also lead to applications in the analysis of experimental data and the development of advanced patient treatments [adapted from Gustavo Deco, personal communication].

of recent computer systems and the deepened knowledge of the brain from experimental research make such models increasingly feasible. We will outline some of these models in later chapters of this book. These models rely on the fundamentals that we introduce in this book, and the feasibility of the study of such models relies strongly on suitable simplification. Explaining the reasons for making such simplifications is therefore at the heart of learning computational neuroscience.

1.3 What is a model?

Modelling is an integral part in many scientific disciplines, and neuroscience is no exception. Indeed, the more complex a system is, the more we have to make simplifications and build model systems that can give us insights into some aspects of the complex system under investigation. The term 'model' frequently appears in many scientific papers and, it seems, describes a vast variety of constructs. Some papers present a single formula as a model, some papers fill several pages with computer code, and some describe with words a hypothetical system. It is important to understand what a model is and, in particular, what the purposes of models are.

To start with, it is important to distinguish a model from a hypotheses or a theory. Scientists develop hypotheses of the underlying mechanisms of a system that have to be tested against reality. In order to test a specific feature of a hypothesis we build a model that can be evaluated. Sometimes we try to mimic a real system by artificial means in order to test this system in different conditions or to make measurements that would not be possible in the 'real' system. A model is therefore a simplification of a system in order to test particular aspects of a system or a hypothesis.

> *Models are abstractions of real world systems or implementations of an hypotheses in order to investigate particular questions or to demonstrate particular features of a system or a hypothesis.*

 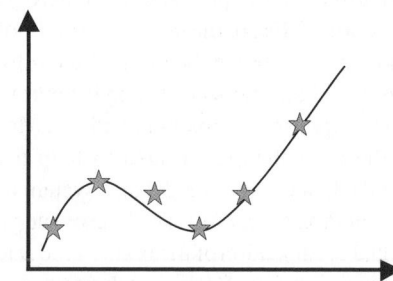

Fig. 1.3 What is a model? On the left is a computer model of a house giving a three-dimensional impression of the design. The right graph shows some data points, for example, from experimental measurements, and a curve is shown that fits these data points reasonably well. The curve can be a simple mathematical formula that fits the data points (heuristic model) or result from more detailed models of the underlying system.

A good example is the use of models in architecture. Small-scale paper models of buildings, or computer graphics generated with sophisticated three-dimensional graphic packages, can be used to give a first impression of the physical appearance and aesthetic composition of a design. A model has a particular purpose attached to it and has to be viewed in this light. A model is not a recreation of the 'real' thing. The paper model of a house cannot be used to test the stability of the construction, and a building engineer uses different models for this purpose. In such models it is important to scale down physical properties of the building material regardless of the physical appearance such as the colour of the building.

1.3.1 Descriptive models

In science we typically represent experimental data in the form of graphs and subsequently seek to describe these data points with a mathematical function. An example of this is the 'modelling' of response properties (receptive fields) of neurons in the lateral geniculate nucleus (LGN), which can be fitted with a specific class of functions called *Gabor functions* by adjusting the parameters within this class of functions accordingly. The Gabor functions are therefore said to be a 'model' of the receptive fields in the LGN. Of course, this *phenomenological model* does not tell us anything about the biophysical mechanisms underlying the formation of receptive fields and why cells respond in this particular way, and such a 'model' therefore seems rather limited. Nevertheless, it can be useful to have a functional description of the response properties of LGN cells. Such parametric models are a shorthand description of experimental data that can be used in various ways. For example, if we want to study a model of the primary visual cortex to which these cells project, then it is much easier to use the parametric form of LGN responses as input to the cortical model rather than including complicated models of the earlier visual pathway in detail.

1.3.2 Explanatory models

As scientists we want to find the roots of natural phenomena. The explanations we are seeking are usually more profound then merely parametrizing experimental data with specific functions. Most of the models in this book are intended to encapsulate processes that are thought of as being the basis of the information-processing capabilities of the brain. These include models of single neurons, networks of neurons, and specific architectures capturing brain organizations. Models are used to study specific aspects of a hypothesis or theory, but also to help to interpret experimental data. A good example of the latter type of model on a system level is a technique from statistics called *structural equation modelling*. The idea of such models is to make an educated guess at the functional organization of the brain in order to deduce effective connectivities in the brain from imaging data. Such models are very useful in analysing experimental data. We will not elaborate on these types of models in this book. Instead we will concentrate on synthetic models that can illuminate the principles of information processing in the brain, which we speculate are the building blocks of brain functions.

A major role of models in science is to illustrate principles underlying natural phenomena on a conceptual level. Sometimes the level of simplification and abstraction is so crude that scientist talk about *toy models*. However, this term may obscure the importance of such models in science. The simplifications made in such models might be necessary in order to employ analytical methods to analyse such models at depth that is not possible in more realistic models. The educational importance of those 'toy models' should not be underestimated, in particular, in demonstrating the principal mechanisms of natural phenomena.

The current state of neuroscience, often still exploratory in nature, frequently makes it difficult to find the right level of abstraction as discussed in the last section. Some models in computational neuroscience have certainly been too abstract to justify some of the claims derived from them. On the other hand, there is a danger in keeping more details than are essential for a scientific argument. Models are intended to simplify and thereby to identify which details of the biology are essential to explain particular aspect of a system. Modelling the brain is not a contest in recreating the brain in all the details on a computer. It would indeed be questionable how such models would add to understanding the functionality of the brain. Models have to be constructed carefully, and reasonable simplification demands high scientific skills. Justifications of the assumptions in models have a high priority in scientific investigations. The purpose of models is to better comprehend the function of complex systems, and simplicity should be a major guide in designing appropriate models.

1.4   Emergence and adaptation

1.4.1   The computer analogy

Standard computers, such as PCs and workstations, have one or perhaps a small number of central processors. Each processor is rather complex with specialized hardware and microprograms implementing a variety of functions such as loading data into registers, adding, multiplying, and comparing data, as well as communicating with external devices. These basic functions can be executed by instructions that are binary data loaded into a special interpreter module. Complicated data processing can be achieved by writing often lengthy programs that are instructions representing a collection of the basic processor functions. Programming a computer means that we have to instruct the machine to follow precisely all the steps that we have figured out beforehand can solve a particular problem. The sophistication of the computer reflects basically the smartness of the programmer.

By contrast, information processing in the brain is very different in several respects. Briefly, it employs simpler processing elements, but lots of them. To explore the information processing in networks of neurons we will mainly employ only very simple abstractions of real neurons, and we will call these fundamental processing units *nodes* to stress this drastic simplification. Nodes can be implemented in hardware or simulated on a standard computer; for the discussions in this book this does not make a difference. We keep the functionality of nodes as simple as possible for the sake of employing lots of them, typically hundreds, thousands, or many more. The employment of many parallel working processors has led to the term *parallel distributed processing* being used in this area. However, with this term one is tempted to think that the processes are independent because only processes that are independent can be processed on different processors in parallel. In contrast to this, a major ingredient of the information processing in the brain is certainly the *interaction of nodes*, and the interaction of nodes is accomplished by assembling them into large networks.

1.4.2   Emergence

In the study of *neural networks* we are interested in understanding the consequences of interacting nodes. It is the interaction that enables processing abilities not present in single nodes. Such capabilities are good examples of emergent properties in *rule-based systems*. *Emergence* is the single most defining property of neural computation, distinguishing it from parallel computing in classical computer science, which is mainly designed to speed up processing by distributing independent algorithmic threads. Interacting systems can have unique properties beyond the mere multiplication of single processor capabilities. It is this type of ability we want to explore and utilize with neural networks. We label these system properties as 'emergent' to stress that we did not encode them directly into the system. To appreciate this better we should distinguish the description of a system on two levels, the level of basic rules defining the system and the level of description aimed at understanding the consequences of such rules.

Scientific explanations have been dominated in the past by the formulation of a set of principles and rules that govern a system. A system can indeed be defined by a set of rules like a game that is defined by rules. In science we make the assumption that natural systems are governed by a finite set of rules. The search for these basic rules, or fundamental laws as they are called in this case, was indeed the central scientific quest for centuries. It is not easy to find such laws, but enormous progress has nevertheless been made. Newton's laws defining classical mechanics or Maxwell's equations of electromagnetism are beautiful examples of fundamental laws in nature. We do not have a theory of the brain on this level, and some have argued that we might never find a simple set of rules explaining brain functions. However, in many scientific disciplines we have begun to realize that, even with a given set of rules, we might still not have a sufficient understanding of the systems. Knowing the rules of the card game 'bridge' is not sufficient to be a good bridge player.

Rules define a system completely and it can therefore be argued that all the properties of the system are encoded in these rules. However, we have to realize that even a small set of rules can generate a multitude of behaviours of the systems that are difficult to understand from the basic rules, and different levels of description might be more appropriate. For example, thermodynamics can describe the macroscopic behaviour of systems of many weakly interacting particles appropriately even though the systems are governed by other microscopic rules. On the other hand, there are *emergent properties* in Newtonian systems that are not well described by classical thermodynamics, for example, turbulent fluids. A deeper understanding emergent properties is becoming a central topic in the science of the 21st century.

1.4.3 Adaptive systems

The importance of emergent properties in networks of simple processing devices is not the only deviation from traditional information processing that we think is crucial in understanding brain functions. An important additional ingredient is that the brain is an example of an *adaptive system*. In our context we define adaptation as the ability of the system to adjust its response to stimuli depending upon the environment. Humans are a good example of systems that have mastered such abilities. This is an area that has attracted a lot of interest in the engineering community, as learning systems, systems that are able to change their behaviour based on examples in order to solve information-processing demands, have the potential to solve problems when traditional algorithmic methods have not been able to produce sufficient results. Adaptation has two major virtues. One is, as just mentioned, the possibility of solving information-processing demands for which explicit algorithms are not yet known. The second concerns our aim to build systems that can cope with a continuously changing environments.

A lot of research in the area of neural networks is dedicated to the understanding of learning in particular networks. Engineering applications of neural networks are not bound by biological plausibility. In contrast to this, we want to concentrate in this book on biologically plausible learning mechanisms that can help us to comprehend the functionality of the brain.

1.5 From exploration to a theory of the brain

The brain is still a largely unknown territory, and exploring it is still a major domain in neuroscience. Recordings with micro-electrodes from single cells contribute significantly to this exploration, and searching for the response properties of neurons is at least as exciting as it must have been for explorers like Marco Polo to discover new lands. With new brain imaging techniques such as fMRI (functional magnetic resonance imaging) we are now able to monitor the living brains of subjects performing specific mental tasks. All this is essential to advance our knowledge of what the brain does. Many data on the brain have been gathered with many different techniques, and we are now slowly entering a new phase in neuroscience, that of formulating more quantitative hypotheses of brain functions. This, in turn, demands some more specific experimental analysis and more dedicated tests of such hypotheses. Consequently, the experimental style is slowly changing. It is increasingly important to formulate alternative hypotheses more precisely and to quantify such hypotheses in such a way that experimental tests can verify or disprove them.

This sounds a lot like quantitative scientific areas such as chemistry or physics. But, can there be a theory of the brain? To fully understand each feature of an individual brain we have to know exactly all of the structural details of the particular brain as well as its precise environment and current state. This is similar to other scientific areas such as physics. To describe completely the physics of an individual aeroplane we have to know the precise location and form of each nut and bolt and all other structural details up to the amount of dirt on the wings and the details of the air it is flying through. As another example, a pot of boiling water consists of lots of individual molecules in a very dynamic state. Measuring all the microscopic details in the last two examples seems impossible. Yet, we have a fairly good understanding of the process of boiling water and why an aeroplane can fly. This wasn't possible overnight but is the result of dedicated scientific research over the last centuries. Thus, once again, the right level of description such as the average behaviour of an idealized gas or the principal mechanisms of flowing air is important. We know today about the essential quantum nature of atomic and subatomic interactions, but a description of Mount Everest on this level is not reasonable. There are geological theories of mountain formation that are more appropriate than employing quantum theory on these questions.

There is no reason to conjecture that the brain cannot be tractable with similar scientific rigour. The brain is certainly more complex than a gas of weakly interacting atoms. However, there are very fundamental questions that we can attack, for example, how a network can store memories that can be recalled in an associative way. We have today a good understanding of the basic mechanisms behind this ability, which is already present in networks of fairly simple processing elements. Another fundamental question is why the brain is relatively stable while still able to adapt to novel environments. Brain theories of this kind are now emerging. It is too early to talk about a theory of the brain, and we might never be able to explain the brain with a few equations. Nevertheless, many researchers are convinced that we will significantly advance our knowledge of how the brain works in the next decades.

Further reading

Patricia S. Churchland and Terrence J. Sejnowski

The computational brain, MIT Press, 1992.

This classic book on computational issues of brain function discusses many topics that are central in computational neuroscience. It also includes discussions of several specific brain processes not discussed here. This is very much recommended reading for everyone interested in computational neuroscience.

Peter Dayan and Laurence F. Abbott

Theoretical neuroscience, MIT Press, 2001.

Two of the leading scientists in computational neuroscience have recently released a book that will become essential for everyone who wants to further their understanding on the theory of neuroscience. This book is very much recommended reading to further understanding of most of the issues outlined in this book.

2 Neurons and conductance-based models

Neurons are specialized cells that enable specific information-processing mechanisms in the brain. We summarize the basic functionality of neurons and outline some of the fascinating biochemical processes on which many of the models in the remainder of the book are based. These include an outline of chemical synapses, the parametrization of the response of the postsynaptic membrane potential to synaptic events, the origin of action potentials and how they can be modelled using elegant differential equations introduced by Hodgkin and Huxley, generalizations of such conductance-based models, and compartmental models that incorporate the physical structure of neurons. The review in this chapter is intended to justify simplifications used in subsequent models and to at least mention some of the neuronal characteristics that may be relevant in future research.

2.1 Modelling biological neurons

In order to describe the networks of neuron-like elements that we think are at the heart of many information-processing abilities of the brain, we first have to gain some insight into the working of single neurons. Important issues for us therefore are the mechanisms of information transmission within single neurons and between neurons, and how the biologically complex neurons can be modelled sufficiently to address the questions we will consider later in this book.

Many of the computational studies on which we will concentrate are based on strongly simplified versions of the mechanisms present in real neurons. The reasons for such simplifications are twofold. On the one hand, such simplifications are often necessary to make computations with large numbers of neurons tractable. On the other hand, it can also be advantageous to highlight the minimal features necessary to enable certain emergent properties in networks. It is, of course, an important part of every serious study to verify that the simplifying assumptions are appropriate in the context of the specific question to be analysed via a model. We will usually label simple neuron-like processing elements as *nodes* to stress their crude approximation to real neurons.

Some of the following review is intended to at least mention some of the sophisticated computational abilities of neurons, which some readers may want to study in more detail using the literature outlined at the end of this chapter. Many advances in the future are likely to be based on understanding the consequences of more sophisticated processes in single neurons on the behaviour of networks that have rarely been investigated so far. We include an outline of some of the computational approaches used to describe single neurons. Such computational approaches are active areas within computational neuroscience in their own right, and we will not follow them in great depth. They are mainly included to allow you to gain some orientation as to what type of model may be useful in specific cases.

2.2 Neurons are specialized cells

Neurons are biological cells and therefore have most of the features of other cells such as a *cell body* (also called *soma*), a *nucleus* containing *DNA*, *ribosomes* assembling proteins from the genetic instructions, and *mitochondria* providing the energy for the cell. In contrast to other cells, neurons are specialized in signal processing utilizing special electrophysical and chemical processes. A lot of progress has been made in unveiling some of those complex processes. A detailed description is beyond the scope of this book, but we will provide here an overview of the basic mechanisms.

2.2.1 Structural properties

There are many different types of neurons, with differences in size, shape, or physiological properties. However, there are many general features of neurons that we will summarize by describing a generic neuron, as illustrated in Fig. 2.1A, while outlining where variations are common. The major structural features are a *cell body* (or *soma*) and root-like extensions called *neurites*. The neurites are further distinguished into *dendrites*, the receiving end of neurons, and one major outgoing trunk, the *axon*. The topology of neurons in different structures of the nervous system can vary considerably, and topological criteria have indeed been used as a scheme for classifying neurons. Some examples of the shapes of neurons are illustrated in Fig. 2.1B–E. Structural properties have computational consequences that must be explored. We outline later in this chapter how we can incorporate cell topologies in specific simulations of single neurons.

2.2.2 Information-processing mechanisms

Neurons can receive signals from many other neurons, typically on the order of 10,000 neurons. Some neurons have many fewer efferents such as some motor neurons in the brainstem, but there are also examples of cells that receive many more inputs such as pyramidal cells in the hippocampus, which have been estimated to receive on the order of 50,000 inputs. The sending neurons contact the receiving neuron at specialized sites, either at the cell body or the dendrites. The English physiologist *Charles Sherrington* named these sites *synapses*. It is known that various mechanisms are utilized to transfer information between the *presynaptic* neuron (the neuron sending the signal) and the *postsynaptic* neuron (the neuron receiving the signal). The general information-processing feature of synapses is that they enable signals from a presynaptic neuron to alter the state of a postsynaptic neuron, and eventually to trigger the generation of an electric pulse in the postsynaptic neuron. This electric pulse, the important *action potential*, is usually initiated at the rooting region of the axon, the *axon-hillock*, and subsequently travels along the axon. Every neuron has one axon leaving the soma, although the axon can branch and send information to different regions in the brain.

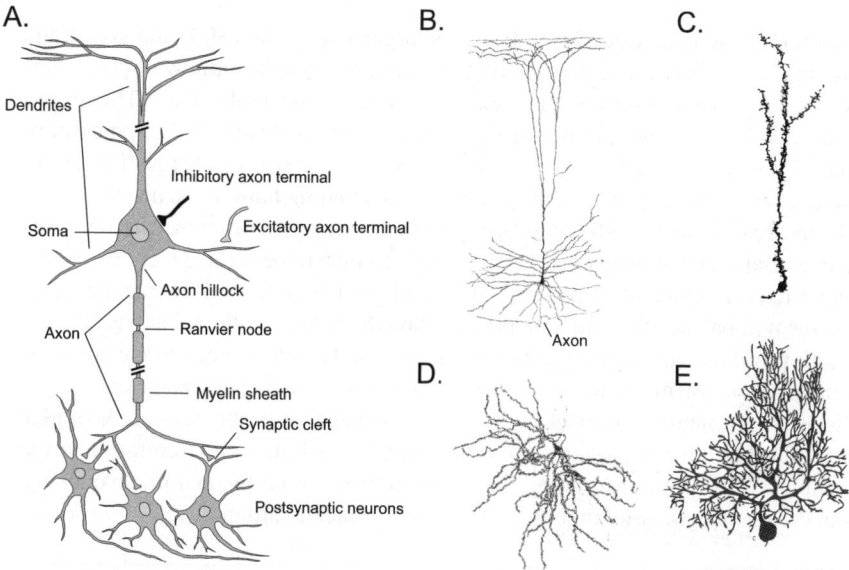

A.

Dendrites

Inhibitory axon terminal

Soma

Excitatory axon terminal

Axon hillock

Axon

Ranvier node

Myelin sheath

Synaptic cleft

Postsynaptic neurons

B.

C.

Axon

D.

E.

Fig. 2.1 (A) Schematic neuron that is similar in appearance to pyramidal cells in the neocortex. The components outlined in the drawing are, however, typical for most major neuron types [adapted from Kandel, Schwartz, and Jessell, *Principles of neural science*, McGraw-Hill, 4th edition (2000)]. (B–E) Examples of morphologies of different neurons. (B) Pyramidal cell from the motor cortex [from Cajal, *Histologie du système nerveux de l'homme et des vertèbres*, Maloin, Paris (1911)], (C) Granule neuron from olfactory bulb of mouse [from Greer, *J. Comp. Neurol.* 257: 442–52 (1987)], (D) Spiny neuron from the caudate nucleus [from Kitai, in *GABA and the basal ganglia*, Chiara & Gessa (eds), Raven Press (1981)], (E) Golgi-stained Purkinje cell from the cerebellum [from Bradly, Berry, *Brain Res.* 109: 133–51 (1976)].

For example, some neurons spread information to neurons in neighbouring areas of the same neural structure with axon branches called *axon collaterals*, while other axons can reach through the *white matter*[2] to other brain areas. At the receiving site the axons split commonly into several fine branches, so-called *arbors* (French for harbours), with *axon terminals* or axon *boutons* (French for buttons) forming part of the synapses to other neurons or muscles.

2.2.3 Membrane potential

Important for the signalling capabilities of a neuron is the ability of a neuron to vary the intrinsic electrical potential of a neuron, the so-called *membrane potential*. The membrane potential, which we shall denote by V_m, is defined as the difference between the electric potential within the cell and its surrounding. The inside of the cell is neutral when positively charged ions, called *cations*, are bound to negatively charged ions, called *anions*. The origin of this potential difference comes from a different concentration of ions within and outside the cell. For example, the concentration of

---

[2] The white matter contains no neuron bodies, only axons and Schwann cells that provide the myelin sheath for the axons as described later.

potassium (K^+) ions is around twenty times larger within the cell compared to the surrounding fluid. This concentration difference encourages the outflow of potassium by a diffusion process as long as the membrane is permeable to this ion.[3] The neuron membrane is, however, not permeable to the anions that typically bind to potassium, and the diffusion process leaves therefore an excess of negative charges inside the neuron and an excess of positive charges in the surrounding fluid of the neuron.

The increasing electrical force resulting from the increased difference in the membrane potential will eventually be strong enough to match the force generated by the concentration difference of potassium within and outside of the cell. Hence, the electrical force will balance the diffusion process, and the neuron settles at an equilibrium state. This process involving only potassium ions would result in a membrane potential of around $E_K = -80$ mV in a typical neuron. Only a very small percentage of K^+ ions have to leave the neuron in order to establish this potential of the neuron. A similar process for other ions, in particular sodium (Na^+), which has a concentration in the surroundings of a neuron exceeding the concentration within the neuron, eventually leads to the *resting potential* of a neuron, which is typically around $V_{rest} = -65$ mV.

2.2.4 Ion channels

The permeability of the membrane to certain ions is achieved by *ion channels*. Ion channels are special types of proteins embedded in the membranes of the cell (see Fig. 2.2). These proteins form pores that enable specific ions to enter or to leave the cell. Common ions involved in such processes within the nervous system are sodium (Na^+), potassium (K^+), calcium (Ca^{2+}), and chloride (Cl^-). The ion channels that drive the resting potential of a cell are usually open all the time (see Fig. 2.2A). It is hence appropriate to call them *leakage channels*. We will shortly meet other types of ion channels that can open and close under various conditions.

The property of information transmission within and between cells is essential for the information processing of single neurons and the nervous system as a whole. It is therefore rewarding to study the mechanisms enabling these capabilities in more detail. We will start in the following section with a description of the major synaptic mechanisms, the chemical synapses, followed by a discussion of the generation of action potential in Section 2.4. In Section 2.5 we will put the pieces together and describe the major functionality of neurons using a compartmental model. We will then discuss briefly some extensions of the basic mechanisms before concentrating on the simplifications that enable us to study the behaviour of large networks of neurons.

2.3 Basic synaptic mechanisms

Signal transmission between neurons is central to the information-processing capabilities in the brain as a whole. There are many known types of transmission mechanisms with a variety of effects on the state of postsynaptic neurons, and some of the details are very important even for models on a system level. One of the most exciting discoveries

---

[3] The potential due to the diffusion process is described by the *Nernst equation* which relates the work that is necessary to compensate for the diffusion gradient to the logarithm of the ratio of the concentrations, the absolute temperature, and some system-specific parameters.

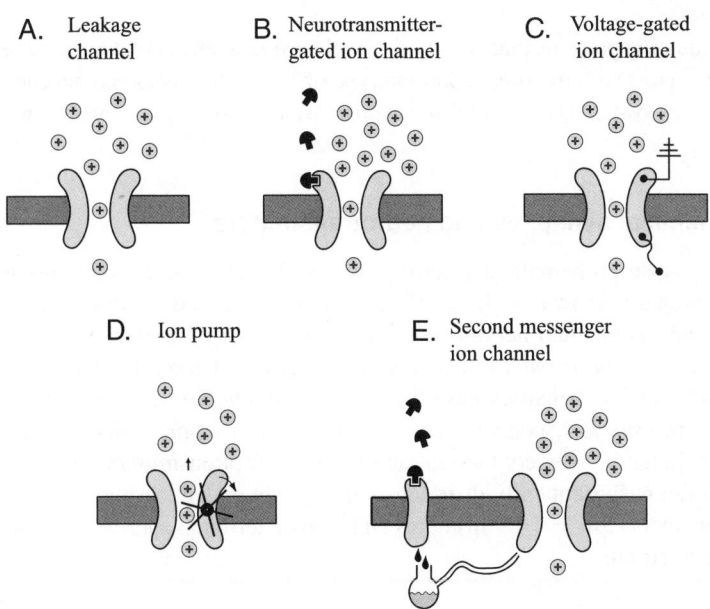

Fig. 2.2 Schematic illustrations of different types of ion channels. (A) Leakage channels are always open. (B) Neurotransmitter-gated ion channel that opens when a neurotransmitter molecule binds to the channel protein, which in turn changes the shape of the channel protein so that specific ions can pass through. (C) The opening of voltage-gated ion channels depends on the membrane potential. This is indicated by a little wire inside the neurons and a grounding wire outside the neuron. Such ion channels can in addition be neurotransmitter-gated (not shown in this figure). (D) Ion pumps are ion channels that transport ions against the concentration gradients. (E) Some neurotransmitters bind to receptor molecules which triggers a whole cascade of chemical reactions in neurons which produce secondary messengers which in turn can influence ion channels.

in neuroscience was that the effectiveness of the transmission can be modulated in various ways. The excitement of this discovery comes from the fact that it allows the brain to adapt to changing circumstances. That not only facilitates the stability of the brain, which is extraordinary for such a complex system in its own right, but is also thought to be the basis of associations, memories, and many other mental abilities. *Synaptic plasticity* will be a central ingredient in most of the models developed in this book. We start here by describing some of the basic mechanisms of the information transmission itself and will later discuss the (largely still unknown) biochemical processes and, in particular, the computational consequences of synaptic plasticity.

Signal transduction within the cell is mediated by electrical potentials as we will discuss in more detail below. These potentials can be generated by a variety of processes, and some neurons are specialized in particular mechanisms. For example, sensory neurons are specialized in converting external stimuli into electrical signals; examples are tactile neurons, which are sensitive to pressure, and photoreceptor neurons in the retina, which are sensitive to light. The information transmission between neurons, with which we are mainly concerned in this book, is also achieved in a variety of ways. For example, so-called *electrical synapses* or *gap-junctions* consist of special conducting proteins that allow a direct electromagnetic signal transfer between two neurons. However, the most common type of connection between neurons in the central nervous system is the *chemical synapse*, which will be our focus for the rest of this section.

2.3.1 Chemical synapses and neurotransmitters

Chemical synapses, schematically outlined in Fig. 2.3A (an electron micrograph of a synaptic terminal is shown in Fig. 2.3B), are the workhorses of neuronal information transfer in the mammalian nervous system. They consist of a specialized extension on the axon, the *axon terminal*, and specific receiving sites on the dendrites. Axon terminals of chemical synapses have the ability to synthesize special chemicals, so-called *neurotransmitters*, concentrated and stored in *synaptic vesicles*. The arrival of an action potential triggers the release of these neurotransmitters. These released neurotransmitters subsequently drift across the *synaptic cleft*, a small gap of only a few micrometers ($1\mu m = 10^{-6}m$) between the axon terminal and the dendrite of the postsynaptic neuron.

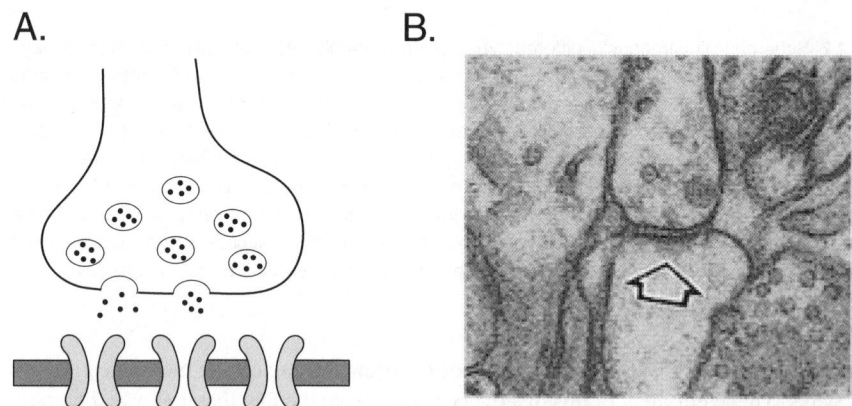

A.   **B.**

Fig. 2.3 (A) Schematic illustration of chemical synapses and (B) an electron microscope photo of a synaptic terminal.

Many different neurotransmitters have been identified as being active in the nervous system. Very common neurotransmitters in the central nervous system include small organic molecules such as glutamate (Glu) or gamma-aminobutyric acid (GABA). Dopamine (DA) is another neurotransmitter that has attracted a lot of attention. Synaptic transmission at neuromuscular junctions, important for initiating muscle movements, is mainly mediated by acetylcholine (ACh), and sensory nerves can utilize a variety of other neurotransmitters. Different synapses and their associated neurotransmitters initiate specific processes in the postsynaptic neurons, which we will outline further below. An important mechanism for fast synaptic transmission, mediated, for example, by the above-mentioned neurotransmitters Glu and GABA, is the binding of these messengers to *neurotransmitter-gated ion channels* in the postsynaptic neuron (see Fig. 2.2B).

Neurotransmitter-gated ion channels are special types of ion channels that open and close under the regulation of neurotransmitters. This is achieved by special proteins that bind specific neurotransmitters and thereby influence the size of the pore that enables the ion flow between the inside and outside of the neuron. This special type of ion channel can therefore regulate the conductance of the membrane through the binding of the neurotransmitter. Without the presence of the neurotransmitter the size of the pore is too small to let ions flow through it. Neurotransmitter-gated ion channels open when neurotransmitters bind to specific proteins of the ion channels, and ions of the right size and shape can flow through them changing the membrane potential of the cell. The response in the membrane potential is called the *postsynaptic potential* (PSP), and we dedicate a later section to its functional description. A larger quantity of neurotransmitter released by the presynaptic neuron can trigger a stronger response in the postsynaptic membrane potential. It is, however, good to keep in mind that this relationship does not have to be linear; twice the amount of neurotransmitter or twice the number of synapses involved in the overall synaptic event does not necessarily increase the membrane potential twofold.

2.3.2   Excitatory and inhibitory synapses

Different types of neurotransmitters and their associated ion channels have different effects on the state of the receiving neuron. An important class of neurotransmitters opens channels that will allow positively charged ions to enter the cell. These neurotransmitters therefore trigger the increase of the membrane potential that drives the postsynaptic neurons towards their excited state. The process (and sometimes simply the synapses) is therefore said to be *excitatory*. The synaptic channels gated by the neurotransmitter Glu are a very common example of such an excitatory synapse. In contrast, other neurotransmitters initiate processes that drive the postsynaptic potential towards the resting potential. Such processes (and sometimes simply the synapses associated with them) are therefore said to be *inhibitory*. A prominent example of inhibitory synapses uses the neurotransmitter GABA. Many of the synapses act in this direct way.

However, not all neurotransmitters influence the postsynaptic potential in this direct way. Other types of receptors and their associated neurotransmitters have attracted much interest recently. Such *non-standard synapses* can influence ion channels in an indirect way, thereby opening the possibility of much more complex synaptic responses to presynaptic transmitter releases. Such non-standard neurotransmitters typically trigger a cascade of chemical reactions that can have a variety of consequences on the functional properties of the neuron. Some of these influences can be described as *modulation*. These neurotransmitters do not open ion channels directly, but can, for example, modulate the effectiveness of standard synapses through a chain of internal biochemical reactions. Some forms of modulatory interaction can be described with a logical *AND function*; for example, only when both types of synapses are active is there a change in the membrane potential of the neuron. We will come back to this interesting form of synaptic interaction at the end of Chapter 4, and we will discuss some possible computational consequences of such nonlinear interactions later in this book.

2.3.3 The time-course of postsynaptic potentials

As mentioned above, the variation of the membrane potential initiated from a presynaptic spike can be described as a *postsynaptic potential* (PSP). In this section we are mainly concerned with the empirical description of these potentials that can be used in simulations of networks as we will see in the following chapters.

Excitatory synapses increase the membrane potential in what is called the *excitatory postsynaptic potential* (EPSP). The change of the membrane potential typically starts with a delay after the firing of a presynaptic neuron. This delay is caused by the time it takes for the neurotransmitter to be released in the axon terminal, the duration of the diffusion process of the neurotransmitter, and the time necessary to open the channels. However, all of these processes are typically very fast (often much less than 1 ms) and will be neglected in most of the models discussed in this book. The principal form of an EPSP after the synaptic delay, in particular the EPSP resulting from non-NMDA (N-methyl-D-aspartate; glutamatergic) receptors, is often described by a function of the form,

$$\Delta V_{\mathrm{m}}^{\mathrm{non-NMDA}} = w \, t \, \mathrm{e}^{-t/t^{\mathrm{peak}}}. \tag{2.1}$$

The symbols ΔV_{m} denotes the change in the membrane potential. The amplitude factor w is a parameter describing the strength of the EPSP (or the efficiency of the synapses as discussed further below). The equation describes the change of the synaptic potential as a function of time t after the synaptic delay. A function of the form $f(x) = x \exp(-x)$ is called an α-*function*, and the Greek symbol α is therefore often taken to stand generally for the functional form of a PSP. The particular function is simply chosen because it fits experimental curves relatively well while being compact. However, we will see shortly that other functions can be used and are sometimes more appropriate. The α-function with the parameter $w = 1$ is illustrated in Fig. 2.4 as a solid line. The parameters of the empirical models have to be fitted to experimental data. The scale of the rise and fall of the PSP in the above parametrization is given by the time constant t^{peak} in eqn 2.1, which corresponds to the time of the peak. Values for such a parameter can vary considerably for different type of synapses. Non-NMDA EPSP can be quite rapid with a rise time of around 0.5 ms.

Inhibitory postsynaptic potentials (IPSP) lower or inhibit the rise of the membrane potential. The functional form of IPSP is frequently also parametrized with the α-function. The difference between an EPSP and an IPSP can then be described by the sign of the parameter w; a positive value indicates an excitatory synapse and a corresponding EPSP, and a negative value for w indicates an inhibitory synapse and a corresponding IPSP. An example of an IPSP is illustrated as a dashed line in Fig. 2.4.

The α-function is not the only possible mathematical description of experimentally measured PSP. For example, the combination of two exponentials, of which two examples with different parameters are shown in Fig. 2.4 as dotted lines, can approximate PSP quite well with the appropriate choice of the parameters. In the figure we have not tried to match the difference of two exponentials with the α-function for better visibility. Note that the functions used for the description of the experimental data are somewhat arbitrary. These empirical (or phenomenological) models are only intended to parametrize experimental findings for convenience without attempting to describing the biochemical and physical mechanisms of ion channels. This is sufficient for most of the discussions in this book. If necessary, more details of the underlying biochemical and physical mechanisms can be incorporated into the model.

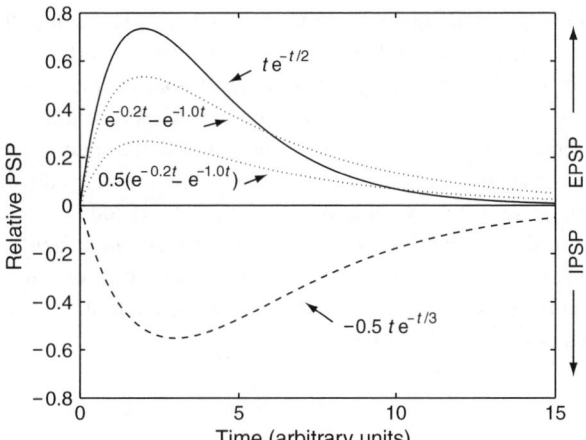

Fig. 2.4 Examples of an EPSP (solid line) and IPSP (dashed line) as modelled by an α-function after the synaptic delay. The dotted lines represent two examples of the difference of two exponential functions with different amplitudes. Note that the time scale is variable and not meant to fit experimental data in the illustration. Indeed, NMDA synapses are often much slower than non-NMDA synapses, often showing their maximal value long after the peak in the effect of the non-NMDA synapses.

NMDA synapses are excitatory synapses, which are typically slower than the fast synapses mentioned above, with a rise time often more than 10 times slower than that of non-NMDA synapses. NMDA synapses are often associated with synaptic plasticity as discussed further in Chapter 7, and the slow time dynamic can facilitate traces that may be utilized in Hebbian learning in recurrent networks. Furthermore, NMDA synapses are an example of transmitter-gated ion channels that are voltage-dependent. An example of a function that has been used to describe the time and voltage dependence of EPSP triggered by this channel is

$$\Delta V_{\mathrm{m}}^{\mathrm{NMDA}} = c(V_{\mathrm{m}})\mathrm{e}^{-t/\tau_1} - \mathrm{e}^{-t/\tau_2} \tag{2.2}$$

which now has two time scale parameters τ_1 and τ_2, and an amplitude that is dependent on the membrane potential. Important for some discussions later in this book is the fact that these ion channels open only after the postsynaptic neurons have been excited, that is, the membrane potential has to be previously increased by other ion channels. The reason for this is that the NMDA channel is blocked by magnesium ions in its resting state. The ions have to be removed before sodium and calcium, the two major ions that can pass through this channel, can enter the neuron.

2.3.4 Superposition of PSP

Electrical potentials have the physical property that they superimpose as the sum of the individual potentials. Many of the models discussed in the literature incorporate the linear superposition of synaptic input. However, it is good to keep in mind that the membrane potential of a neuron depends on several other factors such as the physical properties of the dendrites as we will discuss further in Section 2.5. Also, we just mentioned that some ion channels are voltage-dependent, often with a nonlinear voltage–current relationship. The effect of neurotransmitters therefore depends on the state of the neuron. Finally, there are many sources of nonlinear interactions between synaptic ion channels. For example, a certain type of GABA receptor, called GABA$_a$, is known to have no effect on the membrane potential if the membrane potential is at rest, so it does not reduce the potential further. Instead, it can reduce the effect of excitatory currents. This is an example of nonlinear interactions between excitatory and inhibitory synapses, often referred to as *divisive* or *shunting inhibition*. Nonlinear interactions between synaptic inputs can add greatly to the information-processing capabilities of neural networks. We will discuss some examples in this book, but generally this area is still largely unexplored.

2.4 The generation of action potentials: Hodgkin–Huxley equations

As discussed in the previous section, the resting potential of the neuron can be altered by the release of neurotransmitters from a presynaptic neuron that open specific ion channels and allow ions to leave or enter the neuron. In the following we describe how the change in the resting potential can ultimately trigger the generation of an action potential (spike). A typical form of an action potential, as measured on the giant axon of the squid by *Alan Hodgkin* and *Andrew Huxley*, is shown in Fig. 2.5. It is characterized by a sharp increase (depolarization) of the membrane potential to positive values, followed by a sharp decrease in the membrane potential even undershooting the resting potential of the neuron, called *after hyperpolarization*, before returning to the resting potential of this neuron.

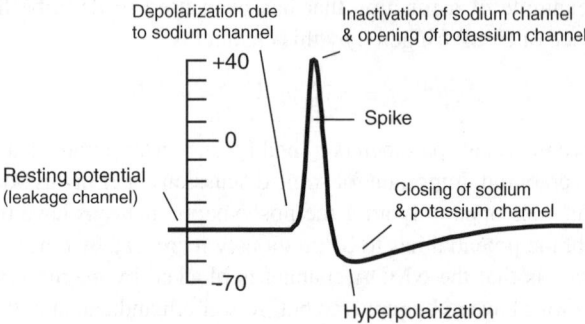

Fig. 2.5 Typical form of an action potential, redrawn from an oscilloscope picture of Hodgkin and Huxley.

It was a major scientific accomplishment by Alan Hodgkin and Andrew Huxley to quantitatively describe the form of the action potential by equations that we will summarize in this section. It is not only the beauty of the description that made them famous, but also the fact that they thereby quantified the process leading to the generation of such action potentials before ion channels were known and long before the details of this process were measured directly. Indeed, their model made specific predictions that could be verified experimentally and by so doing guided further research and many discoveries.

2.4.1 The minimal mechanisms

A least two types of voltage-dependent ion channels and one static ion channel are necessary for the generation of a spike as illustrated in Fig. 2.6. In contrast to the neurotransmitter-gated ion channels discussed in the previous section, voltage-dependent ion channels open and close as a function of the voltage of the membrane (see Fig. 2.2C). A voltage-dependent sodium channel is responsible for the rising phase of the action potential. When neurotransmitter-gated ion channels depolarize the neuron sufficiently, voltage-dependent sodium channels open leading to an influx of Na^+ due to the negative potentials and the lower concentration of Na^+ within the cell. The domination of the sodium channel will therefore bring the membrane potential close to its related sodium resting potential of around $V_m = +65$ mV.

Two contributing processes cause the subsequent falling phase. First, the sodium channels inactivate due to a blockade of the channel by a part of the protein making up the channel at around 1 ms after the opening of the channel. Secondly, a second class of potassium channels opens, leading to an efflux of K^+. In contrast to the voltage-dependent sodium channels, which open nearly instantaneously after crossing the threshold, the voltage-dependent potassium channels open after a delay of about 1 ms after the initial depolarization of the action potential. The dominance of potassium channels will drive the membrane potential close to the potassium equilibrium of around $V_m = -80$ mV, undershooting the normal resting potential of the neuron. Finally, the hyperpolarization of the neuron relative to the resting potential V_{rest} of the neuron causes the voltage-dependent potassium channels to close and the voltage-dependent sodium channels to deactivate, so that the normal resting potential V_{rest} of the neuron will eventually be reinstalled.

2.4.2 Ion pumps

Repeated generation of action potentials results in a repeated efflux of K^+ and influx of Na^+. If the neuron would repeat the process as described so far many times, it would decrease the potassium concentration and increase the sodium concentration within the cell repeatedly, causing the ultimate failure of generating action potentials. Another type of ion channel, not included in Fig. 2.6, is therefore vital for a cell. These specific ion channels are called *ion pumps* and are illustrated schematically in Fig. 2.2D. Ion pumps can transfer specific ions against their concentration level between the inside and outside of the neuron. The price to pay is the large amount of energy necessary for this process. It was estimated to account for about 70% of the total metabolic consumption of a neuron, or about 15% of the total energy consumption in humans when we take into account that the brain accounts for one-fifth of our total energy consumption.

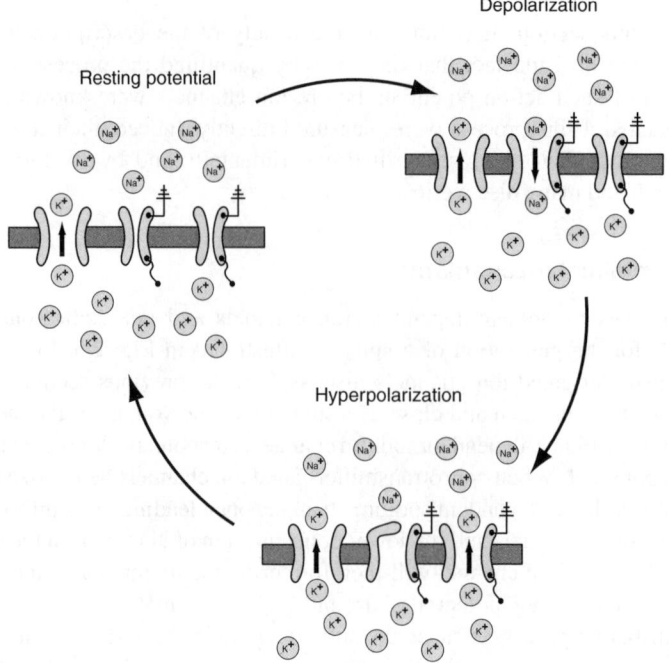

Fig. 2.6 Schematic illustration of the minimal mechanisms necessary for the generation of a spike. The resting potential of a cell is maintained by a leakage channel through which potassium ions can flow as a result of concentration differences between the inside of the cell and the surrounding fluid. A voltage-gated sodium channel allows the influx of positively charged sodium ions and thereby the depolarization of the cell. After a short time the sodium channel is blocked and a voltage-gated potassium channel opens. This results in a hyperpolarization of the cell. Finally, the hyperpolarization causes the inactivation of the voltage-gated channels and a return to the resting potential.

2.4.3 Hodgkin–Huxley equations

Hodgkin and Huxley quantified the process of spike generation with a set of four coupled differential equations, formalizing their measured findings of the giant axon of the squid. The net movement of the ions across the membrane is a current that we shall label with the symbol I_{ion}. The number of open ion channels is proportional to an electric conductance (the inverse of the resistance), which is commonly labelled g_{ion}. In the following we will express the membrane potential relative to the resting potential and label it with V. The relation between this electric potential, the current, and the conductance is given by *Ohm's law*,

$$I_{ion} = g_{ion}(V - E_{ion}). \tag{2.3}$$

where E_{ion} is the equilibrium potential for this channel. This potential is also expressed relative to the resting potential of the neuron and can be represented as a battery (see Fig. 2.7).

As we have discussed before, a major ingredient of the general framework for generating the action potential is the fact that the K^+ and Na^+ channels are voltage-

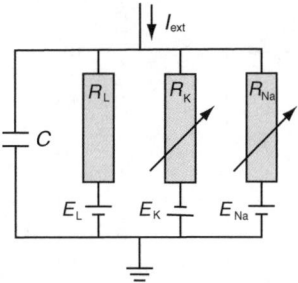

Fig. 2.7 A circuit representation of the Hodgkin–Huxley model. This circuit includes a capacitor on which the membrane potential can be measured and three resistors together with their own battery modelling the ion channels; two are voltage-dependent and one is static.

dependent. Hodgkin and Huxley introduced empirically three dynamic variables, n, m, and h, to describe this voltage dependence and the dynamics of the channels. The variable n describes the activation of the potassium channel, m describes the activation of the sodium channel, and h describes the inactivation of the sodium channel. These variables are included into the formula as modulation factors around maximum conductances, $\bar{g}_{\mathrm{ion}}$, by setting

$$g_{\mathrm{K}} = \bar{g}_{\mathrm{K}} n^4 \tag{2.4}$$

$$g_{\mathrm{Na}} = \bar{g}_{\mathrm{Na}} m^3 h. \tag{2.5}$$

Hodgkin and Huxley chose the dynamics and voltage dependence of the variables appropriately so that the experimental data could be approximated reasonably well. They formulated this dynamic elegantly with a set of three first-order differential-equations, one for each variable. All three differential equations have the same form and can hence be expressed by a single formula,

$$\frac{\mathrm{d}x}{\mathrm{d}t} = -\frac{1}{\tau_x(V)}[x - x_0(V)] \tag{2.6}$$

where x should be substituted by each of the variables n, m, and h. The particular dynamics of the variables and the specific form of the functional dependence of x_0 and τ_x on the voltage were just some choices made by Hodgkin and Huxley. They chose these specific forms and some related parameters only to get a reasonable fit to the experimental data. We will not write down the lengthy expressions for the functions $\tau_x(V)$ and $x_0(V)$ together with all the values of the included parameters that Hodgkin and Huxley chose at this point. Instead, we have included all these details in the program hh.m discussed further in Chapter 12. The voltage dependence of these functions is illustrated in Fig. 2.8. This describes the voltage dependence of the potassium and sodium channels. The leakage channel is static, and the corresponding conductance, g_{L}, is therefore a constant.

The final ingredient that we have to consider is the fact that neurons store electric charges. This is formally expressed as a capacitance C. The model can thus be represented with electronic components as illustrated in Fig. 2.7, consisting of a capacitor in parallel with three resistors, each supplied with their own battery (from the Nernst

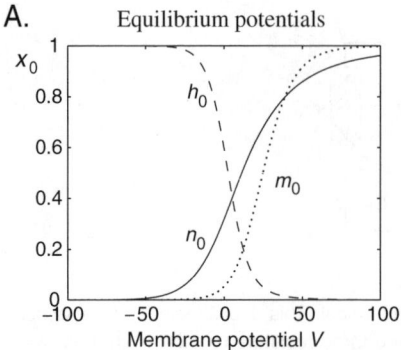
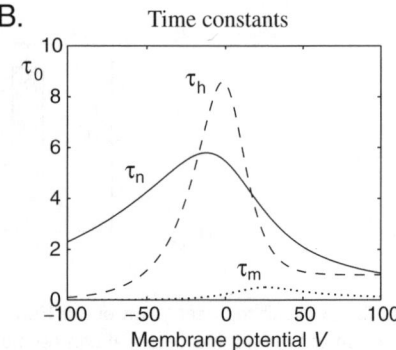

Fig. 2.8 (A) The equilibrium functions and (B) time constants for the three variables n, m, and h in the Hodgkin–Huxley model with parameters used to model the giant axon of the squid.

potential). One of the resistors is constant, whereas the other two can change depending on the state of the system. The combined effects of the different components can easily be expressed by taking the conservation of electric charge into account. This is formalized in *Kirchhoff's law*,

$$C\frac{dV}{dt} = -\sum_{\text{ion}} I_{\text{ion}} + I(t). \qquad (2.7)$$

This formula, a first-order differential-equation, describes the change of the membrane potential with time. $I(t)$ is an external current, for example, the current from the neurotransmitter-gated ion channels. To summarize the four differential-equations of the Hodgkin–Huxley model we substitute the three ionic currents of the basic scheme discussed above into formulas (2.6) and (2.5), resulting in

$$C\frac{dV}{dt} = -g_K n^4(V - E_K) - g_{Na} m^3 h(V - E_{Na}) - g_L(V - E_L) + I(t)$$

$$\tau_n \frac{dn}{dt} = -[n - n_0(V)] \qquad (2.8)$$

$$\tau_m \frac{dm}{dt} = -[m - m_0(V)] \qquad (2.9)$$

$$\tau_h \frac{dh}{dt} = -[h - h_0(V)] \qquad (2.10)$$

We discussed the generation of action potentials by a sodium, potassium, and leakage channels. It should be clear that the mechanism is a result of the simultaneous working of many such ion channels. The conductances in the Hodgkin–Huxley equations are hence the net result of individual ion channels in the membrane. The densities of those ion channels have to exceed a certain threshold in order to be able to generate an action potential. These conditions are usually only fulfilled in the axon membrane, although active action potentials have recently been found in some dendrites. The axon can be excited typically at any location,[4] although *in vivo* neurons typically initiate action potentials in the axon-hillock.

---

[4] As long as it is not *myelinated* as explained below.

2.4.4 Numerical integration

The Hodgkin–Huxley equations can be easily integration numerically. A short introduction to numerical integration techniques is included in Appendix C, and a MATLAB
program for the integration of the Hodgkin–Huxley model, including the parameters
used originally by Hodgkin and Huxley to model the axon of the giant squid, is discussed further in Chapter 12. The results of this numerical integration simulating the
Hodgkin–Huxley model are shown in Fig. 2.9. These simulations demonstrate that a
constant external current with strength parameter $I_{ext} = 10$ leads to a constant firing
of the neuron with a stereotype waveform as shown in Fig. 2.9A.[5] It is remarkable how
the form of this simulated action potential captures the essential details of the action
potentials in real neurons. The *frequency–current (f–I) curve*, plotted in Fig. 2.9B,
reveals two more characteristics of the basic Hodgkin–Huxley axon. First, the onset
of firing follows a sharp threshold around $I_{ext} = 6$. Secondly, there is a fairly limited
range of frequencies when the axon fires, starting at about 53 Hz and increasing only
slightly with increasing external current.

A.   **B.**

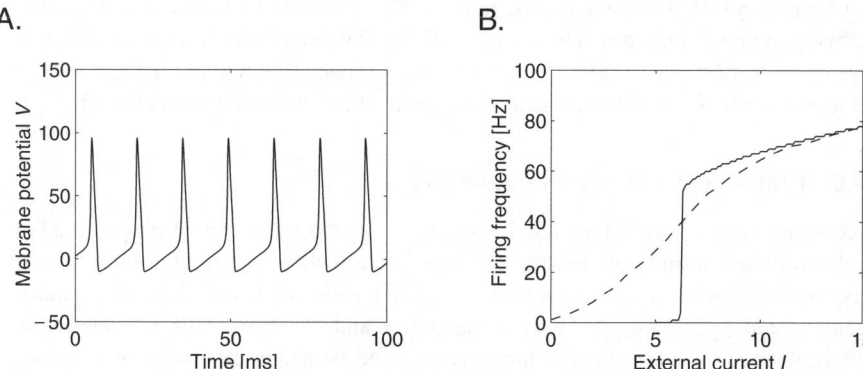

Fig. 2.9 (A) A Hodgkin–Huxley neuron responds with constant firing to a constant external current
of $I_{ext} = 10$. (B) The dependence of the firing rate with the strength of the external current shows
a sharp onset of firing around $I_{ext}^c = 6$ (solid line). High-frequency noise in the current has the
tendency to 'linearize' these response characteristics (dashed line).

2.4.5 Refractory period

There is a maximum firing rate with which any neuron can respond. The reason for this
is that the inactivation of the sodium channel makes it impossible to initiate another
action potential for about 1 ms. This time is called the *absolute refractory period*,
limiting the firing rates of neurons to a maximum of about 1000 Hz. In addition, due
to the hyperpolarizing action potential it is relatively difficult to initiate another spike
during this time period. This *relative refractory period* reduces the firing frequency
of neurons even further. Some very fast spiking frequencies of neurons have been

---

[5] Note that the current parameter has no units at this point as it is only the numerical value we have used
in the program. We can relate this to specific values such as currents measured in nA when we relate the
other parameters in the model to experimental data.

measured in brainstem neurons, with frequencies sometimes exceeding 600 Hz. In contrast, cortical neurons respond typically with much lower frequencies, sometimes with only a few spikes per second. This cannot be explained by the refractory periods alone. It is indeed not obvious why cortical neurons fire with such low frequencies (and very irregularly at the same time), and the reasons behind these facts will be the basis for an interesting discussion later in this book.

For now, it is sufficient to realize that many circumstances influence the firing times of neurons. Among them are the details of charge transmission within the dendrites, interactions between PSP, other types of ion channels, additional mechanisms not included in the Hodgkin–Huxley equations, and the interaction within a network. We will discuss some of these issues in more detail later in this book. Also, a constant external current is physiologically unrealistic in the working brain, and one has to consider more realistic driving currents to get a better feeling for a typical response characteristic of a neuron. To demonstrate how simple alterations in the basic mechanisms can alter the response characteristic of an Hodgkin–Huxley neuron we added the results of some more simulation in Fig. 2.9B shown as a dashed line. We included some high-frequency (1000 Hz) white noise into the otherwise constant external current that is driving the model neuron. The results indicate that such noise has the tendency to 'linearize' the f_I curve resulting in the sigmoidal-shaped function plotted as a dashed line in Fig. 2.9B. We will come back to this point at the end of the next chapter.

2.4.6 Propagation of action potentials

Once an action potential is initiated, it will rapidly increase the membrane potential of neighbouring axon sites and thus travel along the axon with a speed of around 10 m/s. This form of signal transportation with an *active membrane* is loss-free; the signal is regenerated at each point of the axon membrane and the signal will not deteriorate with the distance. This is of major importance as the axon can sometimes be very long (on the order of metres). On the other hand, this form of loss-free signal transportation is much slower than simple current transportation within a conductor, such as in an electrical wire or cable, and also consumes a lot of energy as we discussed above. Nature responded to these problems by covering most of the axon with a sheath of *myelin*. This sheath is provided by another type of cell abundant in the nervous system, so-called Schwann cells. These cells are thought to cover several supporting functions, one of which is the myelination of axons. The myelin sheath changes the electrical properties of the axon to allow fast signal transfer within the axon. However, the myelin sheet is regularly interrupted in so-called *Ranvier nodes*, at which the action potential can regenerate by the active membrane mechanism. Nature combines in this way the advantages of fast and cheaper signal transmission with regenerative amplifiers. It is interesting to note that a large part of the myelination happens after an initial organizational phase of the nervous system, in humans at around their second year of age.

2.5 Dendritic trees, the propagation of action potentials, and compartmental models

We have discussed the effects of neurotransmitter-gated ion channels, leakage channels, and voltage-dependent ion channels on the membrane potential of a neuron. To get a comprehensive view of the state of a neuron we must also include the conductance properties within the neuron and the physical shape of the neuron. In contrast to axons with active membranes able to generate action potentials, dendrites are a bit more like passive conductors,[6] analogous to the long cables for which the physics was worked out in the mid-19th century by *Lord Kelvin* enabling the first transatlantic communication cables. Many researchers have since applied this to neural transmission, for example, *Wilfrid Rall* who has contributed much to the theory and its applications. Taking these missing links into account we can outline the basic idea behind compartmental models. Such modelling alone is an active (and very well developed) area of computational neuroscience. We will only describe the general ideas in the following, which should be sufficient to get an impression of what such often mentioned models are. There are several dedicated publications, some of which are mentioned in the 'Further reading' at the end of this chapter, that give much more detail if further study of such models is desired.

2.5.1 Cable theory

To give you a feeling of some equations involved in compartmental modelling we briefly outline the *cable equation*. This equation describes the spatial–temporal variation of an electric potential along a cable-like conductor driven by an injected current I_{inj} and is given for an idealized one-dimensional linear cable as

$$\lambda^2 \frac{\partial^2 V_m(x,t)}{\partial x^2} - \tau \frac{\partial V_m(x,t)}{\partial t} - V_m(x,t) + V_{rest} = -4\, d\, R_m\, I_{inj}(x,t). \quad (2.11)$$

This is a partial differential equation,[7] second-order in space and first-order in time. The solution of this equation, $V_m(x,t)$, describes the potential of the cable at each location of the cable and how it varies at each location with time. The constant λ describes the physical properties of the cable and has the dimensions of Ωcm. For example, a cylindrical cable of diameter d in an equipotential surrounding has a lambda parameter of

$$\lambda = \sqrt{\frac{dR_m}{4R_i}}, \quad (2.12)$$

where R_m is the specific resistance of the membrane and R_i is the specific intracellular resistance of the cable. A specific resistance is the resistance of a piece of material with

---

[6] Dendrites also have active ion channels, and spikes can indeed be generated in dendrites. Nonlinear effects in dendrites are also important. However, many general features of dendrites can be studied using passive cable equations, and this approximation is also used to outline the principal idea behind compartmental modelling in this section.

[7] The equation includes partial derivatives symbolized by ∂ instead of total derivatives symbolized by d. Partial derivatives only consider explicit dependencies on the parameters and ignore implicit dependence on other parameters. This differential equation is formally a hyperbolic differential equation in case you want to look it up in mathematics books.

constant cross-section divided by the volume of this resistor. Membrane resistance is, of course, much larger than the intracellular resistance, which explains why most of the current flows within the dendritic tree. Note that the lambda parameter changes with the diameter of the cable. The time constant τ_m depends also on the physical properties of the cable and is given by

$$\tau_m = R_m C_m, \tag{2.13}$$

where R_m is the specific resistance of the membrane as above, the inverse of the leakage conductivity, and C_m is the specific capacitance, the capacitance per unit area.

The solutions of eqn 2.11 depend, of course, on the form of the injected current $I_{inj}(x, t)$, and analytical solutions can only be given for some simple examples which are nevertheless instructive.[8] Is is clear that the cable potential should reach a stable configuration (distribution within the cable) if the injected current is not time dependent. For example, we can consider a cable that is clamped at the ends to certain potentials. This steady-state solution of the linear cable equation can be written as

$$V_m(x) = v_1 e^{-\lambda x} + v_2 e^{\lambda x}, \tag{2.14}$$

which you can verify by inserting this solution into eqn 2.11. The parameters v_1 and v_2 have to be adjusted to fit the boundary conditions. For example, if we consider a semi-infinite cable that starts at $x = 0$ with a fixed potential V_0 and extends to infinity thereafter, then the potential decays exponentially with

$$V_m = (V_0 - V_{rest})e^{-\lambda x} + V_{rest}. \tag{2.15}$$

The exponential decay is, of course, a problem for information transmission[9] and the signal in a long cable has to be amplified periodically in order to be able to transport it over large distances.

If we include voltage-dependent ion channels as in the Hodgkin–Huxley equations, we have a voltage-dependent injected current. In this case we have to solve the *nonlinear cable equation,*

$$\lambda^2 \frac{\partial^2 V_m(x,t)}{\partial x^2} - \tau \frac{\partial V_m(x,t)}{\partial t} - V_m(x,t) + V_{rest} = -4\,d\,R_m\,I_{inj}(V_m(x,t),t). \tag{2.16}$$

No general analytical solution can be given for this equation, but we can use numerical integration techniques to solve the equation for particular cases.

2.5.2 Physical shape of neurons

The solution of the cable equations depends on the physical properties of the cable, and it is time to take the physical shape of neurons into account. The first step in solving the cable equations for cables with structures other than a simple homogeneous linear

---

[8] The equation can be integrated numerically, a fact on which we typically rely in practice as discussed later.

[9] In the case of signal transmission we have to solve the cable equation in the presence of a time-varying current. However, the general features of a decaying signal hold also in this case.

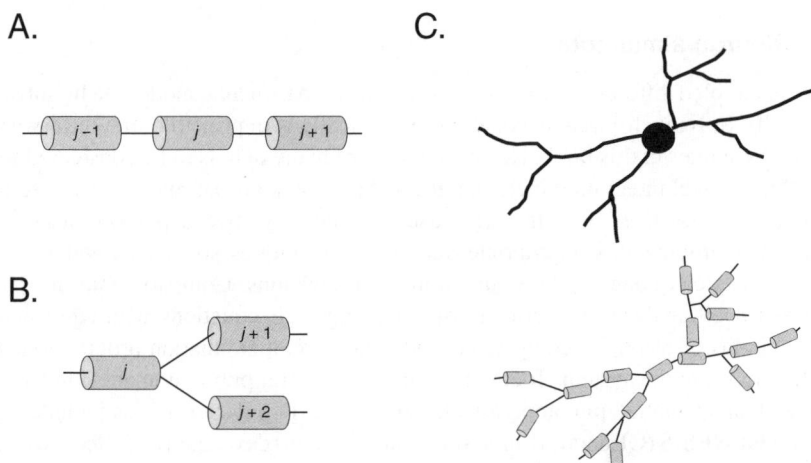

A.

B.

C.

Fig. 2.10 Short cylindrical compartments describing small equipotential pieces of a passive dendrite or small pieces of an active dendrite or axon when including the necessary ion channels. (A) Chains and (B) branches determine the boundary conditions of each compartment. (C) The topology of single neurons can be reconstructed with compartments, and such models can be simulated using numerical integration techniques.

cable is to divide the cable into small pieces (for example, small cylindrical cable), where each piece has to be small enough so that the potential within this unit is approximately constant. We will call this unit a *compartment* (see Fig. 2.10A). Each compartment is governed by a cable equation for a finite cable, which is a first-order differential equation in time as we have seen above. For small compartments we can assume a constant potential within the compartment. We have, however, also to take the boundary conditions into account. This can be done by replacing the continuous space x with a discrete spatial locations labelled with an index, for example, x_j, and replacing the spatial differentials with differences,

$$\frac{\partial^2 V(x,t)}{\partial x^2} \rightarrow \frac{V_{j+1} - 2V_j(t) + V_{j-1}(t)}{(x_{j-i} - x_j)^2}. \tag{2.17}$$

With similar formulas we can also take branching cables (see Fig. 2.10B) into account. A neuron is therefore replaced with a discrete model of small cylinder-like cables as illustrated in Fig. 2.10C. This *compartmental model* is governed by a set of first-order differential equations in time as we have replaced the spatial differentials with difference equations. The number of compartments should, of course, be large to accurately represent a neuron. In practice it depends on the complexity of the neuron how many compartments are necessary to get good approximations of the real neuron. Models with from a few hundred up to several thousand compartments have been considered in the literature to represent single neurons very accurately.

2.5.3   Neuron simulators

The set of coupled difference equations defining a compartmental model can be solved numerically. Solving differential equations numerically is in principle straightforward as we will see later in this book. However, there are many details to be considered for efficient numerical integration including the stability of solutions and efficient use of computational resources. It is therefore useful to employ special purpose-designed software that implements appropriate numerical techniques so that computational neuroscientists can concentrate on the biological questions. Compartmental models have often been analysed numerically by replacing their equations with equivalent representations of electrical components and using circuit simulation programs such as SPICE to solve the system. Today there are also several public-domain simulation packages that specialize in compartmental modelling. Popular examples include the simulator GENESIS (GEneral NEural SImulation System) developed at Caltech, which can be downloaded at `http://www.bbb.caltech.edu/GENESIS/genesis.html`. This program has, in addition to the single-neuron modelling capabilities, the capability of simulating a network of spiking neurons. This program is so far only available for UNIX platforms running the X-Window environment.

   Another simulator called NEURON is by M. L. Hines, J. W. Moore, and N. T. Carnevale from Duke and Yale Universities. This package can be downloaded from the NEURON home page at Duke University (`http://neuron.duke.edu`) and is also available for a Windows platform. The distribution package includes an example of a three-dimensional reconstruction of a pyramidal cell as shown in Fig. 2.11. We injected in the simulation a 20 nA current at $t = 1$ ms for 0.5 ms at the location on the dendrite indicated by the blob and cursor in the graphical outline of the cell. We then measured the response at the soma, which can be seen in the graph window. The passive dendrite passed the current on to the soma with some delay and a more gradual response.

2.6   Above and beyond the Hodgkin–Huxley neuron: fatigue, bursting, and simplifications

The Hodgkin–Huxley equations have taught us the basic concepts of the generation of action potentials, and describe the firing dynamics in the giant axon of the squid with high precision. Today we know more about the functions of ion channels, which can be incorporated in models to get an even better fit of an action potential. Such models are vital for understanding the details of the biochemical mechanisms involved. We will not follow this line of active research much further in this book. Instead, we will concentrate primarily on the more abstract information-processing mechanisms of the nervous system for which this level of description is often sacrificed in favour of understanding global features of the brain. This choice is motivated by the fact that the principal form of an individual spike is rather stereotyped for a given type of neuron, and information transmission within those variations seems not to be the primary mechanism of information processing in the nervous system. The variations in the spike time of neurons and the variety in the dynamics of spiking pattern are, however, essential for information processing in the nervous system. We discuss in this section

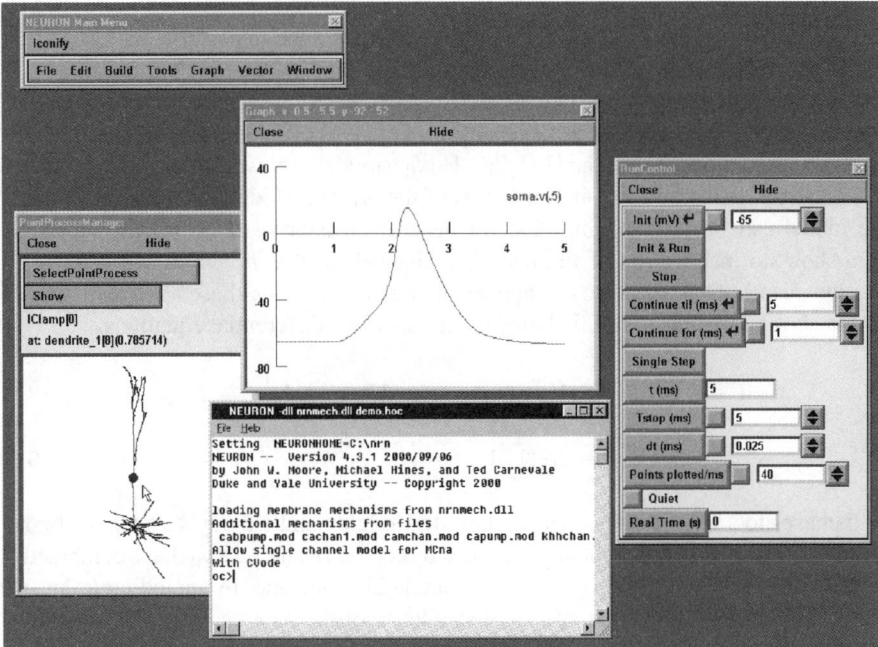

Fig. 2.11 Example of the NEURON simulation environment (see http://neuron.duke.edu). The example shows a simulation with a reconstructed pyramidal cell, in which a short current pulse was injected at $t = 1$ ms at the location of the dendrite indicated by the dot and the cursor in the window on the left.

several mechanisms within a single neuron influencing the spiking characteristics. These include, for example, the effects from specific ion channels such as *spike time adaptation*. Such effects often have to be included explicitly in models if we do not base the neuron models on conductance-based models with corresponding ion channels.

2.6.1 Simplification of the Hodgkin–Huxley model

In contrast to the giant axon of the squid, mammalian neurons have additional types of ion channels that enable a larger complexity of neuronal response. At least a dozen types of ion channels can be involved in the spike generation of human neocortical neurons. The analysis of such high dimensional system is very complex and appropriate simplification are necessary to gain more insight into the principle dynamic mechanisms. Models that reduce the dimensionality of the systems have therefore been advanced since the original work of Hodgkin and Huxley. *Hugh R. Wilson* has shown that the dynamic variety of neocortical neuron responses can be approximated in great detail with a model of only four principle currents. The simplicity of this model allows extended mathematical analysis, and is efficient enough to allow simulations of small networks of such neurons.

The essential mechanism of spike generation by fast K^+ and Na^+ channels has, of course, to be incorporated in such models. We can, however, simplify the Hodgkin–Huxley model by noting that the time constant τ_m is so small for all values of V

(see Fig. 2.8B) that the dynamic variables m, describing the rate of the Na$^+$ channel activation, quickly approaches the corresponding equilibrium value, $m_0(V)$, and can be replaced by it. Furthermore, the rate of inactivation of the Na$^+$ channel is approximately reciprocal to the opening of the K$^+$ channel (see Fig. 2.8A), so that we can set $h = 1-n$. These simplifications reduce the Hodgkin–Huxley model to a two-dimensional system with an action potential very similar to that of the original Hodgkin–Huxley neurons. The model can be further simplified for neocortical neurons. Neocortical neurons often show no inactivation of the fast Na$^+$ channel, so that h can be set to $h = 1$ (or, equivalently, $n = 0$ within our approximations). Following these approximations Wilson showed that the simplified model with only two differential equations,

$$C\frac{dV}{dt} = -g_K R(V - E_K) - g_{Na}(V)(V - E_{Na}) + I(t) \tag{2.18}$$

$$\tau_R \frac{dR}{dt} = -[R - R_0(V)] \tag{2.19}$$

still behaves to a large extent very similarly to Hodgkin and Huxley's system of four differential equations. In the model described by eqns 2.18 and 2.19 we have combined the Na$^+$ and leakage channels into a new single Na$^+$ channel by including a linear term in the description of the voltage dependence of the new channel. The symbol of the only remaining dynamic modulation variable R should signify this variable as describing the recovery of the membrane potential.

2.6.2 Extensions of the Hodgkin–Huxley model

We have so far only expressed the principal mechanisms already described by *Hodgkin* and *Huxley*. In the following we will include two more types of ion channels that seem essential for the more diverse firing properties of mammalian neocortical neurons. The first channel is a cation influx channel with more graded influx characteristics described by a dynamic gating variable T. Such channels, in particular a Ca^{2+} channel, are indeed a major ingredient in mammalian nervous systems. The more graded Ca^{2+} influx is mainly responsible for the more graded response characteristics of neocortical neurons compared to those of the giant axon of the squid. The second channel describes a slow hyperpolarizing current, such as that of a common Ca^{2+}-mediated K$^+$ channel, with a dynamic gating variable H. We will see that this type of channel is essential for the generation of a more complex firing pattern based on spike frequency adaptation. The complete Wilson model of mammalian neocortical neurons includes these types of channels, and is given by

$$C\frac{dV}{dt} = -g_{Na}(V)(V - E_{Na}) - g_K R(V - E_K) \tag{2.20}$$
$$-g_T T(V - E_T) - g_H H(V - E_H) + I(t)$$

$$\tau_R \frac{dR}{dt} = -[R - R_0(V)] \tag{2.21}$$

$$\tau_T \frac{dT}{dt} = -[T - T_0(V)] \tag{2.22}$$

$$\tau_H \frac{dH}{dt} = -[H - 3T(V)], \tag{2.23}$$

with polynomial parametrizations of the voltage dependence of the effective conductances,

$$g_{Na}(V) = 17.8 + 0.476V + 33.810^{-4}V^2 \qquad (2.24)$$
$$R_0(V) = 1.24 + 0.037V + 3.210^{-4}V^2 \qquad (2.25)$$
$$T_0(V) = 4.205 + 0.116V + 810^{-4}V^2. \qquad (2.26)$$

Despite drastic simplifications, such as neglecting the typical inactivation of the Ca^{2+} channels, this model is able to approximate mammalian spike characteristics in great detail. This includes the shape of single spikes as well as all major classes of spike characteristics such as regular spiking neurons (RS), fast spiking neurons (FS), continuously spiking neurons (CS), and intrinsic bursting neurons (IB) by choosing appropriate values of the remaining constants. We will demonstrate such behaviour in the following.

2.6.3 Simulations of the Wilson model

Simulations of the Wilson model can be done with the base program included in the file `wilson.m` printed and discussed further in Chapter 12. All of the following simulations were done with fixed values of the constants $E_{Na} = 50$ mV, $E_K = -95$ mV, $E_T = 120$ mV, $E_R = E_K$, $C = 100\mu Fcm^{-1}$, $g_{Na} = 1$, $g_K = 26$, $\tau_T = 14$ ms, $\tau_R = 45$ ms. The following simulations are also done with a constant external driving current $I_{ext} = 1$ nA. Ignoring the slow calcium mediated potassium channel ($g_H = 0$) and setting $\tau_R = 1.5$ms, $g_T = 0.25$ results in a constantly rapid spiking shown in the upper graph of Fig. 2.12. Such spike trains are typical for inhibitory interneurons in the mammalian neocortex when stimulated with a constant current.

Excitatory neurons typically have a more complex behaviour with slower spike rates and elongated spikes. In the following simulations we therefore use a time constant of $\tau_R = 4.2$ ms. The inclusion of the slow calcium mediated potassium channel ($g_H = 5$) with a slightly smaller calcium conductance ($g_T = 0.1$) results in the spike train shown in the middle graph of Fig. 2.12. The slow hyperpolarizing channel has the important effect of reducing the firing rate after the initial stimulation of the neuron. This *spike rate adaptation* is sometimes called *fatigue*. The firing rates of such regularly spiking neurons are often reduced to about half of their initial value in about 50 ms, and some neurons even cease firing altogether.

Another important class of neurons, showing short bursts interleaved with silent phases, can be generated by increasing the slow hyperpolarizing conductance ($g_H = 9.5$) as well as the calcium conductance ($g_H = 2.25$). The firing response of such a neuron is shown in the lower graph of Fig. 2.12. Such *bursting neurons* typically show a long-lasting *after-depolarizing potential* (ADP) also present in the Wilson model.

Conclusion

Neurons utilize a variety of specialized biochemical mechanisms for information processing and transmission. These include ion channels that allow a controlled influx and outflux of currents, the generation and propagation of action potentials, and the

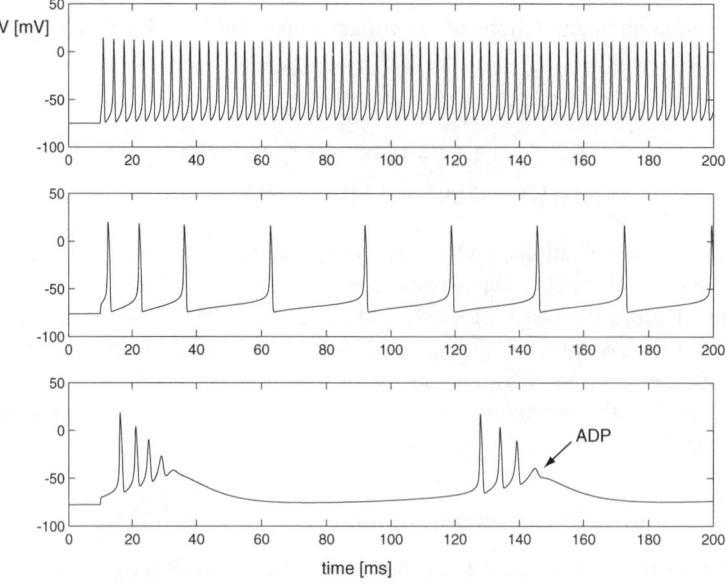

Fig. 2.12 Simulated spike train of the Wilson model. The upper graph simulates fast spiking neurons (FS) typical of inhibitory neurons in the mammalian neocortex ($\tau_R = 1.5$ms, $g_T = 0.25$, $g_H = 0$). The middle graph models regular spiking neurons (RS) with longer spikes ($\tau_R = 4.2$ms, $g_T = 0.1$). The slow calcium-mediated potassium channel ($g_H = 5$) is responsible for the slow adaptation in the spike frequency. The lower graph demonstrate that even more complex behaviour, typical of mammalian neocortical neurons, is incorporated in the Wilson model. The parameters ($\tau_R = 4.2$ms, $g_T = 2.25$, $g_H = 9.5$) result in a typical bursting behaviour, including a typical after-depolarizing potential (ADP) [see also Wilson, *J. Theor. Biol.* 200: 375–88 (1999)].

release of neurotransmitters. Neurons with such basic mechanisms can be modelled with compartmental models that allow a detailed comparison of the mathematical description of such mechanisms with experimental measurements. The basic mechanisms summarized in this chapter are not by any means all of the mechanisms that may be relevant for information processing in the nervous system. For example, some evidence is emerging that some production of proteins in the soma may be utilized to support information processes in neurons, for example, as the formation of long-term memories. Furthermore, the functionality of a neuron is not static. Various parameters can vary, for example, the efficiency with which a spike of a presynaptic neuron influences the membrane potential of postsynaptic neurons. This will be important in subsequent chapters of this book where we will discuss more details of synaptic plasticity.

Further reading

Mark F. Bear, Barry W. Connors, and Michael A. Paradiso
  Neuroscience: exploring the brain, Williams and Wilkins, 1996.
    A very illustrative and basic introduction to neuroscience.

Eric R. Kandel, James H. Schwartz, and Thomas M. Jessell

Principles of neural science, McGraw-Hill, 4th edition, 2000.

A standard reference which still gets better. This is a very good book if you are searching for more details of biophysics, brain anatomy, and specific brain areas.

Christof Koch

Biophysics of computation; information processing in single neurons, Oxford University Press, 1999.

This book is a very comprehensive discussion of the biophysics of single neurons, and is very much recommended reading for many details that we have only sketched on the surface. It also includes several important models that we have not included in our summary.

Christof Koch and Idan Segev (eds.)

Methods in neural modelling, MIT Press, 2nd edition, 1998.

This collection of works from specialist in the corresponding areas is one of the best references for the basic information processing of neurons, including cable theory, active dendrites, ion channel dynamics, and compartmental models. The first edition of this book with different contributions, published in 1989, is also still worthwhile reading.

Gordon M. Shepherd

Neurobiology, Oxford University Press, 3rd edition, 1994.

This is a standard reference in neurobiology. The first nine chapters of the book are in particular dedicated to the description of single neurons, including not only the basic processes sketched in this chapter but also neuromodulators and developmental issues.

Gordon M. Shepherd (ed.)

The synaptic organization of the brain, Oxford University Press, 4th edition, 1998.

A detailed description of the synaptic organization in different brain areas including different cell types and synaptic mechanisms.

Hugh R. Wilson

Spikes decision and actions: dynamical foundations of neuroscience, Oxford University Press, 1999.

A discussion of conductance-based models with MATLAB programs, to a large extent from the dynamical systems point-of-view. I can also very much recommend his recent paper in J. Theor. Biol. 200: 375–88, 1999, on which most of the discussion in this chapter is based.

C. T. Tuckwell

Introduction to theoretical Neurobiology, Cambridge University Press, 1988.

A detailed analysis of single neurons, cable theory, and stochastic modelling. Very detailed and very mathematical.

3 Spiking neurons and response variability

The neuron models described in the last chapter are often numerically too sumptuous to be used in large network simulations. We introduce in this chapter further abstractions that are used often to study the importance of spike timings in networks of such elements. We will then discuss the variability of neuronal response of cortical neurons and outline three different ways in which to include noise in the simulations and the consequences of the influence of noise on the behaviour of neurons. This also gives us the opportunity to mention a few specific forms of probability distribution that are often encountered in computational neuroscience.

3.1 Integrate-and-fire neurons

3.1.1 Stereotyped spike forms

As mentioned in the last chapter, the form of spike generated by neurons is very stereotyped. It therefore seems that this type of information transmission is at least not dominant in a nervous system. This leaves several other types of information representation in the brain, which we will explore in more detail in Chapter 5. We will discuss there the relevance of the timing of the spike for information transmission. To study such questions it is advisable to make further simplifications of biological neurons. For the questions we are going to ask it is not important to describe the precise form of a spike. Only the influence of a presynaptic spike on the postsynaptic membrane potential, which is rather stereotyped and often well described by an α-function as discussed before, is essential for the firing times. We can therefore neglect the detailed ion-channel dynamics within the spike and concentrate entirely on the dynamics leading up to the generation of a spike. This dynamics was driven mainly by the sodium and leakage channels in Hodgkin–Huxley neurons. For simplicity we will neglect the voltage dependence of the sodium channel in the following. This approximation is reasonable as the voltage dependence is only weak for subthreshold neurons. Furthermore, we will not include the effects of frequency adaptation mechanisms as discussed in the last chapter.[10]

---

[10] In principle, there is no problem in including both factors, the voltage dependence of the sodium channel and frequency adaptation in simulations, but it is not essential for the discussions in this chapter, nor for most of the features of neuronal networks that we discuss in the remainder of the book. Also, most current models in computational neuroscience neglect these factors. There are some recent studies that explore the consequences and functional roles of frequency adaptation. Further exploration of the possible effects of these factors on computations would be an interesting research topic.

3.1.2 The basic integrate-and-fire neuron

The modelled membrane potential will be denoted in the following by the symbol u to distinguish these models from the previous models with more detailed description of ion channels. The main effects of the sodium and leakage channels are captured by a simple differential equation of a *leaky integrator*,

$$\tau_{\mathrm{m}} \frac{du(t)}{dt} = -u(t) + RI(t). \tag{3.1}$$

We have here introduced a membrane time constant τ_{m} determined mainly by the average conductances of the sodium and leakage channels.

The input current $I(t)$ is made up of the sum of synaptic currents generated by firings of presynaptic cells. This depends on the *efficiency* of individual synapses, and we describe these by a *synaptic strength value* w_j for each synapse that is labelled by an index j for each postsynaptic neuron. We also call this weight value *connection strength*, *synaptic efficiency*, or *weight value*. In the following we assume that there are no interactions between synapses; we will discuss such interactions later. Without interactions we can write the total input current to the neuron under consideration as the sum of the individual synaptic currents where each stereotyped response of the postsynaptic current is multiplied by a weight value (synaptic efficiency),

$$I(t) = \sum_j \sum_{t_j^{\mathrm{f}}} w_j \alpha(t - t_j^{\mathrm{f}}). \tag{3.2}$$

The function α parametrizes the form of the stereotyped postsynaptic response. We have discussed some examples in Chapter 2, which included the specific mathematical function with the name α-function. The symbol α is here taken to be more general, but this symbol is often used as the specific α-function is a common parametrization of the response function α. The variable t_j^{f} denotes the firing time of the presynaptic neuron of synapse j. The firing time of the postsynaptic neuron is defined by the time the membrane potential u reaches a threshold ϑ,

$$u(t^{\mathrm{f}}) = \vartheta. \tag{3.3}$$

Note that this firing time, which is the firing time of the integrate-and-fire neuron we are describing, has no index and is thereby distinguished from the firing times of presynaptic neurons, which are labelled with an index.

To complete the model we have to reset the membrane potential after the neuron has fired. Several mechanisms can be considered here that will be explored further below. A particularly simple choice is to reset the membrane potential to a fixed value u_{res} immediately after a spike, for example,

$$\lim_{\delta \to 0} u(t^{\mathrm{f}} + \delta) = u_{\mathrm{res}}. \tag{3.4}$$

In addition, we can incorporate an absolute refractory time by holding this value constant for a certain period of time. A more graded refractoriness can be simulated with smoother functional description as the sudden reset of the membrane potential, but most of the models in this book will not depend crucially on these details. Equations (3.1)-(3.4) define the (leaky) *integrate-and-fire neuron* sometimes simply called *IF-neuron*. Such a model neuron is schematically illustrated in Fig. 3.1.

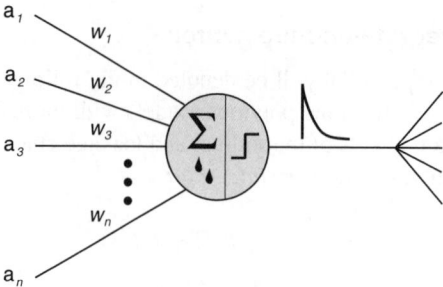

Fig. 3.1 Schematic illustration of a leaky integrate-and-fire neuron. This neuron model integrates (sums) the external input, with each channel weighted with a corresponding synaptic weighting factors w_i, and produces an output spike if the membrane potential reaches a firing threshold.

3.1.3 Response of IF neurons to constant input currents

The dynamic equation for the subthreshold membrane potential, eqn 3.1, is an inhomogeneous linear differential equation that can be solved numerically. It can also be solved analytically for specific examples, which might be instructive and are hence discussed next. We start with an IF-neuron that is driven by a very short input pulse (a very short input current $I(t)$ that is not sufficient to elicit a spike.)[11] Say we start at a situation where the initial potential is equal to zero. We then apply a very short input pulse at $t = 0$ that drives the potential to $u(t = 0) = 1$. After this there is no remaining input current, so that eqn 3.1 becomes a simple homogeneous differential equation for all times $t > 0$ given by,

$$\tau_{\mathrm{m}} \frac{du(t)}{dt} + u(t) = 0. \tag{3.5}$$

The solution of this differential equation is a exponential function that can easily be verified by inserting the result into the differential equation. The membrane potential decays therefore exponentialy after the short external current is applied,

$$u(t) = e^{-t/\tau_{\mathrm{m}}}. \tag{3.6}$$

The time scale of the decay is given by the time constant τ_{m}.

The next example for which can integrate the subthreshold dynamic equations is that of an IF-neuron driven by a constant input current low enough to prevent the firing. This will lead, after some transient time, to a stationary state where the membrane potential will not change further because the input current does not change. When the membrane potential does not change we have

---

[11] Mathematically we write this short input pulse as delta function $I(t) = \delta(t)$ as described further in Appendix A. The delta function is zero for all arguments except one value when the argument is zero. For this argument only the function assumes an infinitely large value. This special mathematical construct is called a functional. It seems hard to imagine that this form can be realized in nature; it would mean having an input pulse that is infinitely strong but has zero duration. However, it is possible to define the delta functional properly as a limiting process of a finite pulse, for example, a Gaussian bell curve, and making the duration smaller while keeping the integral over this function constant. With this limiting process we can solve the differential eqn 3.1.

$$\frac{du}{dt} = 0. \tag{3.7}$$

Inserting this into eqn 3.1 yields the equilibrium potential for large times given by

$$u = RI. \tag{3.8}$$

We made the condition that the input current should be small enough not to elicit a spike. This can now be specified further and is given by $RI < \vartheta$. We only calculated the equilibrium solution of the membrane potential after a constant current has been applied for a long time. The differential equation for constant input can also be solved for all times after the constant current $I_{\text{ext}} = \text{const}$ is applied and is given by

$$u(t) = RI(1 - e^{-t/\tau_m} + \frac{u(t=0)}{RI}e^{-t/\tau_m}), \tag{3.9}$$

which can again be verified by inserting this solution into the differential equation. This solution has two parts: the last term describes the exponential decay of potential at $u(t=0)$ as before, and the first term describes the increase of the membrane potential due to the input current. This solution is illustrated in Fig. 3.2A. The neuron does not fire in this case because we have used a relatively small current, relative to the firing threshold of the neuron, $RI < \vartheta$. In contrast, if the mean external current is larger than the threshold, $RI > \vartheta$, then the neuron fires regularly with a constant interspike interval defined as the time between spikes as shown in Fig. 3.2B.

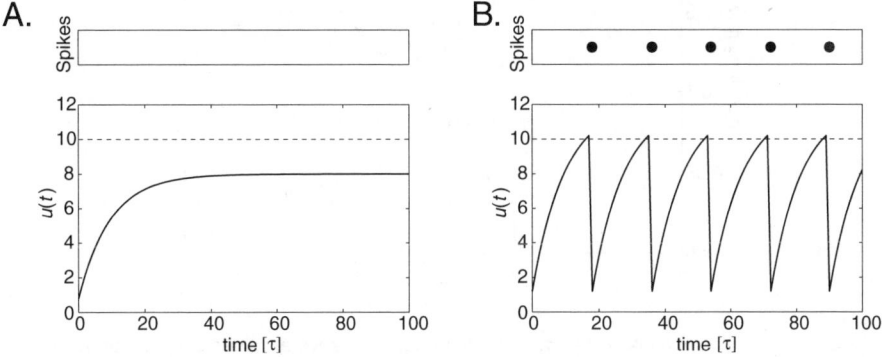

Fig. 3.2 Simulated spike trains and membrane potential of a leaky integrate-and-fire neuron. The threshold is set at $\vartheta = 10$ and indicated as a dashed line. (A) Constant input current of strength $RI = 8$, which is too small to elicit a spike. (B) Constant input current of strength $RI = 12$, strong enough to elicit spikes in regular intervals. Note that we did not include the form of the spike itself in the figure but simply reset the membrane potential while indicating that a spike occurred by plotting a dot in the upper figure.

3.1.4 Gain function

The time between spikes, also called the *first passage time*, can be calculated for a constant input current from the solution 3.9. We can therefore calibrate the time scale so that the last spike is at $t = 0$, at which time the membrane potential is reset, $u(t = 0) = u_{res}$. The time t^f is given by the time when the membrane reaches the *firing threshold*, $u(t^f) = \vartheta$, the solution which is given by

$$t^f = -\tau_m \ln \frac{\vartheta - RI}{u_{res} - RI}. \tag{3.10}$$

The inverse of this time defines the *firing rate*, $\bar{r}$, and is given by

$$\bar{r} = (t^{ref} - \tau_m \ln \frac{\vartheta - RI}{u_{res} - RI})^{-1}, \tag{3.11}$$

where we included an absolute refractory time t^{ref}. This function is the *activation* or *gain* function of this IF-neuron and is illustrated in Fig. 3.3 for several values of the reset potential u_{res} and absolute refractory time t^{ref}. This function quickly reaches an asymptotic linear behaviour after a quick onset with external currents exceeding the threshold value. A threshold-linear function is therefore often used to approximate the gain function of IF-neurons.

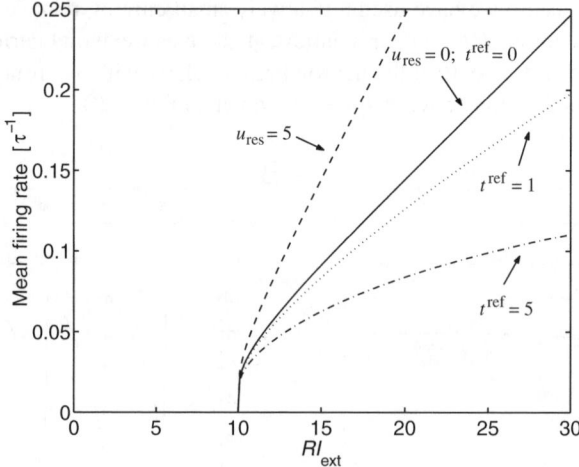

Fig. 3.3 Gain function of a leaky integrate-and-fire neuron for several values of the reset potential u_{res} and refractory time t^{ref}.

The gain function of a neuron can be measured experimentally in a similar way by applying constant external currents of different magnitude to an isolated neuron in a *Petri dish*. Such *in vitro* recordings can be used to examine the principal neuronal parameters that we need to know for biologically faithful simulations.

3.2 The spike-response model

In the last section we integrated the equation describing simple IF-neurons for constant currents driving the neuron, and we calculated the response of the neuron to a very short current pulse. When thinking about information transmission in neurons we are, of course, much more interested in time-varying input currents such as generated by presynaptic spike trains. Actually, we can formally integrate the dynamic differential equation (3.1) for any time-dependent inputs. To show this we utilize our previous result of the response of the neuron to a very short external current pulse. The idea of the solution is to write an arbitrary external current stream $I(t)$ as a collection of many very short pulses modulated by the strength of the input signal at the particular time. The response to each current pulse is an exponential function as we have seen before. The general solution can thus be expressed as a sum over all the exponential responses to very short current pulses. The sum is actually an integral because we have a term for each infinitesimal time step. The time course of a IF-neuron in response to an arbitrary input current $I(t)$ can thus be written as

$$u(t) = R \int_0^\infty e^{-s/\tau_m} I(t-s) \mathrm{d}s. \tag{3.12}$$

The integration variable s stands for all the times that precede the time t, and current pulses at such times influence the membrane potential at time t by an amount that only depends exponentially on the time distance s. This means that more recent spikes have a larger influence on the membrane potential than more distant spikes.

The input current $I(t)$ can be generated by electrodes penetrating the neuron, and we could apply different forms of this current to study the response of the neuron. We are, of course, also interested in the currents generated by presynaptic firing. We therefore develop equations specifically for synaptic inputs in the following. In addition, we have to take into account the resetting mechanisms in the IF-neurons, which we ignored in eqn 3.12. We will account for the response in the membrane potential following a presynaptic spike and the change in the membrane potential following a postsynaptic spike by using two separate terms that we label with ϵ and η, respectively. We therefore split the description of the membrane potential into the components[12]

$$u(t) = \sum_j \sum_{t_j^\mathrm{f}} w_j \epsilon(t - t^\mathrm{f}, t - t_j^\mathrm{f}) + \sum_{t^\mathrm{f}} \eta(t - t^\mathrm{f}). \tag{3.13}$$

We have thereby summed over several ϵ terms, one for each synapse multiplied by the corresponding weight value. The change in the membrane potential described by the ϵ term can depend in general on the last postsynaptic spikes t^f, for example, if the ion channels are voltage-dependent. The main feature encoded in this epsilon term is, however, the influence of the postsynaptic membrane potential on the firing history of the individual presynaptic spikes t_j^f as determined by eqn 3.12. We can substitute into this part the specific form of the input current generated by a synaptic event that we parametrized by the α-function in the last chapter. This can thus be written as

$$\epsilon(t - t_j^\mathrm{f}) = R \int_0^\infty e^{-s/\tau_m} \alpha(t - t_j^\mathrm{f} - s) \mathrm{d}s \tag{3.14}$$

for a synaptic input at synapse j.

---

[12] The membrane potential of the leaky integrate-and-fire model can be expressed as the sum of the individual potentials η and ϵ due to the linearity of the differential equation (3.1).

The second term in eqn 3.13 (the η term) describes the reset of the membrane potential after a postsynaptic spike has occurred. We can write this in a form similar to that of the expression for the synaptic response by including a short negative delta current with the strength of the threshold,

$$RI_{res} = -\vartheta\delta(t - t^f).$$ (3.15)

The integral 3.12 for this special reset current is given by

$$\eta(t - t^f) = -\vartheta e^{-(t-t^f)/\tau_m}.$$ (3.16)

The firing time t^f, written without an index, is again the firing of the postsynaptic neuron given by

$$u(t^f) = \vartheta,$$ (3.17)

in contrast to the presynaptic neuron firing time that have an index, t_j^f. Note that the function η and ϵ can often be given analytically for particular choices of reset mechanisms and α-functions, respectively. Some examples are summarized in Table 3.1. The integrated form of the IF-neurons, defined by eqns 3.17 and 3.13, is called the *spike-response model*. We will use this model to derive descriptions of the average behaviour of populations of neurons in the next chapter.

3.3 Spike time variability

Neurons in the brain do not fire regularly but seem extremely noisy. Neurons that are relatively inactive emit spikes with low frequencies that are very irregular. Also high-frequency responses to relevant stimuli are often not very regular. Single-cell recordings transmitted to a speaker sound very much like the irregular ticking of a Geiger counter when exposed to radioactive material. A histogram of the interspike intervals of one cortical cell is shown in Fig. 3.4A. This prefrontal cell fired around 15 spikes/s without noticeably task-relevant pattern. The interspike interval distribution shows a large variability. The firing pattern of this cortical cell is therefore not well described with the constant interspike intervals generated by the simple IF-neurons studied in the last section. A convenient measure of the variability of spike trains is the *coefficient of variation*, which is defined by the ratio of the standard deviation σ and the mean μ,

$$C_V = \frac{\sigma}{\mu}.$$ (3.18)

This value is $C_V \approx 1.09$ for the cell data shown in Fig. 3.4A. Recordings in the brain often show a high value of variability such as $C_V \approx 0.5–1$ for regularly spiking neurons in V1 and MT as pointed out by Softky and Koch (1993). This is consistent with the finding that spike trains are often well approximated by the Poisson process, which has a coefficient of variation of $C_V = 1$. The regular IF-neuron of the last section has an inconsistent value of $C_V = 0$.

Table 3.1 Examples of spike response functions (right column) describing the membrane potential in response to the corresponding forms of PSPs (left column) used in simulations and analytical models. t^f is the firing time of the presynaptic neuron and t^d is the duration of the α-pulse in the second example. τ^m and τ^s are time constants.

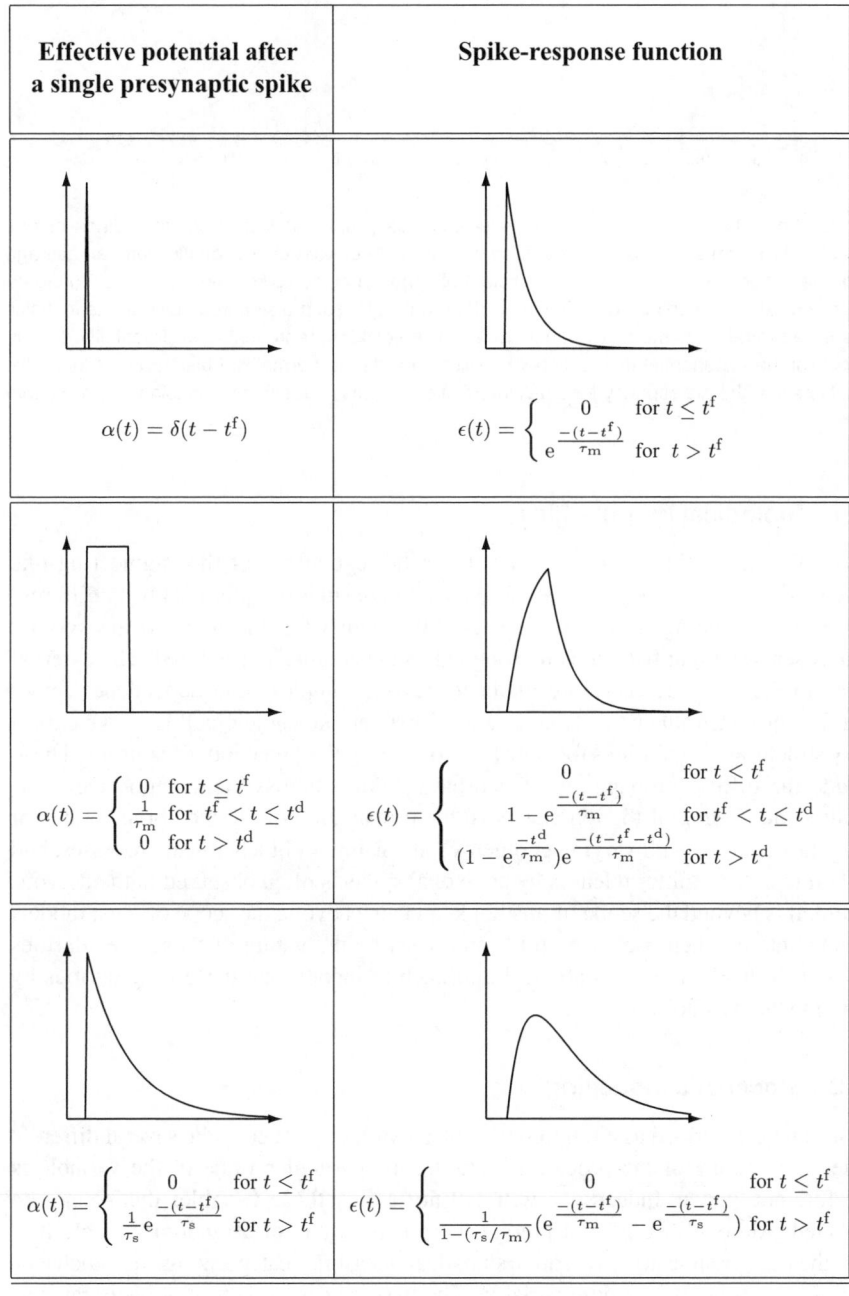

Effective potential after a single presynaptic spike	Spike-response function
$\alpha(t) = \delta(t - t^f)$	$\epsilon(t) = \begin{cases} 0 & \text{for } t \le t^f \\ e^{\frac{-(t-t^f)}{\tau_m}} & \text{for } t > t^f \end{cases}$
$\alpha(t) = \begin{cases} 0 & \text{for } t \le t^f \\ \frac{1}{\tau_m} & \text{for } t^f < t \le t^d \\ 0 & \text{for } t > t^d \end{cases}$	$\epsilon(t) = \begin{cases} 0 & \text{for } t \le t^f \\ 1 - e^{\frac{-(t-t^f)}{\tau_m}} & \text{for } t^f < t \le t^d \\ (1 - e^{\frac{-t^d}{\tau_m}})e^{\frac{-(t-t^f-t^d)}{\tau_m}} & \text{for } t > t^d \end{cases}$
$\alpha(t) = \begin{cases} 0 & \text{for } t \le t^f \\ \frac{1}{\tau_s} e^{\frac{-(t-t^f)}{\tau_s}} & \text{for } t > t^f \end{cases}$	$\epsilon(t) = \begin{cases} 0 & \text{for } t \le t^f \\ \frac{1}{1-(\tau_s/\tau_m)}(e^{\frac{-(t-t^f)}{\tau_m}} - e^{\frac{-(t-t^f)}{\tau_s}}) & \text{for } t > t^f \end{cases}$

A.   ISI-histogram from cell data

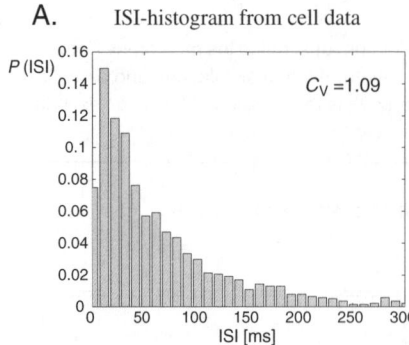

B.   ISI-histogram from Poisson spike train

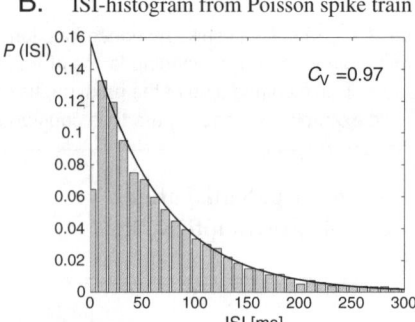

Fig. 3.4 Normalized histogram of interspike intervals (ISIs). (A) Data from recordings of one cortical cell (Brodmann's area 46) that fired without task-relevant characteristics with an average firing rate of about 15 spikes/s. The coefficient of variation of the spike trains is $C_V \approx 1.09$ [data courtesy of Stefan Everling]. (B) Simulated data from a Poisson distributed spike trains in which a Gaussian refractory time has been included. The solid line represents the probability density function of the exponential distribution when scaled to fit the normalized histogram of the spike train. Note that the discrepancy for small interspike intervals is due to the inclusion of a refractory time.

3.3.1 Biological irregularities

Of course, biological networks do not have the regularities of the engineering-like designs of the IF-neurons. We therefore have to consider irregularities from different sources in the biological nervous system. To begin with, the input to the system, such as sensory input from the environment, is never simply structured. The external inputs to the cells are therefore likely to have a complex time dependence rather than the constant value considered so far. Moreover, on a single-cell level we expect many structural irregularities to contribute to an irregular behaviour of neurons. These include the diffuse propagation of neurotransmitters across the synaptic cleft, the opening and closing of the ion channels, the propagation of the membrane potential along the dendrites with varying geometries, the nature of biochemical processes, and the failure of transmitter releases by an axonal spike as often observed in the nervous system. It is beyond the scope of this book, and also beyond the scope of most models in computational neuroscience, to try and describe the nature of these irregularities in detail. Instead we use a statistical approach to incorporate these irregularities by including them as noise.

3.3.2 Stochastic modelling

Noise can be described as a random variable, a variable that can take several different values. The nature of the process generating the particular value of the variable is therefore not known. Indeed, we will call noise only those variables that cannot be predicted. Although we cannot predict the precise value of a random variable it is nevertheless possible to give some statistical measures categorizing the stochastic process generating the random variable. The best we can do is to plot histograms as we have done for the interspike histograms shown in Fig. 3.4A. If we could have an infinite amount of data we could use an infinitesimally small bin width in the histograms. The normalized version of such a histogram, so that the sum or integral over all bin values is one, is called the probability distribution of this random variable. The unique knowledge of the probability distribution of a random variable, more precisely called the *probability density function* (pdf) (see Appendix B), uniquely defines all the statistical measures we could ever hope to measure, such as the mean, the variance, and all the higher moments of the distribution. [13]

3.3.3 Normal distribution

All kinds of probability distributions are possible and many different ones are observable in nature. However, many random processes observed in nature are remarkably well approximated by the Gaussian bell curve illustrated in Fig. 3.5A.

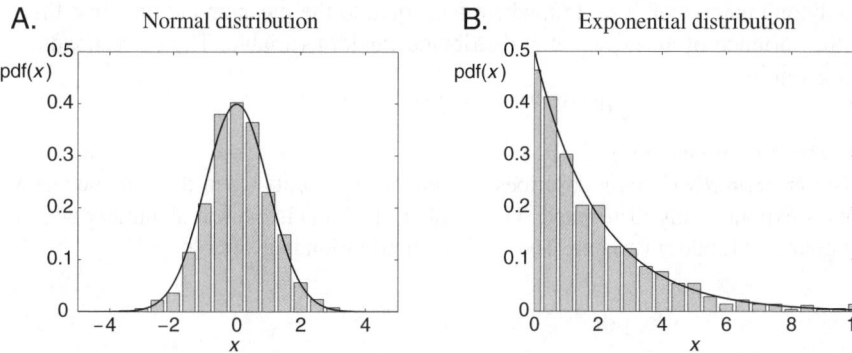

Fig. 3.5 A normalized histogram of 1000 random numbers and the functional form of the corresponding probability distribution functions (pdfs). (A) Random variables from a normal distribution (Gaussian distribution with mean $\mu = 0$ and variance $\sigma = 1$). The solid line represents the corresponding pdf (eqn 3.19). (B) Exponential distribution with mean $b = 2$ (eqn 3.20).

This important probability distribution, given by

$$\mathrm{pdf}^{\mathrm{normal}}(x; \mu, \sigma)(x) = \frac{1}{\sqrt{2\sigma}} e^{(x-\mu)^2/2\sigma^2}, \tag{3.19}$$

is called a *normal distribution* or *Gaussian distribution*, often also denoted by $N(\nu, \sigma)$, where ν is the mean of the distribution and σ the variance. A normally distributed random variable with mean $\nu = 0$ is also called *standard normal distributed* or *white noise*. The reason behind the importance of this particular probability distribution is given by the *central limit theorem* (see also Appendix B). It states that a random variable, which is itself an infinite sum of random variables that all have the same arbitrary probability distribution, has a probability distribution described by the Gaussian

---

[13] The mean is the first moment of the distribution. The variance, which is the second moment of the distribution about the mean, describes how wide the distribution is. The following higher moments describe further characteristics of the distribution such as how symmetric the distributions are around the mean. All moments are required to specify the probability density function uniquely.

bell curve. The central limit theorem assumes a sum of an infinite number of equally distributed random variables. In contrast, physical processes are typically driven by only a limited number of random variables, which, in addition, do not necessarily all have the same probability distribution. Nevertheless, in practice it turns out that the normal distribution is often a good approximation even without fulfilling the precise conditions of the central limit theorem. White noise is therefore often assumed to describe fluctuations around a known mean. However, keep in mind that the normal distribution is not the only distribution observed in nature.

3.3.4 Poisson process

The distribution of the interspike intervals shown in Fig. 3.4 is certainly not well approximated by a normal distribution. Instead they show an exponential decaying tail after a rapid onset determined by the refractory period. The exponential distribution has only one parameter, $\lambda = 1/b$, where b is equal to the mean and at the same time also the variance of an exponential distributed random variable. The corresponding pdf is given by

$$\mathrm{pdf}^{\mathrm{exponential}}(x; \lambda)(x) = \lambda e^{-\lambda x}. \tag{3.20}$$

An example is shown in Fig. 3.5B.

The *Poisson distribution* describes the number of events when the time between events is exponentially distributed. The number of events is a discrete number which is, of course, a random variable. The Poisson distribution is given by

$$\mathrm{pdf}^{\mathrm{Poisson}}(x; \lambda)(x) = \sum_{i=1}^{x} \lambda^i \frac{e^{-\lambda}}{i!}. \tag{3.21}$$

We have characterized the Poisson distribution as the number of events with exponentially distributed interevent times. The exponential distribution and the Poisson distribution can be based on the same process and only count different things in such a process (number of events versus interevent intervals). A *Poisson process* is often applied for generating spike trains when studying nonspecific input to a neural system because the interspike intervals of measured spike trains are often well approximated by an exponential distribution. In contrast, white noise (Gaussian-distributed noise) is often assumed for stochastic processes within neurons. It should always be verified that the scientific arguments derived using such models do not critically depend on the form of noise used in the simulations when no further verifications of the noise models can be given.

3.4 Noise models for IF-neurons

How can we include noise in the neuron models to describe some of the stochastic processes within neuronal responses? For simplicity, we will discuss this for the IF-model, although similar procedures can be applied to other models. We want to determine the stochastic firing times of neurons; in order to do so there are three principal ways in which to include a stochastic component in the IF-model. This is illustrated in Fig. 3.6.

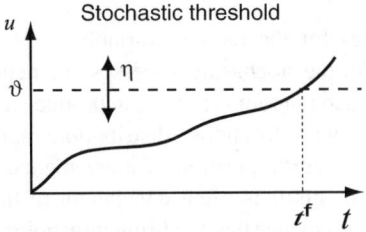

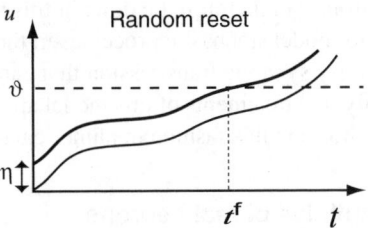

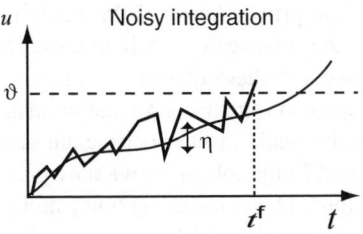

Fig. 3.6 Three different noise models of integrate-and-fire neurons. (A) Stochastic threshold, (B) random reset, and (C) noisy integration [adapted from W. Gerstner, in *Pulsed neural networks*, Maass and Bishop (eds), MIT Press, 1998].

The noise models commonly used for IF-neurons can be summarized as:

1. *Stochastic threshold*: We can replace the threshold that the membrane potential has to pass in order to generate a spike with a noisy threshold

$$\vartheta \to \vartheta + \eta^{(1)}(t) \tag{3.22}$$

2. *Random reset*: We can reset the membrane potential to a stochastic reset potential

$$u^{\text{res}} \to u^{\text{res}} + \eta^{(2)}(t) \tag{3.23}$$

3. *Noisy integration*: The integration mechanisms in the neuron can be noisy such that the leaky integrator of the IF-neurons may be better described by the stochastic differential equation (which is equivalent to saying that the integrator is not noisy but it integrates noisy inputs)

$$\tau_{\text{m}} \frac{du}{dt} = -u + RI_{\text{ext}} + \eta^{(3)}(t) \tag{3.24}$$

With appropriate choices for the random variables $\eta^{(1)}, \eta^{(2)}$, and $\eta^{(3)}$ we can produce equivalent results for the stochastic processes of a neuron, although the same probability distribution for each random variable can produce different results for each noise model. In practice we want to choose distributions appropriate for capturing experimental data, and it is more a question of convenience which noise model to choose. For analytical treatments it is often convenient to use a random threshold model. Although there is less evidence that the firing thresholds of real neurons change over time, this noise model is equivalent to the other noise models and modelling of stochastic processes in neurons can therefore be done in this way. Numerical studies frequently use noisy input to model stochastic processes in the brain. This model has a simple interpretation of noisy synaptic transmission that can indeed be observed in real neuronal systems. Analytical treatments of this model are difficult. It is, however, straightforward to integrate noise in this fashion in numerical studies as shown below.

3.4.1 Simulating variabilities of real neurons

While the choice of the noise model depends only on convenience, the important remaining question is the appropriate choice of the random process, including the probability distribution, and the time scale on which those fluctuations are relevant. We cannot give general answers to these questions as the particular choice depends strongly on the nature of the question and the particular brain area we are investigating. Appropriate choices have to be made to fit experimental data whenever the precise form of fluctuations is relevant. In the following we only give a flavour of the effects of including noise in the IF model by considering noisy input (noise model 3). In the following example we use a normally distributed input current produced by adding white noise to a constant input current,

$$I_{\text{ext}} = \bar{I}_{\text{ext}} + \eta \quad \text{with} \quad \eta \in N(0,1). \tag{3.25}$$

Normally distributed input signals are a good approximation when we consider independent synaptic input from many equally distributed neurons as stated by the central limit theorem. The interspike interval (ISI) distribution of an IF-neuron with mean input current $\bar{I}_{\text{ext}} = 12$ and threshold $\vartheta = 10$ is shown in Fig. 3.7.

This distribution is well approximated by a *lognormal distribution*

$$\text{pdf}^{\text{lognormal}}(x; \mu, \sigma) = \frac{1}{x\sigma\sqrt{2\pi}} e^{\frac{-(\log(x)-\mu)^2}{2\sigma^2}}. \tag{3.26}$$

A fit of the data to this distribution is shown as a solid line in Fig. 3.7. The coefficient of variation of these data is $C_V \approx 0.42$, approaching the lower end of cortical variability of interspike intervals. This is remarkable as we considered only one source of noise. We can, for example, consider this noise as resulting from the noisy internal mechanisms of the integration within the neuron, and could then consider in addition noisy input from the variability in the input spike trains itself. There are several other ways to increase the variability of spikes in the simple IF-neurons, such as using partial and noisy reset after a spike has occurred. This effectively increases the gain in the relation between input current and output spike frequency and has been argued to describe very well the variabilities seen in experiments.

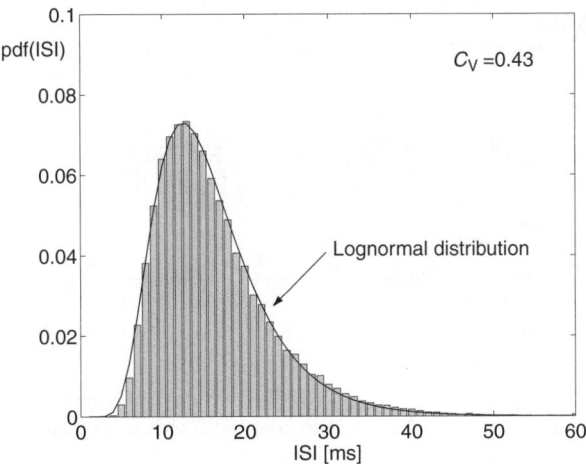

Fig. 3.7 Simulated interspike interval (ISI) distribution of a leaky IF-neuron with the threshold $\vartheta = 10$ and time constant $\tau_{\mathrm{m}} = 10$. The underlying spike train was generated with noisy input around the mean value $RI = 12$. The fluctuation were therefore distributed with a standard normal distribution. The resulting ISI histogram is well approximated by a lognormal distribution (solid line). The coefficient of variation of the simulated spike train is $C_{\mathrm{V}} \approx 0.43$.

3.4.2 Input spike trains

The last example we want to discuss is an IF-neuron that is driven by independent presynaptic spike trains with exponential interspike intervals (Poisson spike trains). Results from such simulations are shown in Fig. 3.8. There we have taken as input 500 Poisson-distributed spike trains with refractory corrections (see Section 12.2.5 for details). The mean firing rate of the input neurons was set to 20 Hz, which was only lowered by a small amount to 19.3 Hz due to the Gaussian refractory time with 2 ms time constant. We took only 500 independent presynaptic spike trains into account as they should represent only the fraction of presynaptic neurons that is active in a particular task (for example, 10% of the inputs to a neuron with 5000 presynaptic neurons). Each presynaptic spike was set to elicit an EPSP in the form of an α-function (eqn 2.1) with amplitude $w = 0.5$ and time constant of 2 ms for all synapses. The synaptic input of all the input spike trains was then large enough to keep the average incoming current, that is, the sum over all EPSPs, larger than the firing threshold of the neuron.

The sum of the EPSPs for the first 1000 ms in this experiment is illustrated by the upper curve in Fig. 3.8A, and the firing threshold is indicated as a dashed line. The average exceeds the firing threshold of the IF neuron, and this in turn results in a regular firing pattern with a high firing rate of around 118 Hz. A constant driving current with a value equal to the mean of the currents produced by many presynaptic spike trains is therefore a good first approximation if we have many inputs to the neuron that produce high postsynaptic responses exceeding on average the threshold of the neuron. The ISI distribution, plotted in Fig. 3.8B, has low variance, and the coefficient of variation of this model neuron is only $C_{\mathrm{V}} = 0.12$. Note that these simulations did not include any noise in the neuron model itself as we discussed before; they only included noise

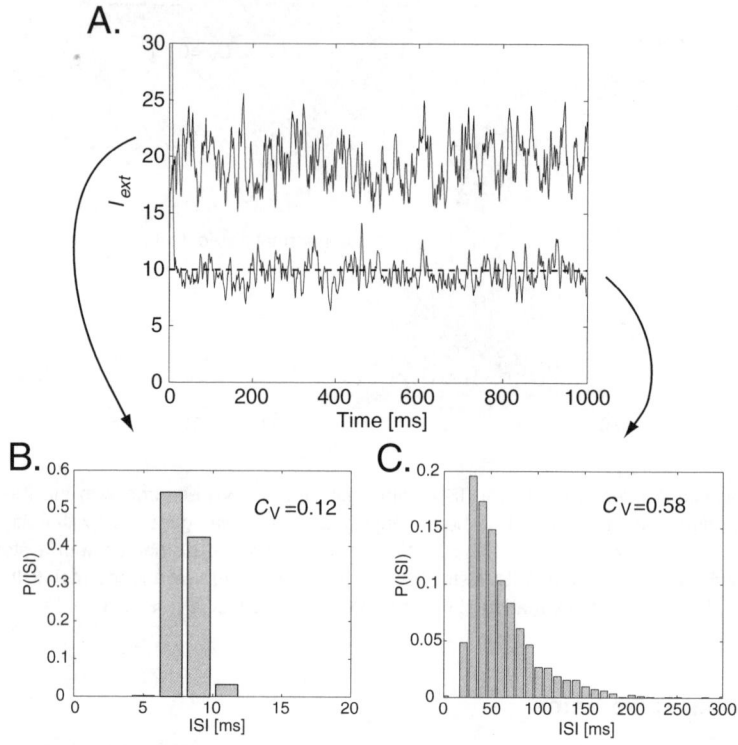

Fig. 3.8 Simulations of a IF-neuron that has no internal noise but is driven by 500 independent incoming spike trains with a corrected Poisson distribution. (A) The sums of the EPSPs, simulated by an α-function for each incoming spike with amplitude $w = 0.5$ for the upper curve and $w = 0.25$ for the lower curve. The firing threshold for the neuron is indicated by the dashed line. The ISI histograms from the corresponding simulations are plotted in (B) for the neuron with EPSP amplitude of $w = 0.5$ and in (C) for the neuron with EPSP amplitude of $w = 0.25$.

in the driving input currents. Noise in the neuron model would further increase the coefficient of variation.

If we lower the effect of each incoming spike by lowering the amplitude of the α-function to $w = 0.25$, then we get an average postsynaptic current, which is shown for the first 1000 ms of the simulation in the lower curve of Fig. 3.8A. The average sum of EPSPs was $I_{\text{ext}} = 9.7$ in this simulation, that is, just below the firing threshold of the neuron. The fluctuation due to the random process of the incoming spike trains is then crucial, and the neuron shows an irregular firing pattern with an average firing rate of 16 Hz. The corresponding ISI histogram is shown in Fig. 3.8C. The coefficient of variation was therefore $C_V = 0.58$, exceeding the lower bounds found in experiments. The question is, of course, how a neuron can adjust the effects of incoming neurons to function in this biologically more realistic regime. It seem rather difficult to achieve the right level of input strength as it seems to require a fine-tuning of the strength parameter by hand. We will discuss this issue further in Chapter 6 where we show that it is possible that synaptic plasticity of the form found in experiments can achieve such fine tuning.

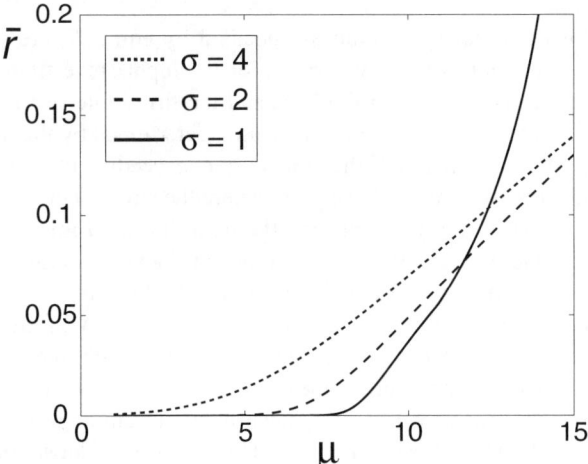

Fig. 3.9 The gain function of an IF-neuron that is driven by an external current that is given a normal distribution with mean $\mu = R\bar{I}$ and variance σ. The reset potential was set to $u_{res} = 5$ and the firing threshold of the IF-neuron was set to $\vartheta = 10$. The three curves correspond to three different variance parameters σ.

3.4.3 The gain function depends on input

A last important point to make is that the activation or gain function of the neuron depends on the variations in the input spike train. This can be seen analytically for an IF-neuron that is driven by a noisy current with normally distributed values. The IF dynamics eqn 3.1 is then a stochastic differential equation as we have argued before in connection with noise model 3. The stochastic differential equations of the leaky IF-neuron can still be solved formally. The first passage time as given in eqn 3.10 is then, of course, a random variable (it is not always the same value). But we can calculate the *mean first passage time* (which is equivalent to the mean interspike interval). These calculations, which we will not review here in detail as it is only important to keep the result in mind, were carried out by *Tuckwell* (see 'Further reading'). He showed that the average firing rate for a stochastic IF-neuron is given by

$$\bar{r} = \left(t^{\text{ref}} + \tau_{\text{m}} \int_{(u^{\text{res}} - R\bar{I}_{\text{ext}})/\sigma}^{(\vartheta - R\bar{I}_{\text{ext}})/\sigma} \sqrt{\pi} e^{u^2} [1 + \text{erf}(u)] du \right)^{-1} \qquad (3.27)$$

where σ is the variance of the Gaussian signal and erf is the error function as described in Appendix B). The firing rate is therefore not only a function of the mean input current, but also depends on higher moments (for example, variance, skewness, etc.) of the distribution describing the input signal. The important result to keep in mind is thus that the mean firing time of an IF-neuron depends on the precise form of the input spike train and not only on the mean firing rate of the inputs, for example,

$$\bar{r} = \bar{r}(\mu, \sigma, ...). \qquad (3.28)$$

Some examples of the mean firing rate of an IF-neuron as a function of the mean firing rate μ of the presynaptic spike train that drives the neuron for different values of the variance of the input spike train as specified by eqn 3.27 are illustrated in Fig. 3.9. The gain function for a low variance of the input spike train has a sharp transition as we have seen before in Fig. 3.3, and the firing rate of the neuron will soon approach the maximal firing rate of the neuron determined by the inverse of the absolute refractory time for means of the input current exceeding the firing threshold. Note also that the effective threshold, the point where the strong increase of the firing rate with external input starts, is lower than the hard threshold imposed on the IF-neuron model. With increasing variance the strong non-linear response is 'linearized', similarly to the linearization by noise in the Hodgkin–Huxley model (see Fig. 2.9). In models that describe average firing rates of neurons, which we will introduce in the next chapter and utilize frequently in the remainder of the book, the dependence of the average firing rate on the structure of the driving signals can be taken into account by activation functions that depend on the mean input current as well as the higher moments of the distribution of the input signals. This is, however, most often neglected in the application of such models in the literature.

Conclusion

We ignored in this chapter most of the detailed processes in the neurons and considered instead simplified neuron models that are designed for the study of information processing in networks of neurons. The simplifications are based on the assumptions that the precise form of the action potential (spike) does not carry information, and that the information is transmitted only by the occurrence of a spike. This assumption is motivated by the stereotyped forms of action potentials, and all the information-processing mechanisms discussed in the remainder of the book rely on this assumption.

A common model of simplified neurons used in network simulations is the integrate-and-fire model or variations thereof. Such models are comprised of a subthreshold leaky-integrator dynamic, a firing threshold, and a reset mechanism. We demonstrated that the variability in the firing times of such model neurons is often much less than the variability in real neurons, and we outlined some noise models that can be incorporated into the deterministic models to increase the variability in neuronal response. However, a fine tuning of neural parameters so that the average presynaptic current is close to the firing threshold of the neuron seems to be necessary to allow a behaviour of the model neurons more closely resembling that of neuron recordings in the brain. We will discuss later mechanisms that could achieve such fine tuning.

Further reading

Wolfgang Maass and Christopher M. Bishop (eds.)

Pulsed neural networks, MIT Press, 1999.

This book includes a clear introduction to the formulation of spiking networks as well as many advanced discussions of networks of spiking neurons. The first introductory chapter by Wulfram Gerstner discusses many models of spiking that we have mentioned in this chapter.

C. T. Tuckwell
Introduction to theoretical neurobiology, Cambridge University Press, 1988.
A detailed analysis of single neurons, cable theory, and stochastic modelling. Very detailed and very mathematical.

Merran Evans, Nicholas Hastings, and Brian Peacock
Statistical distributions, John Wiley & Sons, New York, 3rd edition, 2000.
A useful collection of probability functions. Not an introduction to statistics.

4 Neurons in a network

In this chapter we start exploring networks of neurons. We summarize some anatomical organizations of the neocortex and begin to explore information transmission in chains of simple neurons and random network models. We then discuss population dynamics and introduce simple processing nodes that describe the average firing rate of a population of neurons that we use frequently in the models of the following chapters. The introduction of such nodes is an important step in being able to analyse some important abilities of large networks, and we discuss which domains in the dynamics of networks of spiking neurons are lost with such descriptions. We finally introduce a basic model with the aim of taking into account nonlinear interactions between populations, which may be based on nonlinear interactions between synaptic channels.

4.1 Organizations of neuronal networks

We have discussed some essential features of single neurons in the last two chapters, in particular their ability to integrate presynaptic spike trains and to respond with various firing patterns. There is currently no reason to believe that the higher-order mental abilities of primates are not based on these basic signal-processing mechanisms. However, mental functions are certainly not accomplished by single neurons alone. Instead, we speculate that these functions are an emerging property of specialized networks with many neurons that form the nervous system. The number of neurons in the central nervous system is estimated to be on the order of 10^{12}, and it is certainly demanding to explore such vast systems of neurons. Therefore, rather than trying to rebuild the brain in all its detail on a computer, we aim to understand the principal organization of neuron-like elements and how such structures can support and enable particular mental processes, for example, perception and the learning of motor skills. Integration of neurons into networks with specific architectures seem to be essential for such skills. We will explore the computational abilities and specialities of several principal architectures of neural networks in this book.

A thorough knowledge of the anatomy of the brain areas we want to model is essential for any research that attempts to understand brain functions. Although recent research has revealed many important facts about neural organization, it is still often difficult to specify all the components of a model on the basis of anatomical and physiological data alone, and plausible assumptions have to be made to bridge gaps in the knowledge. Even if we can draw on known details, it might, nevertheless, be useful to make simplifying assumptions that enable computational tractability or the tracing of principal organizations sufficient for certain functionalities. It is beyond the scope of this book to describe all the details of neuronal organization, and more specialized books and research articles have to be consulted for specific brain areas under investigation. We will outline as examples here some principal organizations in the *neocortex*, the outer layer of the *cerebral cortex* often simply referred to as *cortex*, and discuss other large-scale organizations of the brain and some organizations of specific brain areas such as the *cerebellum* and some *subcortical areas* later in this book.

4.1.1 Neocortical organization

Regions of the neocortex are commonly divided into four *lobes* as illustrated in Fig. 4.1, the *occipital lobe* at the rear of the head, the adjacent *parietal lobe*, the *frontal lobe*, and the *temporal lobes* at the flanks of the brain. Further subdivisions can be made, based on various criteria. For example, the German anatomist Korbinian Brodmann identified at the beginning of the 20th century 52 cortical areas based on their cytoarchitecture, the distinctive occurrence of cell types and arrangements, which can be visualized with various staining techniques outlined below. Brodmann labelled the areas he found with numbers as shown in Fig. 4.1. Some of these subdivisions have since been refined, and letters following the number are commonly used to further specify some part of an area defined by Brodmann. *Brodmann's cortical map* is, however, not the only reference to cortical areas used in neuroscience. Other subdivisions and labels of cortical areas are based, for example, on functional correlates of brain areas. These include behavioural correlates of cortical areas as revealed by brain lesions or functional brain imaging, as well as neuronal response characteristics identified by electrophysiological recordings.

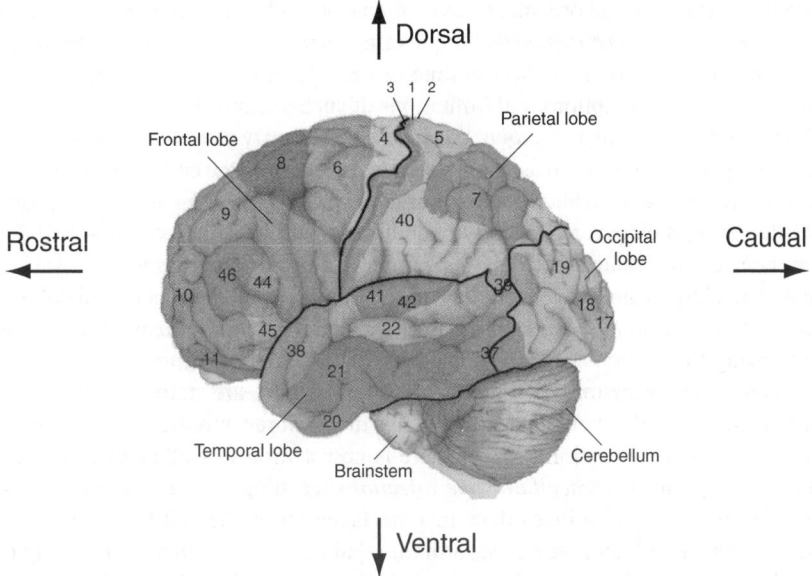

Fig. 4.1 Outline of the lateral view of the human brain including the neocortex, cerebellum, and brainstem. The neocortex is divided into four lobes. The numbers correspond to Brodmann's classification of cortical areas. Directions are commonly stated as indicated in the figure [adapted from Bear, Connors, and Paradiso, *Neuroscience: exploring the brain*, Williams and Wilkins (1996)].

It is, of course, of major interest to establish functional correlates of different cortical areas, a challenge that drives many physiological studies. We might speculate that the diverse functional specialization within the neocortex found with electrophysiological measurements is reflected in major structural differences among the different cortical areas. It is therefore remarkable to realize that this is not the case. Instead it is found that different areas of the neocortex have a remarkably common neuronal organization that includes anatomically distinguishable layers and functionally distinguishable columns as we will discuss in more detail below. The differences in the cytoarchitecture, which have been used by Brodmann to map the cortex, are often only minor compared to the principal architecture within the neocortex, and these variations cannot account solely for the different functionalities associated with the different cortical areas.

The neocortex is different in this respect from some older parts of the brain such as the brainstem where structural differences are much more pronounced. This is reflected in a variety of more easily distinguishable nuclei. We can often attribute specific low-level functions to each nucleus in the brainstem. In contrast to this, it seems that the cortex is an information-processing structure with more universal processing abilities that we speculate enable more flexible mental abilities. It is therefore most interesting to investigate the information-processing capabilities of neuronal networks with the common architecture of the neocortex.

4.1.2 Staining techniques

Knowledge of the neuronal organizations is often derived from studies of stained neo-cortical slices. These experimental techniques are crucial for estimating the parameters on which models are based. A short outline of these techniques is hence appropriate to make clear their assumptions and limitations of such techniques.

Many different staining techniques can be used to identify neurons or parts thereof. Some staining techniques, such as the *Nissl stain*, colour only the cell body and cannot be used to investigate dendritic or axonal organizations. The *Golgi stain*, based on a silver solution, was discovered by the Italian histologist *Camillo Golgi* (1843–1926), and can be used to visualize more parts of the neuron than those accessible by Nissl staining. The Golgi staining method was used extensively by the Spanish histologist *Santiago Ramón y Cajal* (1852–1934) to investigate the circuitry of many brain areas. When viewing illustrations of such stained tissues it is important to know that only a small percentage of neurons, on the order of only 1–2%, are stained by the Golgi staining method, and different neurons can have different receptivities to this stain.

In addition to these traditional dyes there is now a variety of other staining techniques including direct *intracellular dye injections* reaching most parts of a neuron, *anterograde staining* that utilizes dyes that are taken up by the cell body and transported down the axons, and *retrograde staining* that utilizes dyes that are taken up by the terminal endings of axons and transported back to the cell body. The former two staining techniques can be used to identify the projection range of neurons, and the latter is useful to highlight the neurons that project into a particular brain area. Mastering such techniques and applying them carefully to get estimates of neuronal populations and dendritic or axonal organizations is a specialization within neuroscience on which computational neuroscientists rely heavily in order to develop biologically faithful models. The appearance of neocortical slices visualized by different staining techniques is illustrated in Fig. 4.2B.

4.1.3 Common neuronal types in the neocortex

Staining techniques can be used to identify neurons. The most abundant neocortical neurons are categorized into *pyramidal neurons* and *stellate neurons*. Pyramidal neurons, of which an example is given in Fig. 2.1, are by far the most abundant neuronal type in the neocortex, accounting for as much as 75%–90% of the neurons therein. They are characterized by a pyramid-shaped cell body with an axon usually extending from the base of the pyramidal cell body (*basal base*). The dendrites of the pyramidal cell can be divided into a far-reaching dendritic tree extending from the apical tip of the cell body into the upper layer of the cortex (see discussion below), and the basal dendrites with more local organizations. These neurons develop synapses to other neurons, which have asymmetric appearances and are thought to be excitatory.

The two other classes of neuronal types that are common in the neocortex have star-like appearances and are therefore called *stellate neurons*. Dendrites of *spiny stellate neurons* (as well as dendrites of pyramidal cells) are covered with many boutons that are called *spines*. Typically, spiny stellate neurons have an excitatory effect on other cells. In contrast, the *smooth stellate neurons* are not covered with many spines and commonly utilize GABA as their neurotransmitters. They are hence thought to be inhibitory neurons forming mainly symmetric synapses.

It should be clear that the division of neocortical neurons into pyramidal and stellate neurons is only a rough categorization of neuronal types in the neocortex. Other divisions can be made, although the distinction between different classes is often not easy to make on their spatial appearance alone. Some neurons that have been given other names include *Martinotti cells*, which are found mainly in the deeper layers of the neocortex and grow axons that extend commonly into the outer layer, *basket cells*, which are smooth stellate cells that synapse preferentially on the cell body of pyramidal cells, and *chandelier cells*, which synapse preferentially on the initial segment of pyramidal axons.

4.1.4 The layered structure of neocortex

Staining of cell bodies reveals a generally layered structure of the neocortex as illustrated in Fig. 4.2A. The neocortex is historically divided into six layers labelled with Roman numerals from I to VI, although more than six layers, commonly 10 including the white matter, can be identified and are included into the historical labelling scheme by further subdivisions. Layer IV is thereby subdivided into IVA, IVB, and IVC, and layer IVC is further subdivided into layers IVCα and IVCβ. The extent, or thickness, of the layers varies throughout the neocortex up to a point where some layers are difficult to identify if not absent. Examples of stained slices from different areas within the neocortex are shown in Fig. 4.2C. The visual appearance in the stained slices defining the layers is dependent on different populations of cell bodies and neurites. Layer I is easily distinguishable as it is mainly lacking in cell bodies and consists mainly of neurites. The other layers are marked by the domination of different cell types.

The soma of several neuronal types can be found in each neocortical layer, although

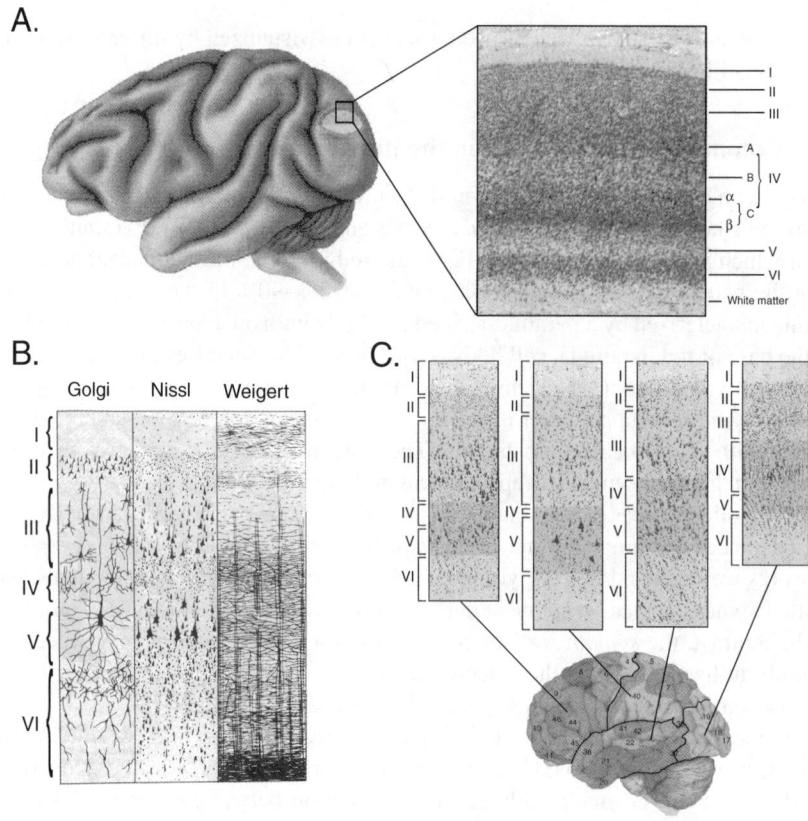

Fig. 4.2 Examples of stained neocortical slices showing the layered structure in neocortex. (A) Nissl stained visual cortex [from Hubble, *Eye, brain and vision*, W. H. Freeman, (1988)] showing cell bodies [adapted from Bear, Connors, Paradiso, *Neuroscience: exploring the brain*, Williams and Wilkins(1996)]. (B) Illustration of different staining techniques [after Brodmann, adapted from Heimer, *The human brain and the spinal cord*, Springer, 2nd edition (1995)]. (C) Different sizes of cortical layers in different areas [adapted from Kandel, Schwartz, and Jessell, *Principles of neural science*, McGraw-Hill, 4th edition (2000)].

the distribution can be used to mark the layers to some extent. As mentioned above, layer I is nearly completely lacking in cell bodies and consists mainly of neurites. Pyramidal cells can be found in most other layers of the neocortex. Layer II and III consist predominantly of small pyramidal cells, although the cells in layer III tend to be larger than those in layer II. Stellate neurons seem, in particular, concentrated around layer IV. In the upper part of this layer (IVA and IVB) one can find a mixture of medium-sized pyramidal cells and stellate cells, whereas the deeper layer (layer IVC) seems to be dominated by stellate neurons. Large pyramidal cells are found predominantly in layer V. A variety of cell types can be found in the deepest layer, layer VI. This includes the above-mentioned Martinotti cells and also cells that have elongated cell bodies and are sometimes used to mark this layer. Such cells are sometimes called *fusiform neurons*.

4.1.5 Columnar organization and cortical modules

The neuronal organizations in the neocortex discussed so far are mainly based on anatomical evidence. There is, in addition, an important functional organization in the neocortex revealed by electrophysiological recordings. These experiments have shown that neurons in a small area of the cortex respond to similar features of an input stimulus. Hubel and Wiesel have investigated such organizations in the *primary visual* (or *striate*) cortex. Neurons in this cortical area respond to visual bars moving in particular directions. More precisely, neurons in a small cortical column perpendicular to the layers and separated by around 30–100 μm respond to moving bars with a specific retinal position and orientation. These regions are called *orientation columns*. Separate from these arrangements are *ocular dominance columns*, cortical columns that respond preferentially to input from a particular eye (see Fig. 4.3A). The relations of orientation columns and ocular dominance columns are illustrated schematically in Fig. 4.3B. Neurons in small columns in other parts of the cortex also tend to respond to similar stimulus features. For example, cortical columns in the somatosensory cortex each respond to specific sensory modalities such as touch, temperature, or pain. The distribution of neurons with specific response characteristics is hence not purely random in the cortex, but there seems to be some form of organization.

Hubel and Wiesel called a collection of orientation columns representing a complete set of orientations a *hypercolumn*. They showed that adjacent hypercolumns in the striate cortex respond to visual input from adjacent retinal areas as illustrated in Fig. 4.3C. Central regions of the visual field are represented by a larger cortical area than peripheral areas. The mapping between the visual field and the cortical representation is therefore not area-preserving; the central visual area is overrepresented, a feature that is called *cortical magnification*. The mapping is, however, *topography-preserving* in that it preserves the topographic relationships between adjacent points. Topography-preserving cortical representations can also be found in other cortical areas. For example, the somatosensory cortex represents tactile input from different body parts with larger cortical areas dedicated to more sensitive areas as illustrated in Fig. 4.3D.

Although we will often simplify cortical organizations to discuss more general aspects of information processing in the cortex, it is conceptually important that neurons in a small areas of the cortex respond to similar sensory stimuli. We call these areas *cortical modules*. For many models in this book it is sufficient to represent the neurons in this area as a single unit as discussed further below. With such models it is then much easier to explore various brain mechanisms, such as the formation of topographic organization as we will do in Chapter 9. In this chapter we continue to outline cortical organizations and discuss some consequences of such organizations on information processing within the cortex.

4.1.6 Connectivity between neocortical layers

The connectivity pattern within the neocortex is, of course, very important for computational models. Neurons in layer IV seem to receive a particularly large number of afferents through the white matter from subcortical and other cortical areas. This layer is therefore often viewed as an input layer. Layer V has many large pyramidal cells

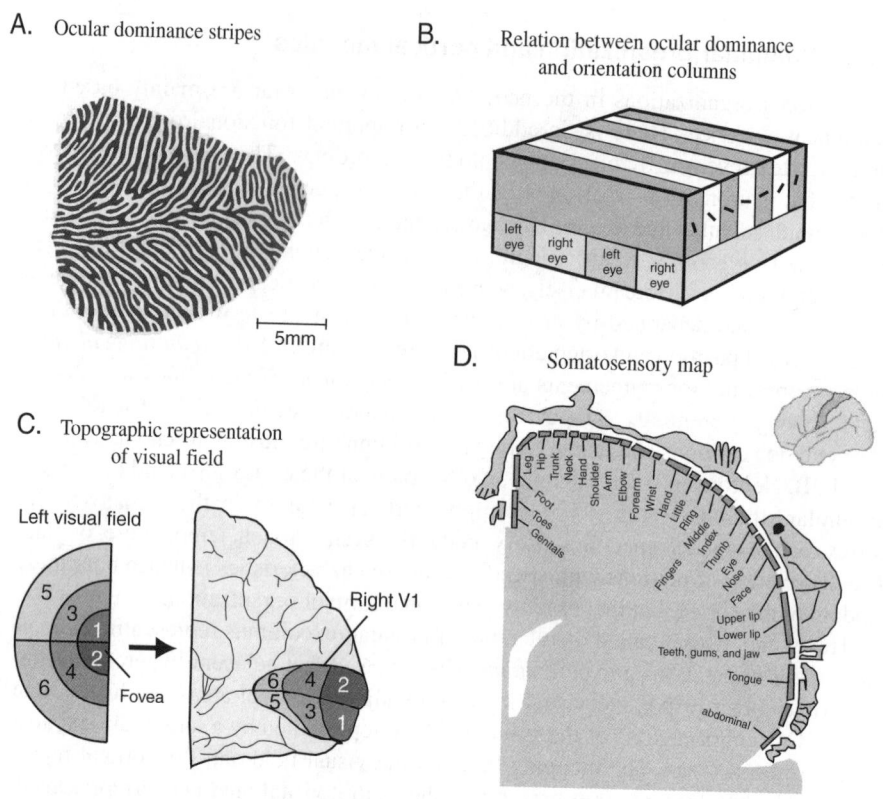

A. Ocular dominance stripes

5mm

B. Relation between ocular dominance and orientation columns

left eye | right eye | left eye | right eye

C. Topographic representation of visual field

Left visual field

5 3 1 2 4 6

Fovea

Right V1

6 4 2
5 3 1

D. Somatosensory map

Leg Hip Trunk Neck Head Shoulder Arm Elbow Forearm Wrist Hand Little Ring Middle Index Thumb Eye Nose Face

Foot Toes Genitals

Fingers

Upper lip
Lower lip
Teeth, gums, and jaw
Tongue
abdominal

Fig. 4.3 Columnar organization and topographic maps in neocortex. (A) Ocular dominance columns [from LeVay, Wiesel, and Hubel, *J. Comp. Neuro.* 191:1–51 (1980)]. (B) Schematic illustration of the relation between orientation and ocular dominance columns. (C) Topographic representation of the visual field in the primary visual cortex. (D) Topographic representation of touch-sensitive areas of the body in the somatosensory cortex [(C) and (D) adapted from Kandel, Schwartz, and Jessell, *Principles of neural science*, McGraw-Hill, 4th edition (2000)].

with axons extending into the white matter. This layer seems therefore to contribute largely to the output of cortical processing. As the white matter is the main pathway between remote cortical areas and in particular between cortical and subcortical areas, it is obvious to suggest that the information flowing through the white matter is, to a large extent, responsible for global information transmission in the brain. In contrast, pyramidal cells in layers II and III are thought to be largely responsible for long-range cortico-cortical tangential connections. Martinotti cells in the deep layers of the neocortex have axons extending into layer I. These could be responsible for information transfer between adjacent cortical modules from which pyramidal neurons in the upper layers receive synaptic input. Stellate neurons, on the other hand, seem to be much more local in their neuritic sphere. The smooth stellate cells are therefore candidates for *inhibitory interneurons*. Their role in the stabilization of cortical processing is an important issue we will discuss in later sections of this book.

An outline of connectivity patterns within a small column of the neocortex is summarized in Fig. 4.4. This scheme is, of course, only a rough approximation of the

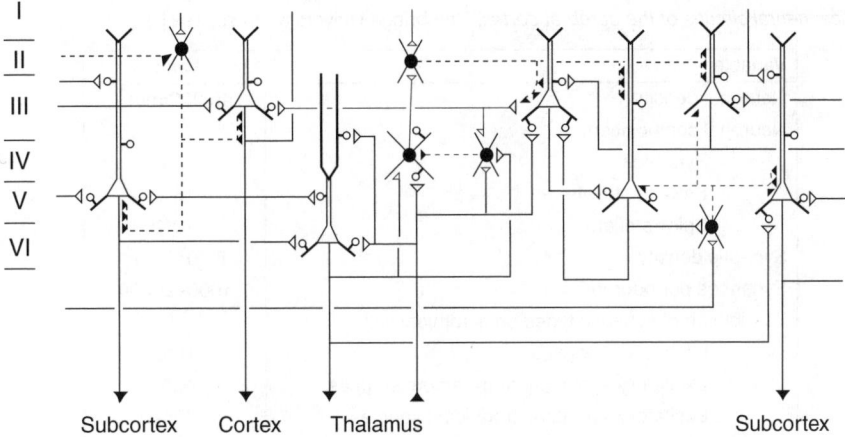

Fig. 4.4 Schematic connectivity patterns between neurons in a cortical layer. Open cell bodies represent (spiny) excitatory neurons such as the pyramidal neuron and the spiny stellate neuron. Their axons are plotted with solid lines that end at open triangles that represent the axon terminal. The dendritic boutons are indicated by open circles. Inhibitory (smooth) stellate neurons have solid cell bodies and synaptic terminal, and the axons are represented by dashed lines. [Adapted from Douglas and Martin, in *Synaptic organization of the brain*, Shepherd (ed.) (1990).]

many details that are known experimentally. More detailed computational studies have still to be performed to understand the functional role of such organizations in more detail.

4.1.7   Cortical parameters

It is not feasible to extract a detailed wiring diagram of the brain. Even an estimation of cortical parameters, such as the number of neurons in a cortical area, the number of connections and their physiological strength, the composition of an area with neuronal types, etc., is often not easy to extract experimentally. In addition, most of the experimental estimations can only be made for particularly favourable cases from which we have to generalize. The generalizations of such experimental studies are often obscured by considerable variations of such numbers within different cortical areas and between different species. Also, the estimation of such parameters varies considerably with different experimental techniques. You might therefore ask yourself how we can build biologically faithful network models of the brain without the necessary experimental support.

The answer is that we have to approach the study of brain networks from different angles. We will consider mainly rather general network architectures and study the general computational capabilities of such networks. These studies reveal, as we will see throughout the book, that many computational abilities of the networks do not depend critically on specific details and are hence present in a large variety of networks within certain classes. Furthermore, we will discuss mechanisms that guide the development and fine tuning of networks to achieve specific computational tasks. We

Table 4.1 Some rough estimates of neocortical parameters (see for example M. Abeles, *Corti-conics: neural circuits of the cerebral cortex*, Cambridge University Press, 1991.)

Variable	Value
Neuronal density	$40,000/mm^3$
Neuronal composition:	
Pyramidal	75%
Smooth stellate	15%
Spiny stellate	10%
Synaptic density	$8\ 10^8/mm^3$
Synapses per neuron	1000–20000
Distribution of synaptic types on pyramidal cell	
Inhibitory synapses	10%
Excitatory synapses from remote sources	45%
Excitatory synapses from local sources	45%
Asynchronous gain (relative synaptic efficiency)	0.003–0.2
Time duration of spike	~ 1 ms
Velocity of spike (myelinated axon of 0.02 mm diameter)	120 m/s
Length of axon	few mm to ~ 1 m
Synaptic cleft	20 nm
Synaptic transmission delay due to diffusion	0.6 ms

therefore approach the study of the brain from the perspective of extracting general principles that guide the organization of the brain as well as revealing the computational consequences of classes of structurally related networks.

To explain brain functions we have, of course, to concentrate on the classes of networks that are consistent with brain networks. Predictions of models therefore have to be tested carefully with experiments on the real brain. Biologically faithful models can also be guided by experimental estimations of general cortical organizations. In Table 4.1 we summarize some rough estimates of neocortical parameters that are good to keep in mind when discussing biologically faithful network models. The values presented only indicate an order of magnitude, which is good to keep in mind when developing very general models, and we will see that it is already instructive to study models with some very crude approximations of cortical organizations. More specific estimations for specific cortical areas that are modelled should, of course, taken into account for more specific studies.

How much of the detail of neocortical organizations is necessary to explain certain brain functions is difficult to assess and has to be considered for each specific question. Some specifics are certainly essential for very detailed explanations, while we can gain a lot of insight into some information-processing principles in the brain from very general organizational principles found in the brain. There are also good reasons to believe that the brain itself has to work within general architectures in contrast to very detailed specific architectures coded, for example, genetically. A mostly accepted hypothesis is that the brain architecture is based on genetically coded organizational principles on which self-organization mechanisms and experience-based learning act to fine-tune the organization to achieve accurate and flexible behaviour.

4.2   Information transmission in networks

It is instructive to consider the general information flow through specific examples of networks and to contrast them with networks with connectivity patterns resembling cortical architectures. We will here follow some thoughts of the neurobiologist *Moshe Abeles* who has analysed cortical connectivity patterns and studied some of the information-processing consequences of such architectures. We start by considering very structured networks such as simple chains of neurons, and consider thereafter random networks with probabilities of synaptic contacts resembling those of cortical organizations.

4.2.1   The simple chain

The simplest network of neurons we can consider is a simple chain of neurons as illustrated in Fig. 4.5A. A spike in the first neuron can propagate through the chain if the strengths of the synaptic connections are sufficient to elicit a postsynaptic spike in each subsequent neuron. Such a mode of information transmission is, however, biologically not reasonable for several reasons. First, a single presynaptic spike is often not sufficient to elicit a postsynaptic spike in the cortex and many other brain areas. At least two or more spikes are often necessary to elicit a postsynaptic spike (see Table 4.1). Secondly, synaptic transmission is lossy. It was estimated that the release of a sufficient amount of neurotransmitters by the arrival of a presynaptic action potential is only generated with a probability not exceeding 0.5 and typically more on the order of 0.25. Thirdly, the death of a single neuron would disrupt the transmission in this chain permanently. Neuronal death is common throughout adult life and is so large that the probability of sustaining complete chains is unrealistically low. The employment of many parallel chains cannot solve the problems realistically as too many redundant chains would be necessary to ensure reliable transmission within the biological parameters mentioned above.

4.2.2   Diverging–converging chains

The restrictive nature of information transmission in simple chains suggests that the firing of a single neuron should not only depend on a single presynaptic neuron but also that a single neuron should transmit a spike to several other neurons. This is reflected in brain networks where a single neuron is contacted by many other neurons and in turn contacts many other neurons. In addition, the convergence and divergence of information flow through the brain enables more interesting information processing as discussed throughout this book. It is therefore instructive to consider *diverging– converging chains* as illustrated in Fig. 4.5B.

   The neurons in each column do not have to be different from neurons in other columns. Rather, the illustration shows the pathways of information flow through time, each column representing a particular step in the information transmission when approximated by a discrete time step scheme. A node that is involved in information transmission at some time when it gets a signal may get recruited again some time later when it gets some feedback from neurons to which it has projected. Neurons in such feedback loops would be repeatedly drawn in this scheme.

   We call the number of neurons in each layer the *width of the chain* and denote it

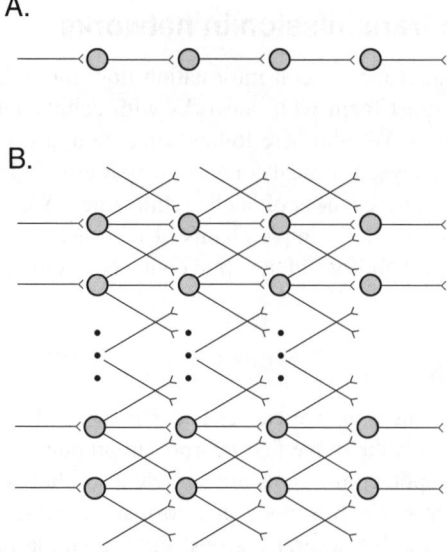

Fig. 4.5 (A) A sequential transmission line of four nodes. Parallel chains are made out of many such sequential transmission lines without connections between them. (B) Diverging/converging chains where each node can contact several other nodes in neighbouring transmission chains.

by N. This number can be equal to the number of neurons in a recurrent network. The number of neurons contacted by each neuron is called the *divergence rate* or *multiplicity* of the chain and denoted by m. In contrast, we will call the number of connections that each neuron receives the *convergence rate* and denote it by C. We will mainly consider the case of equal divergence and convergence rate, $m = C$, for simplicity, although the analysis of networks with different rates is very similar. The case of $N = m = C$ is a *complete transmission chain*, which we will also call a *fully connected network*. A further parameter on which the transmission of spikes will critically depend in such networks is the *synaptic efficiency* or *weight* of the connection, which we will denote with w throughout this book. This is related to the weighting factor of a PSP that we introduced in Chapter 2. The synaptic efficiency times the multiplicity has to be at least on the order of one, $mw > 1$, to enable the transmission of a single spike through this diverging–converging chain.

4.2.3 Immunity of random networks to spontaneous background activity

Cortical neurons typically fire with some background activity that we will assume in the following to be 5 Hz with a variance of 3 Hz. We will also assume that these spike events are independent. Thus, each neuron fires with the probability of $0.005 = 5/1000$ in each time interval of 1 ms. Let us consider a neuron that has 10,000 excitatory dendritic synapses, that is, it receives input from 10,000 spontaneously firing neurons. The sum of independent random presynaptic spikes arriving in each time interval at each single neuron is then normally distributed with mean $\mu = 10000 * 0.005 = 50$ and variance

$\sigma^2 = 0.003$. This follows from the central limit theorem as the sum of N random numbers with mean μ and variance σ^2 is a normally distributed random number with mean $N\mu$ and variance σ^2. The fluctuations of the arriving spikes from the background firing are very small so that at every time step almost exactly 50 spikes will arrive at the postsynaptic neuron. The neuron is immune to the background firing if it has small enough relative synaptic weights. If we consider that a spike arriving at a synapse with weight $w = 1$ would elicit a spike, then the weight would have to be $w < 1/50 = 0.02$ in order for the neuron to be immune against the background firing. We have calculated this value for the average synaptic efficiency when a single presynaptic neuron makes a single synaptic contact with a postsynaptic neuron. In case of multiple synaptic contacts of a single presynaptic neuron with a single postsynaptic neuron we have to divide this value by this multiplicity factor.

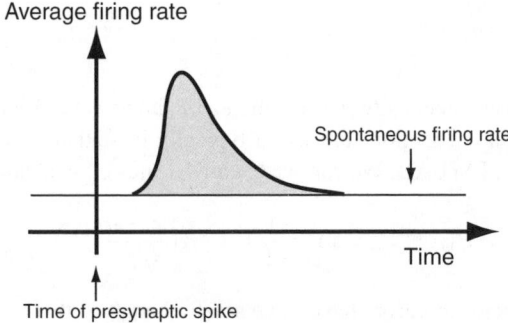

Fig. 4.6 Schematic illustration of the influence of a single presynaptic spike on the average firing rate of the postsynaptic neuron. The delay in the synaptic transmission curve is caused by some synaptic delay, after which, on average, more postsynaptic spikes are generated within a short time window compared to the spontaneous activity of the neurons [adapted from M. Abeles, *Corticonics*, Cambridge University Press, 1991, p. 97].

To compare these values to experimental data we have to consider how to measure the average synaptic efficiency. A way proposed in the literature is to stimulate a presynaptic neuron while recording from the postsynaptic neuron. It is likely that the postsynaptic neuron fires spontaneously, but the influence of the presynaptic spike should be seen in an average increase of the postsynaptic firing rate after many such trials. This is illustrated in Fig. 4.6. The value of the area between the synaptic transmission curve and the spontaneous firing rate of the neuron is called the *asynchronous gain*. This value quantifies the average number of extra spikes that are added to the spikes of a postsynaptic neuron by each presynaptic spike. Reported experimental values of the asynchronous gain of synapses at cortical neurons are often in the range of 0.003–0.2 as mentioned in Table 4.1. The synaptic efficiencies of neocortical neurons could hence indeed be small enough to make cortical neurons to a certain extent immune to the background firing of other neocortical neurons.

4.2.4 Noisy background

We have argued that the variance in the synaptic current from the background firing of other neurons could be very small. However, it is likely that the faulty transmission within the synapses generates a large variability in the effective postsynaptic currents from the background firing. Let us thus discuss the other extreme in which the variance of the sum of presynaptic spikes from the background is on the order of the mean of this firing rate. Taking again the parameters from the previous example but considering now a variance of $\sigma^2 = 50$, we want to calculate again how small the synaptic efficiencies have to be in order for the probability of a postsynaptic spikes generated by the background firing to be less than a certain value p^{BG}. This can be done by calculating the probability of having more than x simultaneous presynaptic spikes. For the Gaussian distribution in our example this is given by

$$P(n^{\mathrm{spikes}} > x) = \frac{1}{\sqrt{2\pi}\sigma} \int_x^\infty e^{-\frac{(y-\mu)^2}{2\sigma^2}} \, dy \tag{4.1}$$

The integral on the right-hand site defines the *error function* (erf) introduced in Appendix B. Values of this functions are listed typically in statistics books, and this is also implemented in MATLAB. We can write eqn 4.1 thus equivalently as

$$P(n^{\mathrm{spikes}} > x) = \frac{1}{2}[1 - \mathrm{erf}(\frac{x-\mu}{\sqrt{2}\sigma})]. \tag{4.2}$$

The synaptic efficiencies therefore have to be small enough so that

$$x = \mu + \sqrt{2}\sigma\mathrm{erf}^{-1}(1 - 2p^{\mathrm{bg}}) \tag{4.3}$$

simultaneous spikes do not elicit a postsynaptic spike. With the values in our example and the probability $p^{\mathrm{bg}} = 0.1$ for the background activity to elicit a spike we get $x \approx 59$, so that the average synaptic efficiencies have to be $w < 1/59 \approx 0.017$. This is still in agreement with experimental values as mentioned before. Synaptic efficiency values only slightly below this value would allow the neuron to be 90% immune against responding to background firing.

4.2.5 Information transmission in large random networks

If we incorporate the stability in response to strongly fluctuating input current from the background activity of other neurons by considering synaptic efficiencies $w < 0.017$, then we have just estimated that we need at least $59 - 50 = 9$ additional presynaptic spikes on top of the average background firing in the cortex to elicit a meaningful postsynaptic spike. Let us consider a large randomly connected network under these conditions, for example, of size 10^{10}, after injecting enough current into 1000 neurons to force them to spike. Each of these 1000 neurons connects to 10,000 other neurons. The spikes are therefore transmitted to nearly $1000 * 10000 = 10^7$ different neurons because the chance of overlaps of receiving neurons is only very small in the randomly connected network. Indeed, we should calculate the probability that a single postsynaptic neuron receives two spikes, one from a separate initially stimulated neuron. Let us select one neuron. The probability that this neuron receives a spike is $10^7/10^{10}$. The probability that this neuron receives two spikes is then $(10^7/10^{10})^2 = 10^{-6}$, a very small probability. It follows that the 1000 initial spikes are, on average, not sufficient to elicit secondary spikes if more than one presynaptic spike is needed to elicit a postsynaptic spike.

The failure of spike transmission through a random network is, of course, a consequence of the small number of connections per neuron relative to the large number of neurons in the network in conjunction with the equal likelihood of each neuron to contact each other neuron in the network. The situation improves in smaller networks. For example, let's consider only a small random network (cortical area) with 1 million (10^6) neurons (each having again 10,000 synapses with other neurons in this set) in which we stimulate 1 neuron; then each neuron has the probability of $10^4/10^6 = 0.01$ of receiving a spike or $1 - 0.01 = 0.99$ of not receiving a spike. If we stimulate 100 neurons, then the probability of not receiving a spike is only $0.99^{100} \approx 0.366$ and the probability of a neuron receiving spikes from two presynaptic neurons is $(1 - 0.366)^2 \approx 0.4$. The probability that presynaptic spikes will elicit secondary spikes, of course, also increases with more focused connectivity patterns as we have to suspect in the cortex (rather than the random connectivities discussed here). Considering a fully diverging–converging chain, we need only to stimulate two of those nodes to elicit a signal transmission because we considered two spikes to be sufficient to elicit a postsynaptic spike.

4.2.6 The spread of activity in small random networks

The probability of contact between two appropriately placed cortical neurons that have large overlaps between the axonal range of the sending neuron and the dendritic tree of the receiving neurons is much higher than we considered in the previous examples. Furthermore, with more specific connectivity, including multiple synapses between two neurons, we can assume higher total synaptic efficiencies between two such closely connected cortical neurons. It is therefore appropriate to assume that only very few active presynaptic neurons can elicit a postsynaptic spike in functionally correlated neurons. It is therefore instructive to study small networks with only a small number of highly efficient synapses. Such networks have been termed *netlets*. Netlets are intended to approximate some aspects of local cortical networks discussed in the following.

Let us first discuss the extreme case where a single presynaptic spike can elicit a postsynaptic spike in all the neurons to which the presynaptic neuron projects efficiently. Let us also assume that the number of these efficiently connected postsynaptic neurons is 10. We consider again a discrete updating scheme with basic time intervals of 1 ms. If we stimulate one neuron at time step $t = 1$ ms then we get 10 spikes at $t = 2$ ms. These 10 spikes will each elicit another 10 spikes so that we have 100 spikes at $t = 3$ ms (as long as the overlap in the target populations is small). We see that the activity of the whole network quickly builds up until all of the neurons are active. The activity of the network becomes smaller if we take some refractoriness into account. For example, we can prevent the model neurons from becoming active in the time step following the time step in which they were last active. This corresponds to an absolute refractory time of 1 ms. We can then expect that around half of the neurons in the netlet are active in any time step because it is likely that a neuron becomes active every second time step.

If we increase the number of spikes that are necessary to elicit a postsynaptic spike by lowering the synaptic weights, then the situation looks a little bit different. A small number of initial spikes will not spread through the network as long as the probability of a postsynaptic neuron receiving the necessary number of spikes simultaneously is low, similar to the arguments we discussed for the large network. However, in small networks we have a larger probability of receiving connections from any other subpopulation of neurons. In addition, if we stimulate initially a relatively large number of neurons in the netlet, then we have a large probability of initiating also secondary and tertiary, etc., spikes. Once we reach this critical number of active spikes we end up again with a quickly spreading activity throughout the network and with nearly half of the neurons active at any time step. Thus, there are two asymptotic modes in such networks: with small initial activity we get an inactive netlet, and with initial large activity we get a nearly maximally active netlet.

4.2.7 The expected number of active neurons in netlets

Random network models such as netlets can be analysed much more systematically. For example, *Anninos* and colleagues have derived a formula that describes the expected fraction of active nodes f in the netlets as a function of the fraction of nodes that are active in the previous time interval. This is given by

$$\langle f(t+1) \rangle = (1 - f(t)) \left(1 - \mathrm{e}^{-f(t)C} \sum_{n=0}^{\Theta-1} \frac{(f(t)C)^n}{n!} \right) \tag{4.4}$$

where the angle brackets indicate the expectation value (the mean if we consider many trials), C is the average number of synapses per neuron, and the firing threshold Θ is the number of presynaptic spikes necessary to elicit a postsynaptic spike. This formula is illustrated for $C = 10$ and various values of Θ in Fig. 4.7. The curves plotted in this figure represent the average expected behaviour when we average over all possible realizations of netlets. Individual netlets with particular realizations of the connectivity pattern can deviate from these curves, but the curves describe what we expect in most cases with only randomly connected model neurons.

The sum of the fraction of active nodes in two consecutive time steps cannot exceed unity with a refractory time of one time step because an active node cannot be active in the following time step. The *saturation* of the network is therefore given by the decreasing diagonal in Fig. 4.7. The other diagonal indicates when the number of active nodes in one time step is equal to the expected number in the next time step. The number of active nodes at this point does not change; this is called a *steady state*. Network activity always increases in netlets for which part of the curve is above the steady state line. The curve for $\Theta = 1$ in Fig. 4.7A starts with a positive slope so that we expect a larger number of active model neurons in consecutive time steps following a small initial set of active nodes. The activity of netlets with parts of the curve below the steady state line always decreases. The activity of netlets with $\Theta > 1$ therefore decreases to zero network activity after small initial activations. The curves for netlets with $\Theta > 3$ never cross the steady state line and therefore always decrease to zero activity.

A.   Without inhibitory neurons

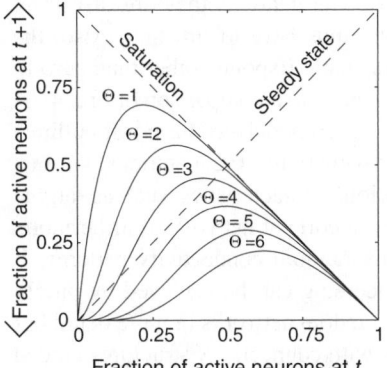

B.   With inhibitory neurons

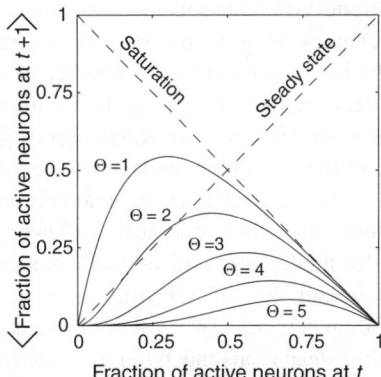

Fig. 4.7 The fraction of active nodes of netlets in the time interval $t + 1$ as a function of the fraction of active nodes in the previous time interval. The different curves correspond to different numbers of presynaptic spikes Θ that are necessary to elicit a postsynaptic spike. (A) Netlets with only excitatory neurons. (B) Netlets with the same amount of excitatory and inhibitory connections [adapted from Anninos et al. , *J. Theor. Biol.* 26: 121–8 (1970)].

The activity of the netlet with $\Theta = 3$ will increase when the initial fraction of activity is larger than about $f = 0.15$. From there it will increase until the activity in the network reaches the second crossing of the steady state line. When the network activity increases further it will be pulled back in the next time step as indicated by the curve. Network activities below this steady state point will increase. This steady state point is therefore an *attractive fixpoint* of the network dynamics. In contrast, the first crossing of the curves with the steady state line is not stable due to the repulsive forces after variations of network activity around this point.

4.2.8   Netlets with inhibition

We have seen that netlets of excitatory neurons have stable fixpoints for low firing thresholds of the model neurons. These fixpoints are near to the saturation of the network with firing rates around 500 Hz. This is certainly in contradiction to cortical activity with much lower firing rates, more realistically in the rage of 5–100 Hz. However, we have so far only considered networks of excitatory neurons. Inhibitory neurons, in particular inhibitory interneurons that respond to the activity in the network, can lower the steady state network activity of netlets. Anninos *et al.* also derived formulas similar to eqn 4.4 for netlets that include inhibitory connections. This formula is included in the MATLAB implementations that we used to plot the curves in Fig. 4.7 (see Section 12.2.6). The behaviour of netlets with the same amounts of inhibitory and excitatory model neurons, both with the same number of synapses, $C = 10$, is shown in Fig. 4.7B. The inhibition does indeed lower the netlet activity of stable steady states. However, the values still exceed typical cortical firing rates.

This discussion shows that it is not obvious that networks should have low firing rates. Even the incorporation of inhibitory model neurons did not improve the situation very much. The situation is worsened in cortical networks if we take into account that inhibitory interneurons become active only after some delay as they are driven by excitatory neurons that receive external input and thus have to fire first. Also, the inhibition does reduce the number of netlets with stable fixpoints other than zero to netlets with only low firing thresholds as can be seen from a comparison of Fig. 4.7A and 4.7B. However, we should not forget that the netlet models we have just outlined are only very simple models with very crude approximations of real neurons. We saw in the previous chapter that real neurons have various characteristics, such as fatigue effects, that can help to achieve lower firing rates in cortical networks. Furthermore, netlet models are still network models with only random connectivity patterns. It is obvious that not much useful information processing can be achieved by purely random networks. We will therefore not consider random networks in more detail but will instead focus much more on network models with connectivity structures guided either by design or by self-organizing principles in most of the remainder of the book.

4.3 Population dynamics: modelling the average behaviour of neurons

In the previous section we only considered models of very naive neurons. It is, of course, possible to use the models of spiking neurons developed in Chapters 2 and 3 in networks with architectures similar to those of the brain. Simulation of networks of spiking neurons of reasonable size has indeed become tractable to a certain extent on recent computer systems. However, even with the increasing power of digital computers we are far from being able to simulate neural systems with spiking neurons on the scale of functional units in the brain, not to mention models on a scale comparable to that of the central nervous system. Furthermore, as discussed in Chapter 1, our aim is not to reconstruct the brain, but rather to extract the principles of its organization and functionality. Many of the models in computational neuroscience, in particular on a cognitive level, are therefore based on descriptions that do not take the individual spikes of neurons into account, but instead describe the average firing rate of neurons.

In this section we discuss the relationship of such models to populations of spiking neurons. The aim is to highlight under what conditions these common approximations are useful and faithful descriptions of neuronal characteristics. It is clear that rate models cannot incorporate all aspects of networks of spiking neurons. However, many of the principles behind information processing in the brain can be efficiently illuminated on the level of networks composed of simple processing elements, and many of the main features of rate models have now been verified in networks of spiking neurons. Many investigations in computational neuroscience have used such models, and, to a large extent, we will follow this path in the remainder of the book.

4.3.1 Firing rates

For the following discussions it is important to recognize that neurophysiological recordings of single cells have been very successful in correlating the firing rates of single neurons with the behavioural perceptions or responses of a subject. It is indeed common in physiological studies to derive an instantaneous firing rate with the help of a sliding window in the spike train. For example, as illustrated in Fig. 4.8A, we can estimate the average temporal spike rate of a neuron with a fixed window of size ΔT,

$$\nu(t) = \frac{\text{number of spikes in } \Delta T}{\Delta T}$$

$$= \frac{1}{\Delta T} \int_{t-\Delta T/2}^{t+\Delta T/2} \delta(t' - t^f)\,dt', \qquad (4.5)$$

where t^f is the firing time of the neuron. This defines for small time windows the *instantaneous firing rate*. The so-called *delta-function* $\delta(x)$, further described in Appendix A, simply contributes a value of 1 to the integral when the argument of this function is zero. The integral therefore describes the sum of the spikes within the time window ΔT. The advantage of this formulation is that it can be used to generalize the firing rate concepts easily to similar definitions including weighted forms of the time window. For example, it is common in physiological studies to use a Gaussian window that does not have the sharp boundaries of the rectangular time window above. The firing rate of a neuron at time t is then defined by

$$\nu(t) = \frac{1}{\sqrt{2\pi}\sigma} \int_{-\infty}^{+\infty} \delta(t' - t^f) e^{(t'-t)^2/2\sigma^2}\,dt'. \qquad (4.6)$$

Information based on average firing rates can be conveyed between neurons when the receiving neuron acts as an integrator with a sufficiently large time constant. We will discuss in the next chapter some estimations of how much information is carried by the average firing rate of neurons compared to the spike timing that could be detected with neuronal integrators with short time constants. We will then see that a lot of information is indeed transmitted by the firing rates of neurons in the brain.

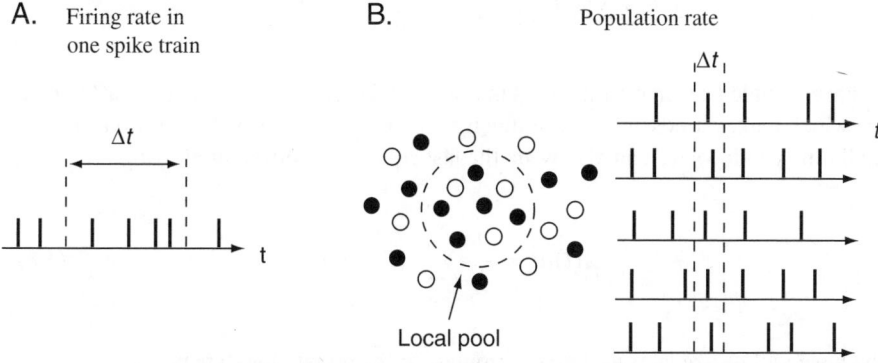

A. Firing rate in one spike train

Δt

t

B. Population rate

Δt

t

Local pool

Fig. 4.8 (A) Temporal average of a single spike train with a time window ΔT that has to be large compared to the average interspike interval. (B) Pool or local population of neurons with similar response characteristics. The pool average is defined as the average firing rate over the neurons in the pool within a relatively small time window [adapted from W. Gerstner, in *Pulsed neural networks*, Maass and Bishop (eds), MIT Press, 1998].

We should note that in physiological estimates of the firing rates of single neurons it is common to include an average over several trials under equal experimental conditions. Such averages over repetitions are often necessary because the firing of a neuron in a single trial is very noisy. This is a valid experimental estimation procedure as it has been found that such averages correlate well with behavioural responses. However, it is also clear that averaging over several trials is not an option for the brain where responses to a single stimulus must be possible. Also, the temporal averaging explained above can only be employed by the brain in a limited way as the time windows of averaging have certainly to be much smaller than the typical response time of the organism. How then can the brain use rate information?

4.3.2 Population averages

A solution can be given if the brain does not rely on the information of a single spike train. Indeed, such reliance would also make the brain very vulnerable to damage or reorganization. We therefore conjecture that there must be a subpopulation or pool of neurons with similar response properties so that the temporal averages measured in the physiological experiments represent spatial averages employed in the brain (see Fig. 4.8B). The neurons in these subpopulations of a cortical module may act in a statistically similar way. This would explain why single-neuron recordings have been so successful in correlating such measurements to behavioural findings. With this conjecture we can identify the experimental estimate of the average rate of a single neuron with the average population activity $A(t)$ of neurons, which include the pool of neurons active in the particular task,

$$A(t) = \lim_{\Delta T \to 0} \frac{1}{\Delta T} \frac{\text{number of spikes in population of size N}}{N}$$

$$= \lim_{\Delta T \to 0} \frac{1}{\Delta T} \int_{t-\Delta T/2}^{t+\Delta T/2} \frac{1}{N} \sum_{i=1}^{N} \delta(t' - t_i^{\mathrm{f}}) \mathrm{d}t'. \tag{4.7}$$

We have included a sum over neurons in a subpopulation in eqn 4.7 (in contrast to eqn 4.5) which makes it possible to use much smaller time windows. In the limit of very small time windows we can also write the last equation in differential form,

$$A(t)\mathrm{d}t = \frac{1}{N} \sum_{i=1}^{N} \delta(t' - t_i^{\mathrm{f}}), \tag{4.8}$$

which we will utilize later to derive formulas for the population dynamics.

It is difficult to verify this conjecture directly with experiments as this would demand the simultaneous recording of many thousands of neurons. Such recording with multiple electrodes is to a large extent beyond a current experimental feasibility, and brain imaging currently average over too many neurons to be able to verify our conjecture. The conjecture is, however, supported indirectly by several known experimental facts, such as the existence of cortical columns where neurons with very similar response are found as discussed earlier in this chapter. In the following we will build on this conjecture to derive models for such a pool of neurons.

4.3.3 Population dynamics in response to slowly varying inputs

To describe the average behaviour of a pool of neurons we divide the population into subpopulations of neurons of the same type with similar response properties. For simplicity we assume further that the neurons in this pool have the same membrane time constant τ_m, the same mean numbers and synaptic efficiencies of afferents, and receive the same input current I^{ext}. The challenge is to derive a compact expression for the dynamics of this pool. A first guess is the dynamics of the leaky integrator similar to the subthreshold dynamics of the individual integrate-and-fire (IF) neurons,

$$\tau \frac{dA(t)}{dt} = -A(t) + g(RI^{ext}(t)). \tag{4.9}$$

The function g is an *activation function* or *gain function* that describes the influence of the external current on the activation of the pool. We will discuss several gain functions in the next section. For now it is sufficient to have a linear function $g(x) = x$ in mind. It is clear that the dynamic eqn 4.9 can only be an approximation of the pool dynamics. The motivation of this description is that, when at any instance in time only a small number of neurons in the pool is firing, then we expect that the population dynamic are mainly characterized by the subthreshold dynamics to slowly varying input currents. The time constant in eqn 4.9 for slowly varying input currents should then be close to the average membrane time constants of the neurons in the pool.

The description given in eqn 4.9 is particularly appropriate when we analyse *asymptotic stationary states* of neuronal networks. *Stationary states* are states that do not change under the dynamics of the system, and by asymptotic we mean that these are the states after some time when the initial transient behaviour levelled off. Saying that the states do not change under the dynamics of the system is mathematically expressed as $dA/dt = 0$. Including this in eqn 4.9 yields

$$A(t) = g(RI^{ext}(t)). \tag{4.10}$$

This is, of course, only true for constant input because the state cannot be stationary with varying input. However, the formula should also hold to a good approximation with slowly varying input. Many information-processing capabilities of the brain have been studied in this limit, which we will also frequently use in some discussions in the later chapters. Due to the importance of these descriptions (eqn 4.9 and 4.10) we will summarize and elaborate these formulations in the next section.

4.3.4 Rapid response of populations

Before proceeding to this summary we want to demonstrate that the dynamic description of population dynamics eqn 4.9 is not always appropriate and indeed breaks down under several circumstances. To demonstrate this we consider a pool of equivalent IF-neurons all with the same time constant $\tau_m = 10$ ms and all receiving the same noisy input $I^{ext} = \bar{I}^{ext} + \eta$ with $\eta = N(0,1)$. A simulation of such a 'network' of independent model neurons (no connections between the neurons) is shown in Fig. 4.9. In these simulations we have switched the external input at time $t = 100$ ms from a low magnitude $R\bar{I}^{ext} = 11$ to a higher magnitude $R\bar{I}^{ext} = 16$. The spike count almost instantaneously follows the jump in the input. The reason for this is that at each instant of time there is always a subset of the neurons that is close to threshold. These neurons can quickly respond to the input, and the other neurons have time to follow quickly. The population rate calculated with the population dynamics in eqn 4.9 follows only slowly this change of input when the time constant is set to $\tau = \tau_{\mathrm{m}}$. The simulation demonstrates that very short time constants, much shorter then typical membrane time constants, have to be considered when using this model to approximate the dynamics of population responses to rapidly varying inputs.

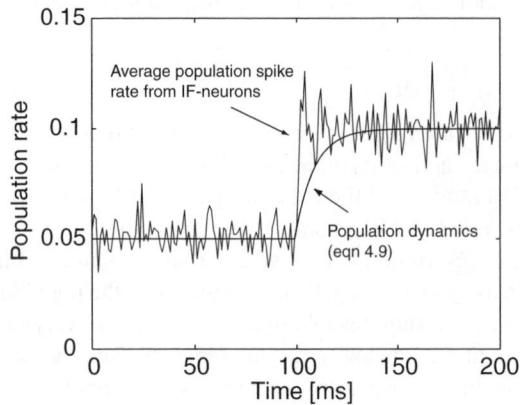

Fig. 4.9 Simulation of a population of 1000 independent integrate-and-fire neurons with a membrane time constant $\tau_{\mathrm{m}} = 10$ ms and threshold $\vartheta = 10$. Each neuron receives an input with white noise around a mean $R\bar{I}^{\mathrm{ext}}$. This mean is switched from $R\bar{I}^{\mathrm{ext}} = 11$ to $R\bar{I}^{\mathrm{ext}} = 16$ at $t = 100$ ms. The spike count of the population almost instantaneously follows this jump in the input, whereas the average population rate, calculated from in eqn 4.9 with a linear activation function, follows this change of input only slowly when the time constant is set to $\tau = \tau_{\mathrm{m}}$.

We have used noise in the models of spiking neurons while leaving any noise out of the node model for the population average. The reason is that the noise is essential for the argument with the spiking neurons. In contrast, we have not included any noise in the population dynamics to outline the average response of the node. Adding noise introduces fluctuations in this curve that may seem to resemble fluctuations in the average of spiking nodes. However, the analysis of the mean response would certainly bring the difference to light. However, adding noise to the population node just to make it look more realistic would be very inappropriate and would only obscure the general argument. In contrast, the noise in a population of spiking neurons might be essential for information processing in the brain.

4.3.5 Advanced descriptions of population dynamics

We stressed that the population dynamics given by eqn 4.9 is only reasonable for slowly varying inputs. A better description of the population dynamics is, of course, of utmost importance and was already addressed many years ago by *Amari*, and *Wilson* and *Cowan*. *Wulfram Gerstner* and *Leo van Hemmen* have more recently advanced such formulations considerably and we will shortly outline here their formulations which is a generalization of the Wilson–Cowan integral equations.

Recall from the last chapter that the membrane potential in the spike response model is given by

$$u_i(t) = \sum_{t^f} \eta(t - t^f) + \sum_j \sum_{t_j^f} w_{ij} \epsilon(t - t_j^f). \tag{4.11}$$

For simplicity, we have here ignored the possible dependence of the ϵ term on the postsynaptic firing. We include new indices i for the postsynaptic neurons as we are now interested in a population of such neurons. We will assume a specific kind of population of neurons in which the neurons react in a similar way to input and have on average a synaptic efficiency

$$w_{ij} = \frac{w_0}{N}. \tag{4.12}$$

We also assume that there is no spike-time adaptation in the neurons, that the total number of neurons in the population stays constant, and that we formally have an infinitely large population of neurons in which only the means of the random variables matter. Using the definition of the population average (eqn 4.8) we can thus express the mean influence of the postsynaptic potential using the rate of the population as

$$u_\epsilon(t) = w_0 \int_0^\infty \epsilon(t') A(t - t') dt'. \tag{4.13}$$

Finally we have to take noise into account, which can be done with either of the noise models mentioned in the last chapter. In general, we can define a probability density, which a neuron fires at time t when it has a membrane potential u and has fired previously at times t^f as $P_u(t|t^f)$. The membrane potential does depend on the population rate as specified in eqn 4.13. The population rate at time t can thus be expressed as the sum (or integral) of all the population rates at previous times multiplied by the probabilities of firing at the corresponding times,[14]

$$A(t) = \int_{-\infty}^t P_u(t|t^f) A(t^f) dt^f. \tag{4.14}$$

This dynamic is exact in the limit of an infinite pool of neurons, and the equation can be solved for particular probability densities $P_u(t|t^f)$ derived from the noise models. It is possible to use such descriptions directly in simulations, though it requires the calculation of large sums which is computationally demanding. It is therefore still desirable to derive approximations of a population dynamics in differential form, and the derivation from the integral equations makes it possible to specify precisely the assumptions that have to be made. *Wulfram Gerstner* has discussed this in great detail (see 'Further reading'). He has shown that, in the *adiabatic limit*, in which the cell assembly is responds only slowly to slowly varying inputs that do not cause specific collective phenomena of the neurons in the cell assembly (such as synchronization

---

[14] See Gerstner in *Neural Computation* 12:43–89, 2000, and references therein, for a more detailed derivation.

or phase locking), the population dynamics can be approximated with differential equations in the form of eqn 4.9 with a gain function of the form

$$g(x) = \frac{1}{t^{\text{ref}} - \tau \log(1 - \frac{1}{\tau x})},$$ (4.15)

where t^{ref} is an absolute refractory time. This result tells us that the average gain function of the population is similar to the gain function of the single IF-neuron (compare eqn 3.11), although only in the adiabatic limit. This gain function 4.15 is plotted in Fig. 4.10A with an absolute refractory time of $t^{\text{ref}} = 1$ ms and a time constant of $\tau = 5$ ms.

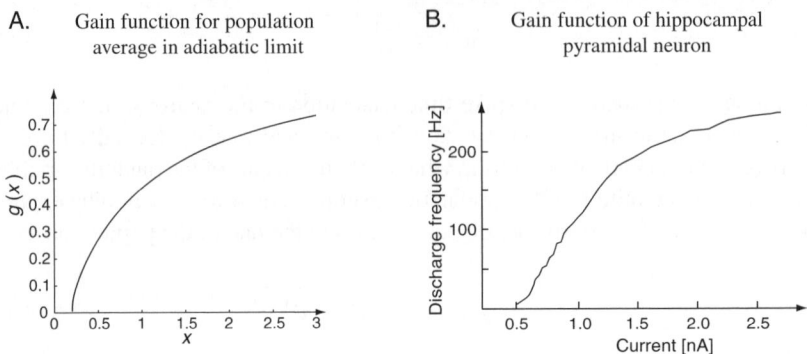

A. Gain function for population average in adiabatic limit

B. Gain function of hippocampal pyramidal neuron

Fig. 4.10 (A) The gain function of eqn 4.15 that can be used to approximate the dynamics of a population response to slowly varying inputs (adiabatic limit). (B) Examples of physiological gain functions from a hippocampal pyramidal cell. The discharge frequency is based on the inverse of the first interspike interval after the cell started to respond to rectangular current pulses with different strength [redrawn from Lanthorn, Storm, and Anderson, *Exp. Brain Res.* 53: 431–43 (1984)].

Some gain functions of single neurons measured electrophysiologically have shown similar response characteristics. An example of a gain function measured from a hippocampal pyramidal cell is shown in Fig. 4.10B. The figure shows the instantaneous firing rate (discharge frequency) of such a neuron in response to a 1.5 s rectangular current stimulus with different amplitudes measured in nanoamperes (nA). The instantaneous firing rate was thus derived from the inverse of the first interspike interval. The neuron quickly adapted to the stimulus (not shown in the figure) so that instantaneous firing based on subsequent spikes would be much smaller. Keep in mind that we are comparing two different definitions of gain functions in Fig. 4.10A and 4.10B. The similarity of the forms could be the result of the particular way in which the measurements are done. Gain functions are typically measured in experiments with isolated neurons, which could account for the similarity of this response to the average response of a population to slowly varying inputs. Such experimental measurements, however, neglect possible interactions *in vivo* that could alter the effective gain function considerably. It is thus important to keep in mind the different interpretations of gain functions used in the following models. Most often we will consider population models with slowly varying inputs.

4.4 The sigma node

Warren McCulloch and *Walter Pitts* published a seminal work in 1943 in which they studied simple binary units that generate an output only when the summed input reached a certain threshold. Such units can be associated with the threshold firing characteristics of single neurons. However, the McCulloch–Pitts nodes do not include the generation and reset of spikes, and a better interpretation is that they describe the response of a population of neurons as discussed in the last section. The binary case then corresponds to the first approximation of a population response in that the cell assembly is either inactive (low firing rate) or active (high firing rate). We discussed in the last chapter the fact that neurophysiological estimates of firing rates have strong behavioural correlates, and we will discuss in the next chapter some experimental evidence that suggests that firing rates indeed carry large amounts of the information processed in the central nervous system.

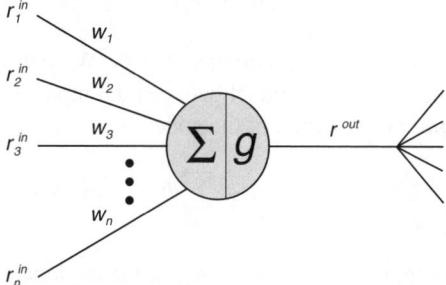

Fig. 4.11 Schematic summary of the functionality of a sigma node most commonly used in networks of artificial neurons. Such a node weights the input value of each channel with the corresponding weight value of that channel and sums up all these weighted inputs. The output of the node is then a function of this internal activation.

Rate models are well established and have contributed considerably to computational neuroscience (as well as to the area of *artificial neural networks* and *connectionist modelling*). Due to the importance of rate models in the literature and in many of the discussions in the following chapters we will summarize and elaborate on the basic units of rate models in the remainder of this chapter. Note that the study of rate models does not imply that the timing of spikes is irrelevant; quite the opposite, we have already discussed one example in the previous section where spike times play a critical role for fast response characteristics that can not be captured with rate models. More examples will follow later in this book, in particular when we discuss spike time-dependent synaptic plasticity in Chapter 6. However, studying rate models can illuminate many of the information-processing mechanisms in the brain.

4.4.1 The minimal neuron model

The minimal unit in a rate model, which is a simple generalization of the McCulloch–Pitts unit, can be summarized as receiving input, doing some local processing, and distributing a subsequently generated output to other units. We will call these functional units *nodes* to remind us of the various interpretations with regard to their neural basis, either as a strong simplification of a single neuron, or, more realistically, as a unit that reflects the average behaviour of a statistically coupled neuronal group that we called *subpopulation* or *pool* in the last section. The most basic type of a node, used most frequently in connectionist and neural network modelling, is called a *sigma node*. Such a node is schematically summarized in Fig. 4.11. The sigma node has several input channels and generates only one output, which is distributed to many other nodes (subpopulations of neurons). The values for each input are most often represented by real numbers, although other values such as binary inputs are possible. We will denote these input values as r_i^{in} to remind us that they are related to rate values of other neuronal groups. The subscript i indexes the particular input channel.

The functionality of the sigma node can be divided into three simple steps:

1. The input value r_i^{in} of each channel is multiplied by the weight value w_i for this channel, which results in a quantity often referred to as *net input*

$$h_i = w_i r_i^{\text{in}} \tag{4.16}$$

2. The node then sums up all the weighted inputs. We call the resulting value, which corresponds to the dot product between the input vector and the weight vector, the *net input* or *activation* of the node:

$$h = \sum_i h_i \tag{4.17}$$

3. It calculates the output r^{out} as a function g of the *net input*

$$r^{\text{out}} = g(h) \tag{4.18}$$

We can also summarize the function of a sigma node using the compact formula

$$r^{\text{out}} = g\left(\sum_i w_i r_i^{\text{in}}\right) \tag{4.19}$$

Other types of nodes are possible and introduced later. However, this type of node describes the most basic functions of neuronal groups and is therefore at the centre of most investigations into the modelling of neural networks.

4.4.2 Common activation functions

We call this function g the *activation function* in this book. This function also represents the *gain function* discussed in Chapter 2. This terminology is appropriate when comparing the node to a single neuron. This function is sometimes also called *transfer function*, which is appropriate when we think of this function as the net effect of various processes within the neuronal group that transform a given input to the specific output of this neuron assembly. This function can therefore have various forms, of which some of the most frequently used are illustrated in Table 4.2. The first example is the *linear function* g^{lin} relating the sum of the inputs directly to the output. Linear units can often be treated analytically and can provide us with many insights into the working of networks. Many other useful activation functions can also be approximated by

Table 4.2 Examples of frequently used activation functions and their basic implementation in MATLAB.

Type of function	Graph	Mathematical formula	Matlab implementation
Linear		$g^{lin}(x) = x$	`X`
Step		$g^{step}(x) = \begin{cases} 1 & \text{if } x > 0 \\ 0 & \text{elsewhere} \end{cases}$	`floor(0.5*(1+sign(x)))`
Threshold-linear		$g^{theta}(x) = x\,\Theta(x)$	`x.*floor(0.5*(1+sign(x)))`
Sigmoid		$g^{sig}(x) = \dfrac{1}{1+\exp(-x)}$	`1./(1+exp(-x))`
Radial-basis		$g^{gauss}(x) = \exp(-x^2)$	`exp(-x.^2)`

piecewise linear functions. A simple example of such a nonlinear function is the *step function* g^{step}. This function only returns two values and therefore produces binary responses. This activation function was used by McCulloch and Pitts, and a node with this activation function is therefore called a *McCulloch–Pitts node*.

The third example, the *threshold-linear function* g^{theta}, is similar to the linear functions except that it is bounded from below. Limiting the node activities from below is sensible as negative activity has no biological equivalent. This activation function is indeed a good approximation to the gain functions of IF-neurons as we have seen in the last chapter. Besides the lower limit on the firing rate it is also biologically sensible to limit the maximal response of the node, because the firing rate of single neurons and neuronal groups is limited by the refractory times of the neurons. A simple realization of such a activation function is a combination of the threshold-linear function and the step function. However, a more frequently used activation function in this class is the *sigmoid function* g^{sig} illustrated as the fourth example. This function bounds the minimal and maximal response of a node while interpolating smoothly between these two extremes. This type of activation function is most frequently used in neural network and connectionist models for reasons that become apparent in the next chapter. It also approximates the gain function of noisy IF-neurons well as we have seen in the last chapter. The last function in the table is very different to the former in the sense that it is a non-monotonic function. The example shown is a *radial basis function*, one of a group of general functions that are symmetric around a base value. The particular example g^{gauss} is the famous Gaussian bell curve.

Suffice it to say that these are only examples of possible activation functions. We have only outlined the general shape of those functions while it should be clear that those shapes can be modified. The functions often include parameters with which some characteristics of the functions can be changed such as the slope and offset of a function. For example, we can generalize the sigmoid function in the table with

$$g^{\text{sig}} = \frac{1}{1 + \exp(-\beta(x - x_0))}, \tag{4.20}$$

where an increasing value of the parameter β increases the steepness of the curve, and x_0 shifts the curve along the abscissa. Keep in mind that these are not the only possible forms of activation functions. Networks of nodes with different activation functions have different characteristics and it is worth exploring the dependence of network characteristics on the specific activation functions. However, it is also interesting to note that many of the information-processing capabilities of networks of such nodes do not depend very critically on the precise form of the activation function. Several types of activation function lead to similar network abilities in the sense that we do not have to fine-tune the functions in order to demonstrate network abilities.

4.4.3 The dynamic sigma node

The above description of the sigma node (eqn 4.19) does not include a time dynamic other than the single response to an external input. For slowly varying external input we can describe the population average with a leaky integrator dynamic as discussed in the last section. We will here derive this equation again from the viewpoint of generalizing the discrete sigma node to continuous dynamics.

Let's assume that the functions of this discrete sigma node, the weighting and summation of the inputs and the generation of the output summarized in eqn 4.19, take a fixed length of time. This sets our basic time scale, which we label with τ. To resemble the previous functionality of the sigma node it is reasonable that the continuous node should integrate only a fraction of the input in each smaller time step Δt. If we set this fraction equal to $\Delta t/\tau$ we ensure that the node still integrates the same amount of input as in the case of one big time step as described by eqn 4.19. Of course, the node has to remember the values it had before to which the new input is added. However, the memory of the previous values generates a small problem. If this node never forgets the values it has integrated before, then the net input will continue to grow without bound as long as input is applied. To solve this problem we use a leaky integrator that loses some of its current state over the time. We can achieve this by introducing a *forgetting factor*. To ensure some stability of the node we set this forgetting factor to acts on the same time scale as the integration of new input. Taking all these considerations together we can write

$$h(t + \Delta t) = (1 - \frac{\Delta t}{\tau})h(t) + \frac{\Delta t}{\tau} \sum_i w_i r_i \tag{4.21}$$

Note that we get back the functionality of the simple node (eqn 4.19) if we make one big time step of $\Delta t = \tau$. For this time step the node has forgotten its previous state and just sums up all the weighted input.

At this level of description we still have a discrete node because we still use small discrete time steps Δt (in units of τ). The next step is therefore to make the time steps smaller and smaller. To do this we first rewrite eqn (4.21) as

$$\tau \frac{h(t + \Delta t) - h(t)}{\Delta t} = -h(t) + \sum_i w_i r_i \tag{4.22}$$

and perform the limit $\Delta t \to 0$. In this limit the differences become differentials. The time-dependent equation for a continuous node is hence

$$\tau \frac{\mathrm{d}h(t)}{\mathrm{d}t} = -h(t) + \sum_i w_i r_i \tag{4.23}$$

while the output of the node at time t is still given by eqn 4.18. This is exactly the equation for the population dynamics we have discussed in the last section (eqn 4.9), and we derived this equation here as a generalization of the discrete dynamics given by eqn 4.19. Remember, however, that the leaky integrator dynamics for population averages is only appropriate under certain conditions such as slowly varying inputs.

4.4.4 Leaky integrator characteristics

The leaky integrator dynamics are often used in computational neuroscience, and it is therefore useful to summarize some of the main behaviour of a leaky integrator. Equation 4.23 without external input is a simple homogeneous first-order differential equation. This can easily be solved and the result is the exponential function,

$$h(t) = h(0)\, \mathrm{e}^{-\frac{t}{\tau}} + h^{\mathrm{rest}}. \tag{4.24}$$

Hence, the activation of the node without external input decays exponentially on a time scale of τ towards the resting activation h^{rest} as illustrated in Fig. 4.12. An input current $I^{\mathrm{in}} = \sum_i w_i r_i$, changes this behaviour. A positive net input slows down the exponential decay or even increases the amount of activation. A negative net input decreases the activation or slows down the recovery from a sub-resting-level activation towards the resting value.

4.4.5 Discrete formulation of continuous dynamics

We have worked our way from the discrete formulation of the sigma node (eqn 4.21) to a sigma node with a continuous time dynamic (eqn 4.23) by approaching the limit of infinitesimally small time steps. Alternatively, we could have started with the continuous dynamics (eqn 4.9) and changed it to a formulation with discrete steps. Note that this is what we have to do when simulating a continuous dynamic on a digital computer. The specific form of eqn 4.21 is, however, only one of many possible forms that would yield the same dynamic in the limit of infinitesimally small time steps (continuum limit). An alternative formulation would be, for example, to incorporate our knowledge of the exponential response to short inputs in the form

$$h(t + \Delta t) = (1 - \mathrm{e}^{-\frac{\Delta t}{\tau}})h(t) + \mathrm{e}^{\frac{\Delta t}{\tau}} \sum_i w_i r_i, \tag{4.25}$$

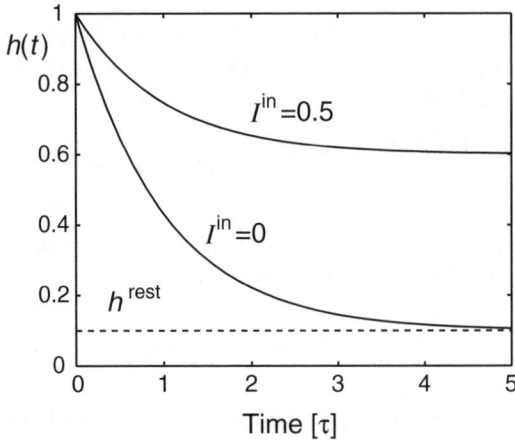

Fig. 4.12 Time course of the activation of a leaky integrator node with initial value $h = 0$. In the lower curve no input I^{in} was applied leading to an exponential decay. The upper curve corresponds to a constant input current $I^{\text{in}} = 0.5$. The resting activation of the node was set to $h^{\text{rest}} = 0.1$.

which can be viewed as taking time steps on a logarithmic scale into account. Formally, eqn 4.25 is another method for the numerical integration, and we can employ many other methods. The discrete formulation of eqn 4.9) in the form of eqn 4.21 is known formally as the *Euler method* and is outlined in Appendix C.

Each discrete model can evolve differently; only in the limit of infinitesimally small time steps do we recover the unique continuous model. If we are concerned with approximating the unique continuous model we have to make sure that the differences due to the finite time steps are negligible. It is well known that numerical integrations with the Euler method can be unstable, and advanced integration methods such as higher-order *Runge–Kutta* methods, outlined in Appendix C, are more commonly used in numerical integrations. We must always verify that our numerical experiments do not critically depend on the time steps used in the simulations when we are interested in the behaviour of a system with continuous time dynamics. Many numerical integration methods include an *adaptive time step algorithm* that varies the time step in small enough values so that the numerical results only vary within a given error bound. Some of the programs discussed in Chapter 12 utilize such advanced integration techniques.

4.5 Networks with nonclassical synapses: the sigma–pi node

The sigma node introduced in the last section is certainly a very rough abstraction of a real neuron. However, we will frequently see in the following chapters that nodes with the above-outlined simple functionality are sufficient to outline important information-processing principles in the brain. To approximate a neuron with a summation node is also strongly motivated as electrical potential such as PSPs generated in the dendrites superimpose. This formulation, however, ignores, besides other things, possible interactions of different ion channels that are thought to be important for some forms of information-processing mechanisms in the brain and that are utilized in some of the models described later in this book. We outline in this section some of the possible sources that lead to nonlinear interactions between input channels and introduce a simple description with nodes that describe average firing rates.

4.5.1 Logical AND and sigma–pi nodes

An example of a strong nonlinear interaction between two ion channels is a spiking neuron with a firing threshold that requires at least two spikes in some temporal proximity to each other. A single spike alone, in this case, cannot initiate a spike. Only if two spikes are present within the time interval, on the order of the decay term of EPSPs or the time constant of the leaky integrator, can a postsynaptic spike be generated. This corresponds to a logical AND function. We can represent such an AND function with a multiplicative term of two binary presynaptic terms, r_1 and r_2, with values one, indicating that a presynaptic spike at a particular synapse is present, or zero, indicating that no presynaptic spike is present. Thus, the dependence of the activation of such a node from two inputs is described by a multiplicative term such as $h = r_1 r_2$.

We can generalize this idea for models describing average firing rates. Let us consider two independent presynaptic spike trains with average firing rates r_j^{in} and r_k^{in}, respectively. The average firing rate determines the probability of having a spike in a small interval. The probability of having two spikes of the two different presynaptic neurons in the same interval is then proportional to the product of the two individual probabilities. This forms the basis of a simple model of nonlinear interactions between synapses in a rate model that includes nonlinear interactions between synapses. The activation of a node i in this model is given by

$$h_i = \sum_{jk} w_{jk} r_j^{in} r_k^{in}, \qquad (4.26)$$

where the weight factor w_{ik} describes now the overall strength of two combined synaptic inputs. The activation of postsynaptic neurons in this model depends on the sum (mathematically depicted by the Greek letter $\sum$) of multiplicative terms (mathematically depicted by the Greek letter $\prod$), and such a node is therefore called a *sigma–pi node*.[15] The model can be generalized to other forms of nonlinear interactions between synaptic channels by replacing the product of the presynaptic firing rates with some function of presynaptic firing rates, for example, $r_j^{in} r_k^{in} \rightarrow g(r_j^{in} r_k^{in})$, but the simple multiplicative model already incorporates essential features of nonlinear ion channel interactions that are sufficient for most of the models discussed in this book. Note that we can also view such interactions as modulatory effects of one synaptic input on the other synaptic input.

---

[15] The product of four terms $x_1 x_2 x_3 x_4$ can be written with this shorthand notation as $\prod_{i=1}^{4} x_i$. We have not included this notation in our equation as we have only three factors in the product and the notation would actually make the formulas look more cluttered.

4.5.2 Divisive inhibition

Equation 4.26 is an example of two nonlinear interacting excitatory channels. An analogous model of an interaction between an excitatory synapse and an inhibitory synapse can be written as

$$h_i = \sum_{jk} w_{ik} r_i^{\text{excitatory}} / r_k^{\text{inhibitory}}. \tag{4.27}$$

This is called *divisive inhibition*. It models the effects of *shunting inhibition* that we mentioned in Chapter 2. Inhibitory synapses on the cell body often have such shunting effects (see Fig. 4.13A), whereas inhibitory synapses on remote parts of the dendrites are better described by subtractive inhibition. It is possible that synapses in between have effects that interpolate between these two extremes.

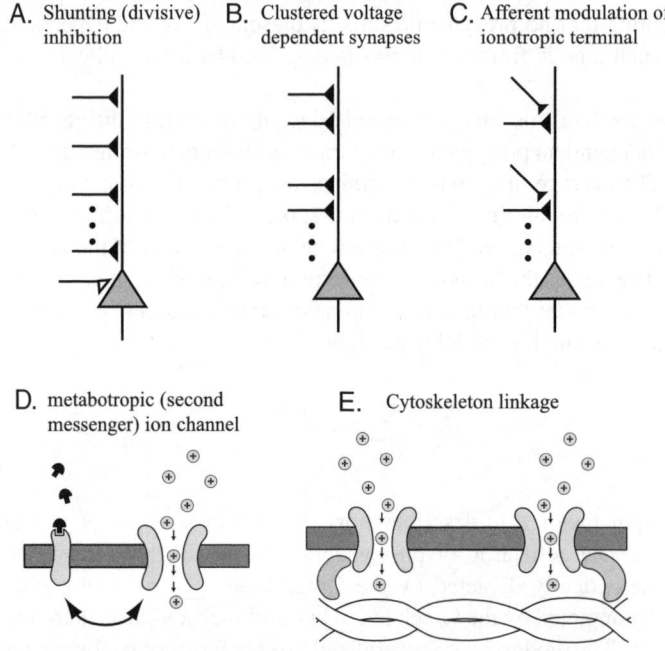

A. Shunting (divisive) inhibition   B. Clustered voltage dependent synapses   C. Afferent modulation of ionotropic terminal

D. metabotropic (second messenger) ion channel   E. Cytoskeleton linkage

Fig. 4.13 Some sources of nonlinear (modulatory) effects between synapses as modelled by sigma–pi nodes. (A) Shunting (divisive) inhibition, which is often recorded as the effect of inhibitory synapses on the cell body. (B) The effect of simultaneously activated voltage-gated excitatory synapses that are in close physical proximity to each other (synaptic clusters) can be larger than the sum of the effect of each individual synapse. Examples are clusters of AMPA and NMDA type synapses. (C) Some cortical synaptic terminals have nicotinic acetylcholine (ACh) receptors. An ACh release of cholinergic afferents can thus produce a larger efflux of neurotransmitter and thereby increase EPSPs in the postsynaptic neuron of this synaptic terminal. D. Metabotropic receptors can trigger intracellular messengers that can influence the gain of ion channels. (E) Ion channels can be linked to the underlying cytoskeleton with adapter proteins and can thus influence other ion channels through this link.

4.5.3 Further sources of modulatory effects between synaptic inputs

Interactions between synapses can result from a variety of other sources in biological neurons. An obvious source is voltage-dependent synapses that are in physical proximity to each other as illustrated in Fig. 4.13B. A good example are NMDA receptors that are blocked for low membrane potentials. If the membrane potential is raised by an EPSP from another non-NMDA synapse in its proximity, then it is possible that the blockade is removed so that the combined effect of a non-NMDA synapse together with the NMDA synapse is much larger than the sum of the activation through each synapse alone. The physical proximity is necessary as the EPSP in a passive dendrite decays with distance, so that the nonlinear effects are largest for nearby synapses. The nonlinear effects can also reach larger distances between synapses with local active dendrites that can activate more remote dendrites through active spike propagation.

There are also examples known of a direct influence of specific afferents on the release of neurotransmitters by presynaptic terminals. This is illustrated schematically in Fig. 4.13C. An example is cholinergic afferents that emit acetylcholine (ACh). This can bind to ACh receptors at a presynaptic terminal that opens calcium channels. The excess of calcium in the presynaptic terminal then triggers an enhanced release of neurotransmitters by a presynaptic action potential. This is an example of the modulation of an ionotropic synapse. A more diffuse modulation can be achieved with metabotropic receptors (see Fig. 4.13D). Such receptors initiate intracellular chemical reactions leading to second messengers that can influence the efficiency or gain of different ion channels. A more direct linkage between ion channels is possible through adapter proteins that link ion channels to the underlying cytoskeleton in a neuron (Fig. 4.13E). There is still more research necessary to understand all the details of such mechanisms. We are mainly concerned here with the principal potentials of such modulatory effects between presynaptic neuron activities, and we will employ such mechanisms in some models discussed in later chapters.

Conclusion

The brain does display characteristic neuronal organizations, and we started to study some properties of networks of spiking neurons in this chapter, in particular the spread of neuronal activities through random networks of simple neuron models. We saw that transmission of information in such networks is not likely to be reliable without diverging–converging chains, and that a sensible activity in random recurrent networks is not easy to achieve. It is thus evident that further mechanisms are crucial in the brain, in particular the more guided connectivity pattern. In the following chapters we will explore some of those mechanisms, in particular the self-organization of synaptic efficiencies and more specific organizations within the brain.

We also introduced the sigma node, which is aimed at modelling the average firing rate of populations of neurons with similar response properties. Such nodes will be the basic processing unit in many models discussed in this book. We argued that it is possible to take into account nonlinear interactions within a neuronal pool by introducing sigma–pi nodes that not only sum the incoming current but can also depend on different currents in a multiplicative way. Sigma nodes or sigma–pi nodes are the basic units in rate models common in the literature. Such rate models are well motivated, although it should be clear that rate models are not able to account for a variety of possible effects in populations of neurons such as synchronization and fast response characteristics of neuronal populations.

Further reading

Moshe Abeles
Corticonics: neural circuits of the cerebral cortex, Cambridge University Press, 1991.
This book includes an introduction to neocortical anatomy and neurophysiology relevant for modellers and explains how some estimations of important parameters are extracted from experiments. It continues to discuss several consequences of information transmission in such networks, including the proposal of so-called synfire chains. The book also contains many good exercises employing probability theory.

Edward L. White
Cortical circuits, Birkhäuser, 1989.
In case you need more details on cortical cell types and organizations.

Wolfgang Maass and Christopher M. Bishop (eds.)
Pulsed neural networks, MIT Press, 1999.
Wulfram Gerstner describes population dynamics in more detail in chapter 10.

Wulfram Gerstner
Population dynamics of spiking neurons: fast transients, asynchronous states, and locking, in *Neural Computation* 12: 43–89, 2000.
This is an important paper in which the integral equation for the population equation is derived and in which Wulfram discusses specific effects that are not covered by other simplified rate models.

Christof Koch
Biophysics of computation: information processing in single neurons, Oxford University Press, 1999.
Christof Koch has promoted for many years the importance of nonlinear interactions between ion channels, and descriptions of many related biophysical mechanisms can be found throughout the book.

Paul S. Katz (ed.)
Beyond neurotransmission: neuromodulation and its importance for information processing, Oxford University Press, 1999.
A collection of contributions by many specialist in the area of neuromodulation. This is not an introductory text, and includes a lot of biochemistry. However, some introductory parts are very interesting to read. See in particular the general introduction by Paul Katz (chapter 1) and the outline of several subcellular mechanisms of neuromodulation by Elizabeth A. Jones and Leonard K. Kaczmarek (chapter 3).

5 Representations and the neural code

In this chapter we discuss how information is represented in the brain. Neuroscientists have long wondered if some messages are transmitted by the specific firing patterns of neurons. We discuss issues concerning the neural code with the help of information theory in this chapter. Specifically, we contrast spike-pattern codes with rate codes and discuss the relevance of precise spike timing. Besides this temporal coding we will also discuss spatial representations of information in populations of neurons, including encoding and decoding of stimuli with distributed representations in neural populations. Finally, we will estimate the sparseness of neural representations in the inferior-temporal cortex.

5.1 How neurons talk

5.1.1 Cortical chatter

The activities of single neurons can be measured extracellularly by guiding a microelectrode close to the cell body as illustrated schematically in Fig. 5.1. The characteristic waveforms in these signals can be used to identify the occurrence of single spikes from these measurements, and the signal can be amplified and monitored with an oscilloscope and loudspeaker. Should you have the opportunity to observe a cell recording you should take it; it is very instructive to hear the chattering of a cortical neuron. The irregular ticking from the speaker has fascinated neuroscientists for many decades.

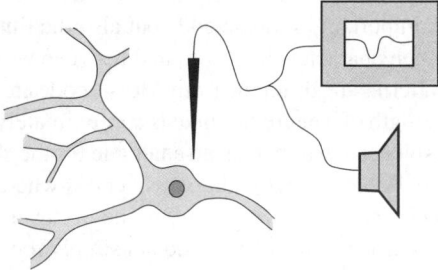

Fig. 5.1 Schematic view of an extracellular recording of neuronal activity using a microelectrode.

Regular spiking of neurons would not be that interesting because we can predict, in such a case, the next firing of the neuron with high precision. The occurrence of a regular spike would therefore not convey any information; only deviation from regular spiking conveys information. Just think about how surprised you would be if

Table 5.1 International Morse code of short (.) and long (-) electrical signals, a variation of a code invented by Samuel Morse around 1835. The length of time for a 'dash' signal is set to be three times as long as a 'dot' signal. The time between signals within characters is equal to one dot signal, equivalent to three dot signals between characters, and the time gap between words is equal to seven dot signals.

A	. -	K	- . -	U	. . -	4	. . . . -
B	- . . .	L	. - . .	V	. . . -	5	. . . . .
C	- . - .	M	- -	W	. - -	6	- . . . .
D	- . .	N	- .	X	- . . -	7	- - . . .
E	.	O	- - -	Y	- . - -	8	- - - . .
F	. . - .	P	. - - .	Z	- - . .	9	- - - - .
G	- - .	Q	- - . -	0	- - - - -	Fullstop	. - . - . -
H	. . . .	R	. - .	1	. - - - -	Comma	- - . . - -
I	. .	S	. . .	2	. . - - -	Query	. . - - . .
J	. - - -	T	-	3	. . . - -		

the predicted spike would not occur. The example of a regularly spiking neuron shows us that we need irregular spiking if we want to convey information via the spiking pattern of a neuron. Information in the spike train is, of course, only useful if it is functionally relevant, that is, if it is not just random noise. Neuroscientists therefore search for functional correlates in the firing patterns of neurons.

5.1.2 Morse code and neural code

The ticking of a speaker driven by a cortical neuron makes us wonder what the neuron is 'saying'. Indeed, we might expect some information in the firing pattern. It is therefore instructive to compare the firing of a neuron with the way in which messages were transmitted in early telegraphic transmissions. A telegram was transmitted by electrical signals over wires using *Morse code*, which specifies the correspondence between the pattern of electrical signals and characters as listed in Table 5.1. With the code table we can decipher a message transmitted to us via this code. Not only are the length of the individual signals important in Morse code but also the times between signals, which are different between characters in a word and between words.

Neuronal spiking patterns are different from Morse code in many respects. For example, the individual length of a neuronal spike is approximately constant due to the stereotyped form of the spike, so that there is no analogue for the short and long signal in the Morse code. It is possible to devise alternative codes where only one length of a signal, for example, a dot, and the timing between the occurrences of a dot is used. However, it is clear that neuronal spike patterns do not simply represent an alphabet as in the case of the Morse code. Our aim in *deciphering the neural code* is understanding the relevance of particular node activities for the representation, transmission, and processing of information in the nervous system.

5.1.3 The firing rate hypothesis

Neuroscientists try to decipher the neural code by searching for reliable correlations between firing patterns and behavioural consequences or correlations between sensory stimuli and neural activity patterns. One of the first scientists to use microelectrode recordings in the 1920s, after sufficient amplification of the small electrical signals became available through vacuum tubes, was the English physiologist *Edgar Douglas Adrian*. And one of the first effects he recognized was that the number of spikes of a neuron often increases with an effective stimulus for this neuron. This corresponds to an increase of the instantaneous firing rate as defined in the last section (see eqns 4.5 and 4.6). An increasing firing rate is easily detectable and is mainly used by neurophysiologists to search for stimuli that 'drive' a neuron.

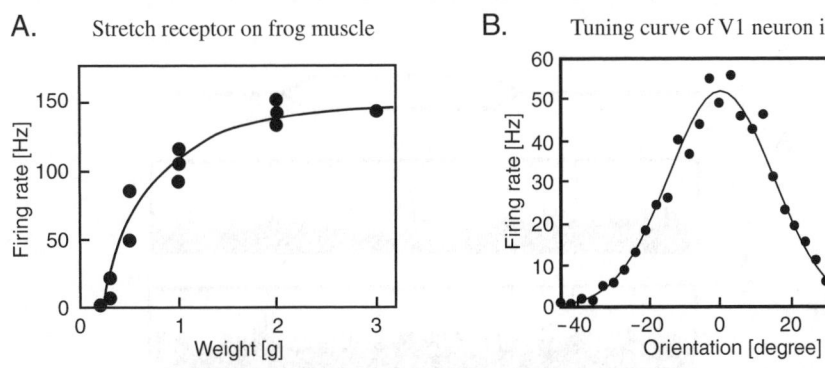

Fig. 5.2 (A) Data from Adrian's original work showing the firing rate of a frog's stretch receptor on a muscle as a function of the weight load applied to the muscle [redrawn from Adrian, *J. Physiol. (Lond.)* 61: 49–72 (1926)]. (B) Response (tuning curve) of a neuron in the primary visual cortex of the cat as a function of the orientation of a light stimulus in form of a moving bar [data from Henry, Dreher, and Bishop, *J. Neurophys.* 37: 1394–1409 (1974)].

An example of a rate code explored by Adrian is that of the stretch receptor on the frog muscle that increases with increasing weights on the muscle as shown in Fig. 5.2A. In general, firing rates of sensory neurons typically increase notably in a short time interval following the presentation of an effective stimulus to the recorded neuron. The set of stimuli that reliably increases the firing rate of neurons is called the *receptive field* of the neuron. An example of the response of a sensory neuron in the primary visual cortex, called the *tuning curve* of the neuron, is shown in Fig. 5.2B. Simple cells in this visual area respond to orientations of bars, and the orientation tuning curve can be fitted reasonably well with a Gaussian function. The information represented in the primary visual area must also be conveyed to later visual areas as the firing rate of sensory neurons can often be correlated to the sensory awareness of a stimulus. The *firing rate hypothesis* has since dominated much research in neurophysiology.

5.1.4 Correlation code

Information processing in the brain usually includes a variation of the firing rate. This is not very surprising as we expect a variation in the number of spikes with varying input currents. However, we want to ask if only the firing rate is used in the brain to convey information. A rare example where the firing rate does not show the relevant information is shown in Fig. 5.3. The figure displays the response of two neurons in the primary auditory cortex to a 4 kHz tone with an amplitude envelope shown at the top. Figure 5.3B shows the average firing rate at 5 ms intervals over many trials for each neuron respectively. No significant variation of the firing rate can be seen in either of the neurons. However, some stimulus-locked variation in the relative spiking of the two neurons can be seen when plotting the rate of spikes from one neuron that occurred within a fixed short interval of the spiking of the other reference neuron (Fig. 5.3C). This rate indicates the probability of co-occurrence of the spikes of the two neurons. We will see next that this rate of spikes from different presynaptic neurons in close temporal proximity is the relevant rate for a receiving neuron that is a leaky integrator.

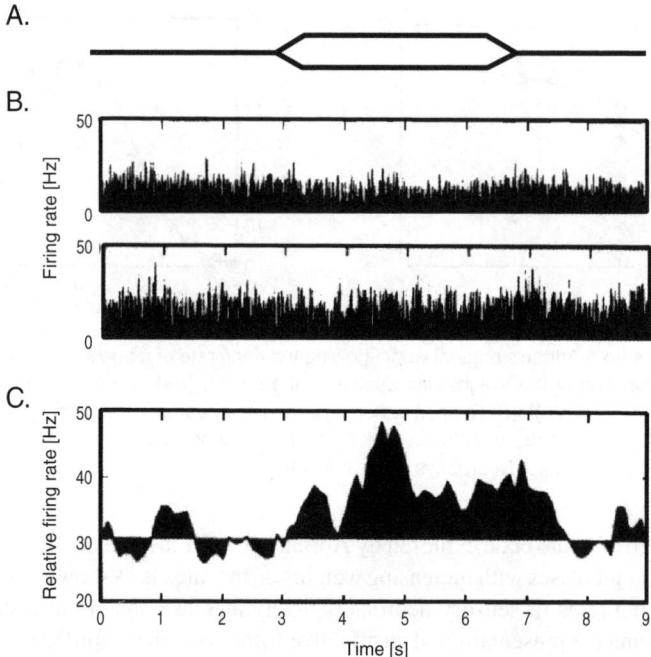

Fig. 5.3 An example of the response of some neurons in the primary auditory cortex that do not show significant variations in response to the onset of a 4 kHz tone with the amplitude envelope shown in (A). (B) Average firing rates in 5 ms bins of two different neurons. (C) Spike-triggered average rate that indicates some correlation between the firing of the two neurons that is significantly correlated to the presentation of the stimulus [from DeCharms and Merzenich, *Science* 381: 610–13 (1996)].

5.1.5 Integrator and coincidence detector

We mentioned in the last chapter that the determination of the firing rate of one sending neuron by a receiving neuron requires a rather long time window of integration in the receiving neuron. A *perfect integrator* is illustrated in Fig. 5.4A, which integrates the spikes of two presynaptic spike trains as indicated at the bottom. The membrane potential accumulates the synaptic currents triggered by the presynaptic spikes and therefore counts the number of presynaptic spikes since the last reset of the membrane potential. This can be contrasted to a leaky integrator, illustrated in Fig. 5.4B. In this case the membrane potential decays after an increase from the input currents triggered by the presynaptic spikes, and the neuron becomes sensitive to the relative timing of spikes of different presynaptic neurons. Such a leaky integrator with a small time constant can be used as a *coincidence detector*.

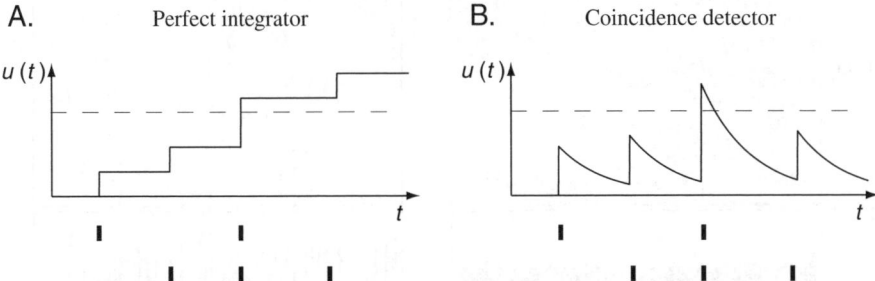

A.   Perfect integrator   **B.**   Coincidence detector

Fig. 5.4 Schematic illustration of (A) a perfect integrator and (B) a leaky integrator that can be utilized as coincidence detector. In this example the membrane potential $u(t)$ integrates short current pulses of the two spike trains shown at the bottom.

A neuron generates a spike after the membrane potential reaches a threshold at which time the membrane potential would have to be reset (not included in Fig. 5.4). With the thresholds indicated by dashed lines in Fig. 5.4, both neurons, the integrator and the coincidence detector, would fire at the same time. However, the reason for the firing would be very different in the two cases. In the first case, that of the perfect integrator, the neuron would fire because four presynaptic spikes had occurred since the last firing of the neuron. This neuron would also fire at this time if one of the two last simultaneous spikes would have occurred at an earlier time. In contrast, the spike of the leaky integrator only occurs because of the occurrence of two simultaneous presynaptic spikes. With a higher firing threshold, larger then the effect of the sum of two simultaneous EPSPs, it is also possible to employ such a leaky integrator neuron to detect the coincidence of more than two spikes. The time window of coincidence depends on the decay constant (membrane constant) and the time course of the EPSPs. In real neurons this should be of the order of a few milliseconds or less.

5.1.6   How accurate is the spike timing?

The transmission of spikes is not very reliable, and we expect a lot of variability in neuronal response in the neural system as discussed in Chapter 3. Another demonstration of spike time variability is shown in Fig. 5.5 for responses of a cell in the middle temporal area (MT) that responds to movements of visual stimuli. The top graph of Fig. 5.5A shows the response of the neuron in several trials to a stimulus with constant velocity as indicated at the bottom. The middle graph shows the trial average. The data indicate a quite reliable initial response of the neuron. After this initial response the neuron still fires rapidly; however, the times of these spikes are rather different in each trial. Some people search for a neural code in the firing pattern of neurons in response to a constant pattern. The data in Fig. 5.5A already indicate that there might not be much information in the continuing firing pattern of the neuron due to the enormous variability. In the next section we will outline how such a statement can be quantified.

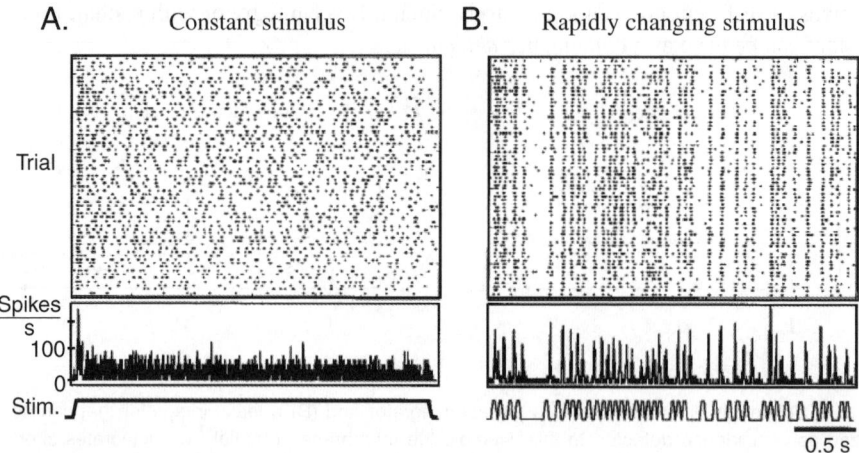

Fig. 5.5 Spike trains (top) and average response over trials (middle) of an MT neuron to a visual stimulus with either constant velocity (A) or altering velocity (B) as indicated in the bottom graph [adapted from Buračas *et al.*, *Neuron* 20: 959–69 (1998)].

Data like those shown in Fig. 5.5A lead to the impression that neuronal spike times are very variable and imprecise. This has, however, to be viewed in relation to the experimental situation. Figure 5.5B shows the response of the same neuron to a fast varying stimulus. The response of the neuron follows the variations in the stimulus rapidly with little jitter. Each neuron does not always respond with a spike to the changing velocity of the stimulus, and some spikes still occur in between the stimulus changes. But the majority of spikes follow the changing stimulus.

We can speculate about the nature of the information processing using the somewhat exaggerated scenario that the firing patterns of neurons code elaborate messages that are processed in the nervous system. We study this hypothesis in this chapter with the help of information theory. Information theory allows us to quantify the amount of information we can gain from messages. We can use this to compare the information in the spike pattern of a spike train with the information that is present in the firing rate of spike trains. We will see that most often (though not always) a large amount of information is carried by the firing of only a few spikes, which makes the firing rate highly relevant. Thus, our conclusion about the nature of information processing in the brain will somewhat revise the above neural code hypothesis. We suggest a revised view in which sensory events elicit a cascade of neuronal activity through a neuronal network. Information processing is thus achieved mainly by the recruitment of neurons along processing pathways. In the rest of the book we will concentrate mainly on information processing through rapid propagation of neuronal activity through neuronal networks.

5.2 Information theory

5.2.1 Communication channel

We talked very casually about information and how information can be transmitted by signals over telegraph wires or in neural systems. It is useful to define information more formally so that we can quantify the amount of information that is or can be transmitted in different systems. This was done in 1948 by *Claude Shannon* who studied the transmission of information over a general *communication channel* as shown in Fig. 5.6. A message x_i, one of several possible messages of a set X from an information source, is converted into a signal $s_i = f(x_i)$ that can be transmitted over the communication channel to a receiver. The receiver converts the received signal r_i back to a message $y_i = g(r_i)$, one of a set Y of possible output messages, that can be interpreted by the destination receiver. Shannon included a noise input η in the channel that is intended to capture faulty transmissions, for example due to physical noise in the transmission channel or faulty message conversion. In the noiseless case the receiver receives the sent signal $r_i = s_i$, which can be converted into the original message when the receiver does the inverse function of the encoding done by the transmitter, that is, $g = f^{-1}$. Note that we can consider several noise models such as additive noise, for example, $r = s + \eta$, or multiplicative noise, for example, $r_i = s_i * \eta$, where η is a random variable.

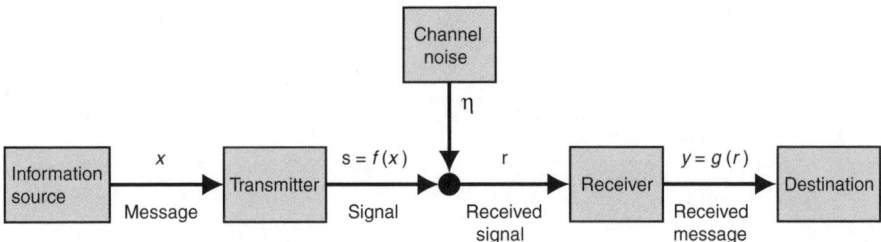

Fig. 5.6 The communication channel as studied by Shannon. A message x is converted into a signal $s = f(x)$ by a transmitter that sends this signal subsequently to the receiver. The receiver generally receives a distorted signal r consisting of the sent signal s convoluted by noise η. The received signal is then converted into a message $y = g(r)$ [adapted from C. Shannon, *The Bell System Technical Journal* 27: 379–423 (1948)].

5.2.2 Information gain

What is the amount of information we gain when receiving a message y_i? This depends highly on our expectations. For example, if I tell you that the first name of 'Shannon' is 'Claude' then you gain no information if you have read carefully the last paragraph where it was already mentioned. If you thought it was either 'Albert' or 'Claude' you gain some information. You gain more information from my message if you thought it was one of four possible choices with equal likelihood. Information depends on the set of possible messages and their frequencies, and a measure of information gain should therefore be a function of the probability $p_i = P(y_i)$ of a message. Furthermore, independent information should be additive, that is, if I tell you that the middle name of Claude Shannon is 'Elwood', then you gain a bit of information independent of and additional to the information I gave you previously. The probability of independent messages is the product of the probabilities of the individual messages. We thus need a function (of the probabilities) that has the property

$$f(xy) = f(x) + f(y). \tag{5.1}$$

The logarithm has these characteristics, and we therefore define the information gain when receiving a message y_i as

$$I(y_i) = -\log_2(p_i). \tag{5.2}$$

The minus sign makes the information positive as the probabilities are less than 1. Note that we use the logarithm with basis 2 and no additional proportionality constant, which is just a convention. This convention defines the units of one *bit*. If we have two possible states with equal likelihood, then the transmission of one state allows us to gain 1 bit of information.

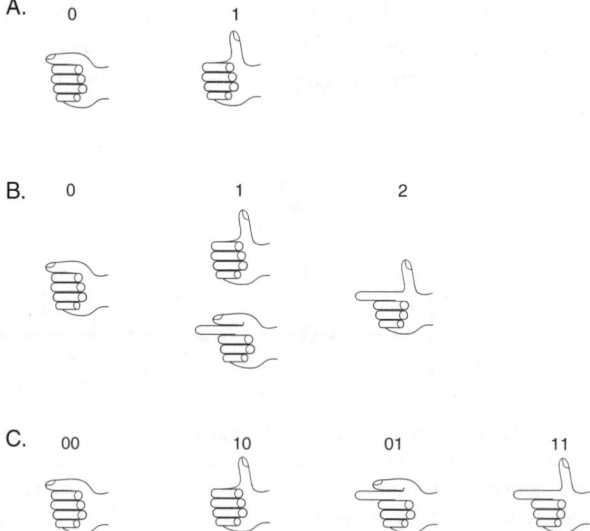

Fig. 5.7 Example of hand signals used to transmit messages. (A) Only two positions of the thumb are used to encode two possible messages '0' or '1'. (B) Two fingers are used to encode three messages by the total amount of extended fingers. (C) Two fingers are used with a different code, in which the position of each of the two fingers used for the signalling is essential.

Let us illustrate the concept further with information transmitted by hand signals as illustrated in Fig. 5.7. In Fig. 5.7A we illustrate a case where we use only the thumb in two possible locations, either horizontal or up, to transmit a message, say '0' and '1'. If both cases are equally likely, that is, $P(0) = P(1) = 1/2$, then we receive an information gain of

$$I = -\log_2(0.5) = 1 \tag{5.3}$$

bit of information. If we use two fingers with two possible locations and the code of the total number of fingers to transmit information as illustrated in Fig. 5.7B, then we can transmit three different messages. The gain of information from each signal is therefore $I = -\log_2(1/3) = 1.585$ bits with equally frequent signals. In the code illustrated in Fig. 5.7B it does not matter if the pointing finger is extended or if the thumb is up. In contrast, in Fig. 5.7C we have changed the code to transmit a message with two fingers where now the position of the finger matters. We can then transmit four different messages, and each equally likely signal would convey 2 bits of information.

We demonstrated the information gain with discrete events that had a certain probability distribution. Discrete events are also most appropriate when thinking about message transmission with a finite alphabet. However, it is mathematically often more convenient to write the following formulations in continuous form. The information gain is thus often written in the form

$$I(x) = -\log_2 P(x). \tag{5.4}$$

Note that this formulation can still be used for discrete events. The probability function $P(x)$ then has non-vanishing values only for a set of discrete values x. If we consider a continuous set X of events x then we have to take the corresponding probability density function $p(x)$ into account (see also Appendix B). The probability of an event x is then an infinitesimally small quantity $P(x) = p(x)\mathrm{d}x$, and the information gain for one of those events goes to infinity. However, averages of continuous sets are still defined as we will see in the following, and it is easy to replace integrals with sums (and vice versa) in the formulas below.

We have so far only considered the cases where a message conveyed a definitive statement. In contrast, if I tell you that the first name of Shannon is 'Claude' with 90% probability out of four possibilities, then there is still some uncertainty left which we have to take into account. We have then to distinguish the probability of an event before the message was sent, called the *a priori* or *prior* probability $P^{\mathrm{prior}}(x)$, and the probability of an event after taking the information into account, which is called the *posterior* probability $P^{\mathrm{posterior}}(x)$. The information gain is then defined by

$$I(x) = -\sum_x P^{\mathrm{posterior}}(x) \log_2 \frac{P^{\mathrm{prior}}(x)}{P^{\mathrm{posterior}}(x)}. \tag{5.5}$$

In the previous examples we knew the precise answer after the message was received, so that $P^{\mathrm{posterior}}(x) = 1$. We also wrote the prior probability simply as $P^{\mathrm{prior}}(x) = P(x)$. Equation 5.4 is therefore only a special case of the more general formula, eqn 5.5.

5.2.3 Entropy, the average information gain

We have used equally likely signals in the previous numerical examples. It is, however, often the case that not all messages are equally likely, and the amount of information gain is therefore different for different messages. For example, let's go back to the example illustrated in Fig. 5.7A and consider the case where the thumb is up in 3/4 of all messages sent out and only horizontal in 1/4 of the cases. When we receive a signal of thumb horizontal, then we gain $I(0) = -\log_2(1/4) = 2$ bits of information, whereas the signal of 'thumb up' conveys only $I(1) = -\log_2(3/4) = 0.415$ bits of information. The average amount of information in the message set of the information source is

$$S(X) = - \sum_i p_i \log_2(p_i), \tag{5.6}$$

which is called the *entropy* of the information source. The entropy is a quantity of the message set and is not defined for an individual message. The entropy of the set of possible received messages $S(Y)$ is defined in a similar way. The average amount of information gained by the transmission of a signal in the previous example is $S = 1/4 * 2 + 3/4 * 0.415 = 0.8113$. The entropy is a measure of the average information gain we can expect from a signal in a communication channel, though it is good to keep in mind that individual messages can have larger and smaller information gain values. Some rare events might indeed convey a very large amount of information.

The entropy of a message set with N possible messages that are all equally likely is

$$S(X) = - \sum_{i=1}^{N} \frac{1}{N} \log_2\left(\frac{1}{N}\right) = \log_2(N). \tag{5.7}$$

The entropy is hence equivalent to the logarithm of the number of possible states for equally likely states. This is a useful fact to keep in mind. More generally, the entropy measures the effective number of states by weighting the logarithm of the number of states with the frequency of their occurrences.

We can formulate the entropy for continuous distributions in an analogous way. We again write the set with capital letters, for example, X, and examples from this set with lower case letters, for example, x. The entropy of a set X with probability density function $p(x)$ is then defined as

$$S(X) = - \int_X \mathrm{d}x p(x) \log_2 p(x). \tag{5.8}$$

For example, the entropy of a Gaussian-distributed set with mean μ and variance σ^2,

$$p(x) = \frac{1}{\sqrt{2\pi}\sigma} e^{-\frac{(x-\mu)^2}{2\sigma^2}} \tag{5.9}$$

is given by

$$S = \frac{1}{2} \log_2(2\pi e \sigma^2). \tag{5.10}$$

This entropy does depends on the variance, not the mean. This is intuitively clear as the average information gain is only a relative quantity and information is transmitted by variations of a signal around its mean value.

5.2.4 Mutual information

So far we have only considered noise-free signal transmission and receivers that invert precisely the encoding of the transmitter ($g = f^{-1}$). This corresponds to perfect transmission and unambiguous decoding. However, in all practical applications of information transmission, including information transmission in neural systems, we must take noise into account. We have then to distinguish the received signal y from the sending signal x. The question we are most interested in is what a received signal y can tell us about an event x. Indeed, we will later attempt to reconstruct an event, such as a sensory input to the neural system, from the firing pattern of neurons. For now we want to ask what is the average information gain from what a message y tells us about the set of events that could happen. We therefore have to consider conditional probabilities such as $P(x|y)$ (or corresponding density functions). The average information gain with such conditional probabilities over both sets is

$$\int_y \mathrm{d}y\, P(y) S(X|y) = \int_X \int_Y \mathrm{d}x\mathrm{d}y\, p(y)p(x|y) \log_2 \frac{p(x|y)}{p(x)}. \qquad (5.11)$$

We can express the conditional probabilities with joint distributions,

$$P(x|y) = \frac{P(x,y)}{P(y)}, \qquad (5.12)$$

so that we can also write eqn 5.11 as

$$I^{\mathrm{mutual}} = \int_X \int_Y \mathrm{d}x\mathrm{d}y\, p(x,y) log_2 \frac{p(x,y)}{p(x)p(y)}. \qquad (5.13)$$

This quantity is called the *cross-entropy* or *mutual information*. It is symmetric in the arguments (which motivated the term mutual) and describes the average amount of information that can be gained by receiving a message y when a signal x is sent (or vice versa). We can also rewrite the mutual information as difference between the sum of the individual entropies $S(X)$ and $S(Y)$ and the entropy of the joint distributions $S(X, Y)$,

$$I^{\mathrm{mutual}}(X, Y) = S(X) + S(Y) - S(X, Y). \qquad (5.14)$$

The joint entropy in the case of independent events in X and Y is $S(X, Y) = S(X) + S(Y)$. The mutual information in this case is equal to zero,

$$I^{\mathrm{mutual}}(\text{all } y \in Y \text{ independent of all } x \in X) = 0, \qquad (5.15)$$

which reflects the fact that received messages are independent of sent messages. However, for information transmission we want messages such that the received message and the sent message are dependent. Mutual information not equal to zero then reflect some correlation of the events in X and Y. The mutual information in a communication channel describes the average amount of information we can expect to flow between input and output events.

5.2.5 Channel capacity

An amount of information that can be transmitted by a change in the signal is proportional to the number of states the signal can have. For example, in Fig. 5.7A we have considered only two allowed positions of the thumb and could therefore send one bit of information in each time interval in which the thumb can change. We can increase the number of states the thumb can take as illustrated in Fig. 5.8. There we allowed for three positions $(0°, 45°, 90°)$ and we could therefore transmit $I = \log_2 1/3 \approx 1.6$ bit of information at every time step with equally likely messages. We could increase the number of thumb states further until it will be difficult for the observer to distinguish the differences. There is therefore a limiting factor given by the resolution of our visual system and the noise introduced by the motor system of the sender. It is intuitively clear that the amount of information that can be transmitted over a noisy communication channel is limited by the ratio of the signal to noise or, to be more precise, by the ratios of the variance of these quantities as the information is conveyed by changes in signals (see eqn 5.10).

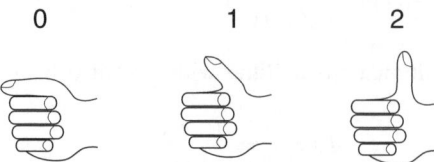

Fig. 5.8 Example of hand signals using the thumb in three different states to transmit messages.

We can calculate the mutual information of a channel with noise for particular noise models and probability distributions for the input signal and the noise. It is common to consider additive Gaussian noise as well as Gaussian-distributed input signals. It turns out that the amount of information transmitted in such a Gaussian communication channel is the best we can achieve. The amount of information that can be transmitted through a Gaussian channel is called the *channel capacity*. The information we can transmit through a noisy channel is therefore always less than the channel capacity, which is given by

$$I \leq \frac{1}{2} \log_2 \left(1 + \frac{\langle x^2 \rangle}{\langle \eta_{\text{eff}}^2 \rangle}\right), \tag{5.16}$$

where $\langle x^2 \rangle$ is the variance of the input signal and $\langle \eta_{\text{eff}}^2 \rangle$ is the effective variance of the channel noise, considering the noise as transmitted itself as signal through the channel. The ratios $\langle x^2 \rangle / \langle \eta_{\text{eff}}^2 \rangle$ is called the *signal to noise ratio* (SNR). It is interesting to note that the information transmission with given noise in the transmission channel can be increased by increasing the variability of the input signals. It seems that the high variability of spike trains, as discussed in Chapter 3, is therefore particularly suited for transmission in noisy neural systems. There are indeed several possible mechanisms that can increase the variability of neuronal response, some of which we will discuss in Chapter 8.

5.3 Information in spike trains

5.3.1 Entropy of a spike train with temporal coding

It is very instructive, and important for the comparisons with experimental data outlined below, to calculate the maximum entropy of a spike train with temporal coding up to a certain precision. To do this we introduce small time bins Δt, small enough that one time bin can contain at most one spike. We can then view the spike train as a binary string with 0 in a time bin corresponding to no spike in this time interval and 1 in a time bin in which a spike has occurred. An example is illustrated in Fig. 5.9. With small time bins we can assume a binary string with many more 0s than 1s, indeed many more than illustrated in the figure. We will also assume that the events in the bins are independent. Dependence would lower the entropy as mentioned before, and we are here interested in estimating the maximaum information carried in a spike train.

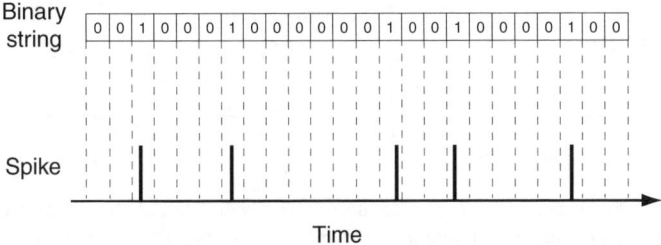

Fig. 5.9 Representation of spike train as binary string with time resolution Δt.

The firing rate r of the spike train fixes the number of 1s and 0s in the spike train of length T. We can then calculate the number of possible spike patterns with the fixed number of spikes in the sequence of length T. The logarithm to base 2 of this number is the entropy of the spike train. This can thus be calculated to be

$$S = -\frac{T}{\Delta t \ln 2}\left[r\Delta t \ln\left(r\Delta t\right) + \left(1 - r\Delta t\right)\ln\left(1 - r\Delta t\right)\right]. \tag{5.17}$$

For time bins much smaller than the inverse firing rate, that is, $\Delta t \ll 1/r$, this is approximately equal to

$$S \approx Tr \log_2\left(\frac{e}{r\Delta t}\right). \tag{5.18}$$

The term $N = Tr$ is the number of spikes in the spike train, so that the average entropy per spike is given by

$$\frac{S}{N} \approx \log_2\left(\frac{e}{r\Delta t}\right). \tag{5.19}$$

This is the maximum entropy per spike of a spike train for a given timing precision Δt when we can assign to each individual spike pattern a unique message or sensory event. It is maximum in the sense that we considered only the noiseless case, and noise would diminish the information gain of each message.

The maximum entropy per spike is plotted for three different firing rates in Fig. 5.10A. What is remarkable is that the entropy per spike is for the parameters shown always larger than one bit. The reason for this is that the interspike intervals are long so that the occurrence of a spike becomes highly relevant. Also, we have normalized the entropy to the number of spikes, which can be a bit misleading; the absence of a spike can also bear information, and the measure of the entropy takes this into account. We have only normalized the spike train entropy to the total number of spikes to make it easier to compare the entropies of different spike trains and coding schemes.

Figure 5.10A illustrates that the maximal entropy decreases with increasing precision interval Δt (decreasing time precision). This is because the number of patterns

A. Maximum entropy with spike code

B. Maximum entropy with rate code

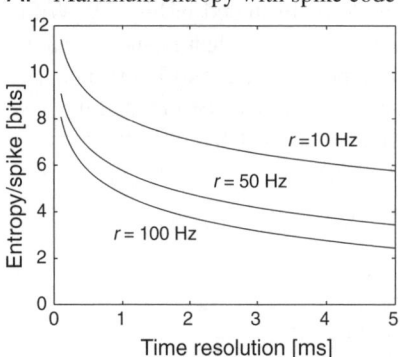

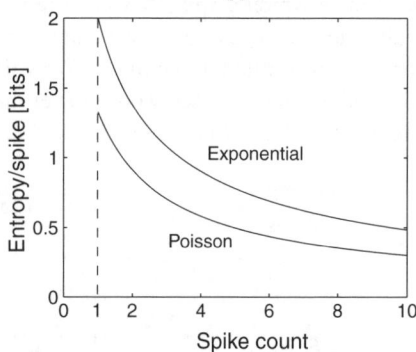

Fig. 5.10 (A) Maximum entropy per spike for spike trains with different firing rates as a function of the size of the time bins, which sets the resolution of the firing times. (B) Maximum entropy per spike for spike trains from which the information is transmitted with a rate code. The two different curves correspond to two different spike statistics of the spike train, a Poisson and an exponential probability of spike counts. Spike trains with exponential spike distributions can convey the maximum information with a rate code for fixed variance.

that can be represented by the spike train decreases with fewer time intervals in a spike train of length T. The entropy per spike also decreases with increasing firing rate because the individual spike is then not so surprising and bears less information. We should also keep in mind that the transmission of the information contained in a spike train is lowered by noise in the transmission process and by spike correlations. However, the maximum entropy as calculated above gives us the scale to which we can compare measurements of information transmission in the brain.

5.3.2 Entropy of a rate code

We can compare the entropy of a spike train above, in which we considered all possible spike patterns (which is therefore a time code), with other coding schemes that do not employ all possible spike patterns such as a rate code. In the rate code only the number of spikes n in a certain interval T is relevant. The entropy of observing N spikes in a time interval T can be calculated by applying the definition of the entropy,

$$S(N;T) = -\sum_n P(n) \log_2 P(n)$$

$$= -\frac{1}{\ln 2} \sum_n P(n) \ln P(n) \tag{5.20}$$

where $P(n)$ is the probability of observing n spikes in the time interval T. For example, let us calculate the entropy for a *Poisson spike train* of length T. The probability of observing n spikes, when we expect $N = rT$ spikes, is given by the Poisson distribution,

$$P(n) = \frac{N^n e^{-N}}{n!}. \tag{5.21}$$

Inserting this probability into the log part of formula 5.20 we get

$$S(N;T) = -\frac{1}{\ln 2} \sum_n P(n)(n \ln N - N - \ln (n!)). \qquad (5.22)$$

We can approximate the last term with Stirling's formula

$$\ln (n!) \approx (n + \frac{1}{2}) \ln n - n + \frac{1}{2} \ln (2\pi). \qquad (5.23)$$

At this point we should remember the definition of the expectation value $E(x)$ of a quantity x,

$$E(x) = -\sum_x P(x)x, \qquad (5.24)$$

and the expectation value of n is just $E(n) = N$. The entropy of a Poisson spike train with rate code is therefore

$$S(N;T) \approx \frac{1}{2} \log_2 N - \frac{1}{2} \log_2 2\pi. \qquad (5.25)$$

We divide this value again by the number of spikes so that we get values for the entropy per spike. This is plotted in Fig. 5.10B. The entropy of the rate code is, of course, much smaller than the entropy of a spike train in which we allowed the precise spike timing to be relevant. The length of time over which we have to integrate to achieve a particular spike count depends on the firing rate. To get a spike count of 2 we have to integrate on average for 40 ms if a neuron fires with mean firing rate of 50 Hz.

We calculated the entropy of a spike train with rate coding for the example of a Poisson spike train. Spike trains with other probability distributions of spike counts can have different entropies within the rate code. We can ask what would be the distribution of the spike count with fixed mean and variance that would result in the maximum entropy. The answer is that the entropy under these conditions is maximized by an exponentially distribution. An exponential distributed spike train

$$P(n) \propto e^{-\lambda n} \qquad (5.26)$$

with average firing rate $r = 1/[T(\exp(\lambda) - 1]$ has an entropy of

$$S(N;T) = \log_2 (1 + N) + N \log_2 (1 + \frac{1}{N}). \qquad (5.27)$$

The entropy per spike for this distribution is also shown in Fig. 5.10B. We can see that the entropy of such spike trains is moderately larger than the entropy for the Poisson spike train and reaches a value of precisely 2 bits/spike in the limit where a single spike carries all the information. The average information in a rate code can therefore be reasonably high so that the information that flows through the brain even with this code can be enormous.

5.3.3 Difficulties in measuring entropy

Measuring entropies is a data hungry task. This comes partly from the fact that we have to estimate probability distributions. To do this we have to observe many instances, from which we can, for example, construct a histogram. The histogram gives us the numbers of the frequencies with which events occur within a certain resolution determined by the bin size of the histogram. We can often get some reasonable estimations of the probability function if we have at least some prior knowledge about the form of the probability function. We can then fit the histograms with the expected function to fix the unknown parameters to values consistent with the experimental observations. Often, however, we do not know *a priori* the form of the distribution functions, and suitable guesses from the data are common. We have then to check that the results do not depend critically on the choice of distribution function.

A further difficulty is that information depends on the logarithm of probabilities. This means that events with small probabilities have a large factor in the entropy, and reliable measurements of rare events do demand a large amount of representative data. To get probability distributions from frequency histograms we have to normalize the histograms so that the sum over all data becomes one (see Appendix B),

$$\int \mathrm{d}x P(x) = 1. \tag{5.28}$$

Due to this normalization procedure we tend to overestimate the information content of the events measured if we miss out some rare events with high information content. Estimations of entropy therefore tend to overestimate the true entropy. This is a potential danger for suitable interpretations of experimental results. For example, a large entropy per spike, much larger than, for example, 2 bits, would rule out rate codes and would therefore indicate that temporal structures in the spike trains are used to transmit information. An overestimation of entropies from experimental data can obscure such conclusions. Some methods have been developed to compensate for such systematic shifts of small data sets, but such methods cannot, of course, completely compensate for real data.

5.3.4 Entropy measured from single neurons

The calculation of the full mutual information between two neurons entails the determination of the firing probabilities of each individual neuron as well as the joint firing probability. It is, however, simpler to measure only one neuron and the probability distribution of the response of this neuron to sensory stimuli s. We have then only to calculate the conditional probability $P(y|s)$ and can consider the quantity

$$I^S(s) = \int_Y \mathrm{d}y P(y|s) \log_2 \frac{P(y|s)}{P(y)}. \tag{5.29}$$

This quantity looks like an information measure. However, we have used conditional probabilities in this measure so that this quantity does not necessarily have the additivity feature that we demand from information measures. We therefore call this measure the *stimulus-dependent surprise*. The average over this quantity still corresponds to the mutual information $I^{\mathrm{mutual}}(Y, S)$ between the stimulus set and neuronal response. Examples for the stimulus-dependent surprise per spike are shown in Fig. 5.11 for two visual neurons in the inferior-temporal cortex that respond to faces. The average firing rate of the responding neurons shown in Fig. 5.11A was 54 Hz. Responses with rates around this firing rates bear little surprise value. Some responses with high firing rates for this neuron carried high surprise values of around 2 bits per spike. However, the average surprise value in the responses of this neuron to all 20 faces in the stimulus set was only 0.55 bits. This value is not surprisingly high and does not suggest the need of a code within the firing pattern of these neurons. Another example is shown in Fig. 5.11B. The surprise values are again lowest around the mean firing rates of the neuron, which is higher than that of the neuron shown on the left. This neuron did carry the largest surprise values for small firing rates and carried an average surprise value similar to that in the previous example.

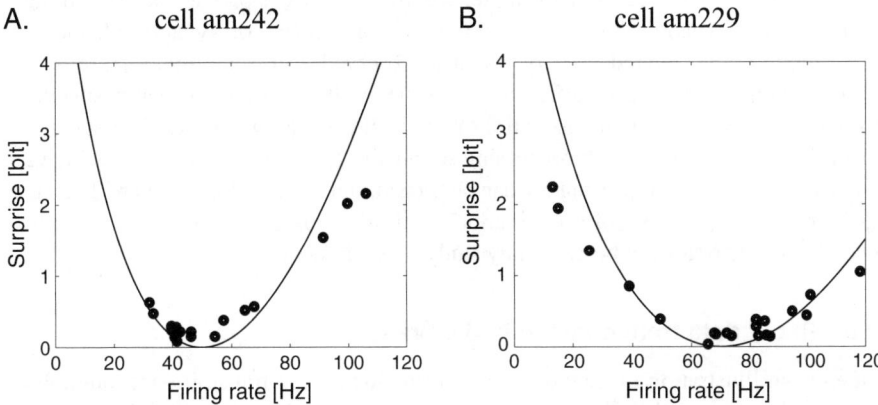

Fig. 5.11 Stimulus-dependent surprise for two cells in the inferior temporal cortex that respond to faces. The curve is 5 times the universal expectation for small time windows (eqn 5.30) [data from Rolls, Treves, Tovee, and Panzeri, *J. Comp. Neuro.* 4: 309–33 (1997)].

The values for the surprise rate were calculated for 500 ms spike trains during the presentation of the faces. Several studies have also shown that much of the information about a stimulus is transmitted in much smaller time intervals. For very small time intervals, so that at most one spike can occur during this time, there is not much freedom for a coding scheme so that we can expect a universal curve, which is given by

$$I^s(\Delta t \ll 1) = r^s \log_2\left(\frac{r^s}{r}\right) + \frac{r - r^s}{\ln 2}, \qquad (5.30)$$

where r^s is the specific response rate for stimulus s, and r is the average firing rate over all stimuli. This universal surprise curve, divided by the average firing rate r and multiplied by a factor of 5, is plotted in Fig. 5.11 as a solid line. We scaled the curve up because the experimental data were obtained in large time windows for which the values can be higher. What is however striking is that the experimental data follow approximately the universal curve for small time windows. This is another hint that elaborate codes of spiking pattern might not be employed by neurons in the central nervous system.

5.3.5 Spike-dependent network processing in the brain

So far there is little evidence that spike patterns are utilized for information processing in the brain. This does not mean that the timing of a spike does not matter. Quite the opposite, the emission of spikes, in particular the temporally close emission of spikes from connected neurons, can stimulate waves of neural activity in the brain. The reason for this is evident from the leaky integrators characteristic of neurons that give more weight to synchronous presynaptic events compared to the same number of spikes distributed over a long period of time. There are also examples where precise spike timings are essential. For example, we will see in Chapter 8 that the timing of pre- and postsynaptic spikes is essential for some forms of synaptic plasticity. Some temporally correlated activity was also observed even for remote neurons in the brain such as neurons in different hemispheres. It is so far unclear if specific information is coded by such correlations or if such correlations only reflect specific temporally related information-processing streams in the brain. In any case we have seen that vast amounts of information can be transmitted in short times even with a rate code (representing the rate of a population), and many insights into the information processing of the brain can be gained by studying such models.

5.3.6 Rapid data transmission in the brain

A interesting illustration of how quickly information can be transmitted through the brain was demonstrated by *Simon Thorpe* and colleagues. They showed that human subjects are able to discriminate the presence or absence of specific object categories, such as animals or cars, in visual scenes that are presented for very short times, as short as 20 ms. The percentage of correct manual responses, which consisted of releasing a button only when an animal was present in a complex image that was presented for only 20 ms, is shown for 15 subjects in Fig. 5.12A plotted against the mean reaction time for each subject. The experiment shows some trade-off between reaction time and recognition accuracy, but the important point to note here is the high level of performance for such short presentations of the images.

The ability to recognize objects with these short presentation times is not the only astonishing result in these experiments. The authors also recorded skull EEGs during the experiments. The event-related potential, averaged over frontal electrodes, is shown in Fig. 5.12B, separated for image presentations with and without animals. The average response is not different for the first 150 ms, but becomes markedly different thereafter. The response of the frontal cortex therefore already indicates a correct answer after 150 ms. This is remarkable because for such categorization tasks we know that neural activity has to pass through several layers of brain areas. Each neuron in the processing stream necessary for the categorization path must thus be able to process and pass on information in time intervals of the order of only 10–20 ms or so.

We discussed in the last chapter the fact that noisy populations of neurons can respond very quickly to changes in the input currents that we speculate are at the core of this fast processing. Furthermore, processing in the brain involves many neurons, and the information-processing capabilities of networks seem to be essential in brain processing. It is possible to distribute information in networks and to transmit small amounts of information via local processes while still processing information with the

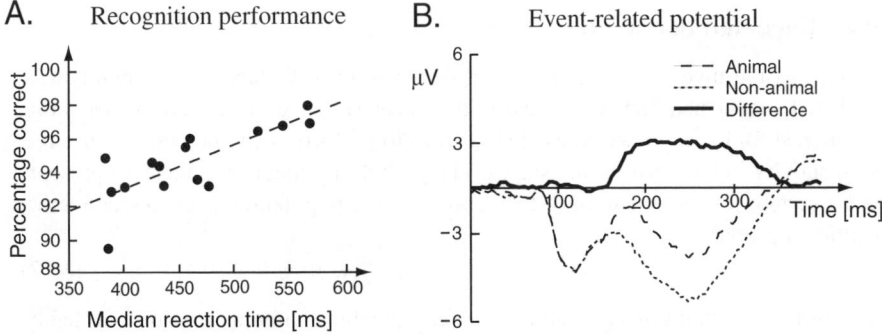

Fig. 5.12 (A) Recognition performance of 15 subjects who had to identify the presence of an animal in a visual scene (presented for only 20 ms) versus their mean reaction time. (B) Event-related potential averaged over frontal electrodes across the 15 subjects [redrawn from Thorpe, Fize, and Marlow, *Nature* 381: 520–22 (1996)].

network on a scale that is necessary to process whole visual scenes. It is therefore important to get some more quantitative estimates of the way in which information is represented in the brain. We will focus on this subject in the remainder of this chapter.

5.4 Population coding and decoding

5.4.1 Encoding

We expect that information about the external world and the processing of information is distributed in the brain. In this and the next section we will discuss this important issue further. In the previous section we discussed some temporal coding schemes, how a stimulus could be represented in the firing pattern of a neuron. In the following we also consider spatial distribution of information in a population of neurons. Reading a coded message is easy if the coding scheme is known. Here we will talk about *encoding*, the specific way in which a stimulus is coded in a neuron or a population of neurons, and in which *decoding*, how a message can be read from the responses of a neuron or a population of neurons. We consider the coding (encoding and decoding) here in a probabilistic framework due to the variability of the neuronal responses to the same presentation of a stimulus. We can then express the encoding of a stimulus pattern in terns of the response probability of neurons in a population with a conditional probability distribution of neuronal responses,

$$P(\mathbf{r}|s) = P(r_1^s, r_2^s, r_2^s, ...|s), \tag{5.31}$$

where s is a certain stimulus and r_i^s stands for the stimulus-specific response of neuron i in the population. We will here mainly consider stimulus-specific firing rates as the response, although other response quantities could be applied in the same framework. Note that we are again using the symbol P as either meaning the probability for discrete random variables or the probability distribution function in the case of continuous random variables.

5.4.2 Bayesian decoding

Decoding is the inverse of encoding, that is, we want to deduce what stimulus was presented from the neuronal responses of a neuron or a population of neurons. Decoding is of interest for brain processes as the brain has to perform such operations to use the neural activity in later processing stages. The probability that a stimulus was present, given a certain response pattern of the neurons in the population, is expressed by the conditional probability

$$P(s|\mathbf{r}) = P(s|r_1^s, r_2^s, r_2^s, ...).\tag{5.32}$$

If we know this conditional probability we can say which stimulus was most probably present given a certain response of the neuron population. It is obvious to choose this most likely stimulus as answer to the decoding problem, for example,

$$\hat{s} = \arg\max_s P(s|\mathbf{r}),\tag{5.33}$$

where $\hat{s}$ is our estimation of the stimulus and $\arg\max f(x)$ is a function that selects the argument x that maximizes the expression $f(x)$. We can estimate the conditional probabilities $P(\mathbf{r}|s)$ from the recording of the cells as stated above, but we need here the conditional probability $P(s|\mathbf{r})$. However, these two conditional probabilities are related through the identity called *Bayes's theorem*[16]

$$P(s|\mathbf{r}) = \frac{P(\mathbf{r}|s)P(s)}{P(\mathbf{r})}.\tag{5.34}$$

To use Bayes's theorem we need to know, in addition to the conditional probabilities $P(\mathbf{r}|s)$, the probabilities of the stimuli $P(s)$ and the probabilities of the responses $P(\mathbf{r})$. There are potential brain mechanisms to estimate such probabilities, although this would demand long learning procedures on many examples. It is therefore interesting to investigate alternative estimation procedures that are more easy implemented. A standard procedure in statistics is the *maximum likelihood estimate* which is based on Bayes's theorem. Bayes's theorem tells us that the conditional probabilities $P(\mathbf{r}|s)$ and $P(s|\mathbf{r})$ are related. If $P(s)$ is a constant then maximizing $P(s|\mathbf{r})$ is equivalent to maximizing $P(\mathbf{r}|s)$. We therefore view the probability $P(\mathbf{r}|s)$ as function of s, which is called the *likelihood function*, and define a maximum likelihood estimate as

$$\hat{s}_{\mathrm{ML}} = \arg\max_s P(\mathbf{r}|s).\tag{5.35}$$

Frequently, the logarithm of the likelihood function is maximized because the probability is often expressed by products of probabilities of independent variables so that the function to be minimized becomes the sum of the independent variables.

There are several other methods of estimating parameters of the distributions from conditional probability distributions $P(s|\mathbf{r})$. The importance of the maximum likelihood estimate is that for a large amount of units it is an *unbiased estimate*, which

---

[16] In the literature there is sometimes the reference (and critique of) on the so-called *Bayesian view*. This refers to the use of probability theory to describe experiments with underlying mechanisms that are thought to be deterministic in nature. Bayes's theorem is an identity within probability (or set) theory and is not questioned by statisticians.

is an estimate for which the expectation value (mean) of the estimate is equal to the correct parameter, $E(\hat{s}_{ML} = s)$. Furthermore, the variance of this estimate is optimal in the sense that it approaches the minimum possible variance given by the *Cramér–Rao bound*,

$$E\left((\hat{s}_{ML} - s)^2\right) = \frac{1}{I_F},\qquad(5.36)$$

where I_F is the Fisher information

$$I_F = -\int p(\mathbf{r}|s)\frac{d^2}{dx^2}\ln p(\mathbf{r}|s).\qquad(5.37)$$

The maximum likelihood estimate can therefore provide some reasonable estimates with limited data sets.

5.4.3 Decoding with response tuning curves

If we do not know the conditional probability distribution $P(\mathbf{r}|s)$ we cannot use the maximum likelihood estimate or some other common estimates in statistics directly. The best we can then do is to make appropriate assumptions and constrain them with experimental data. For example, we have seen in the first section that tuning curves $r_i = f_i(s)$ of neurons can be measured reasonably well (see Fig. 5.2B). Tuning curves are often well approximated with Gaussian functions as illustrated in Fig. 5.13A. Note that we cannot determine the feature value of the stimulus from the firing rate of one neuron unambiguously because a certain firing rate can be the result of two different stimuli. This ambiguity is, however, removed in a population code. A second neuron with a shifted tuning curve, illustrated in Fig. 5.13B, can resolve the ambiguity. In the illustrated example the firing rate of the neuron is higher in the case of the true stimulus corresponding to the right solution compared to the alternative. Note also that the accuracy of the decoding can be different for different values of the stimulus feature value s. A good resolution can be achieved where the tuning curve changes most with respect to the stimulus feature value s, around the largest slope of the tuning curve, $\max(f'(s))$, not at the maximum. However, with noisy data we also have to take the signal to noise ratio into account, which might be highest at the centre of the receptive field.

In the following we will assume that the response fluctuations of the neurons around this average response profile are independent, so that the conditional probability $P(s|\mathbf{r})$ can be written as the product of the conditional probabilities of the individual neurons,

$$P(\mathbf{r}|s) = \prod_i P(r_i|s).\qquad(5.38)$$

Without further knowledge of the individual conditional probability distributions it is reasonable to assume Gaussian distributions

$$P(r_i|s) = \frac{1}{\sqrt{2\pi}\sigma_i}e^{-(r_i-f_i(s))^2/2\sigma_i^2}.\qquad(5.39)$$

Taking the logarithm of this expression and viewing it as a function of s defines the *log-likelihood function*. The maximum likelihood estimator can thus be extracted in this situation from minimizing the expression,

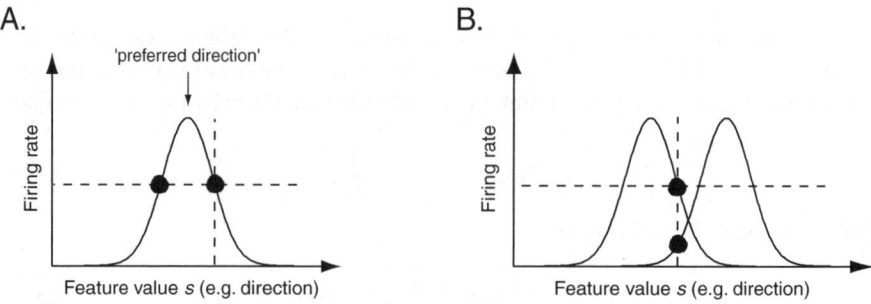

Fig. 5.13 Gaussian tuning curves representing the firing rate of a neuron as a function of a stimulus feature. (A) A single neuron cannot unambiguously decode the stimulus feature from the firing rate. (B) A second neuron with shifted tuning curve can resolve the ambiguity.

$$\hat{s} = \min \sum_i \left(\frac{r_i - f(s)}{\sigma_i} \right)^2, \tag{5.40}$$

which is a least square fit of the data points r_i (measured firing rates) to the expected firing rates $f(s)$ of stimulus s. Without noise in the tuning curve we can recover the stimulus exactly from the firing rates of the population of neurons.

5.4.4 Population vector decoding

We will now outline a simple method, called *population vector decoding*, which could easily be implemented in the brain. To demonstrate this method we consider again a set of neurons with Gaussian tuning curves. In Fig. 5.14 we illustrate a system with eight neurons that respond to the moving directions of a stimulus. The neurons have Gaussian tuning curves,

$$r_i = f(s) = e^{-(s-s_i^{\text{pref}})^2/2\sigma_{\text{RF}}^2}, \tag{5.41}$$

with the same width σ_{RF} (receptive field size) and have equally distributed preferred direction with centres s_i^{pref} every 45 deg. We will compare a system with two different widths of the receptive fields, $\sigma_{\text{RF}} = 10$ deg shown in Fig. 5.14A and $\sigma_{\text{RF}} = 20$ deg shown in Fig. 5.14B. We might intuitively think that sharper tuning curves (that is, smaller receptive fields) lead to more accurate decoding. This is, however, not the case as we will now see. The principal reason for this can be seen in the firing rate pattern of the eight nodes in response to a specific stimulus pattern as shown in the second row of Fig. 5.14. There we plotted as an example the noiseless response of the neurons to a stimulus at 130 deg, indicated by the horizontal dashed line. The neurons with preferred direction close to this stimulus value respond heavily. However, we also get some reasonable response, which helps in the decoding process, from some of the neurons in the population with wide receptive fields.

To decode the stimulus value for the firing pattern of the population we multiply the firing rate of each neuron by its preferred direction and sum the contributions of all the neurons,

$$\hat{s}_{\text{dir}} = \sum_i r_i s_i^{\text{pref}}. \tag{5.42}$$

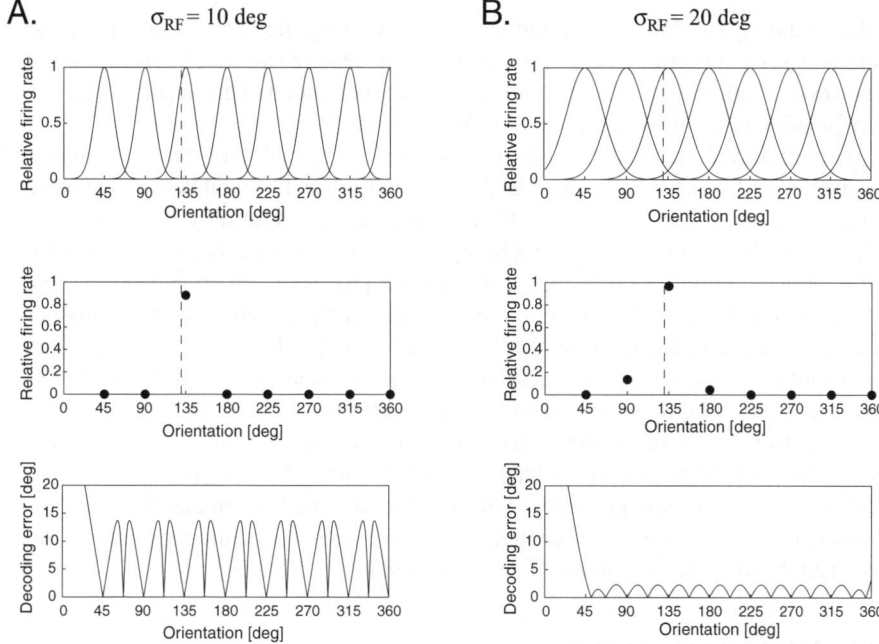

Fig. 5.14 (A) Gaussian tuning curves representing the firing rate of a neuron as a function of a stimulus feature. (B) Example of the firing rates of the 8 neurons in response to a stimulus with direction 130 deg. (C) Decoding error when the stimulus is estimated with a population code.

For a stimulus that coincides with a preferred direction for one neuron, and with small sizes of the receptive fields so that the firing rates of the other neurons can be neglected, we get an estimate that is r_i times the preferred direction of the node that is firing. We will therefore have to apply some further normalization.

In our example the stimulus values are real values for the direction of the moving object. However, we can apply this method to higher-dimensional stimulus presentations, so that the preferred stimulus of a neuron is a feature vector s_i^{pref}. The estimate 5.42 is then a vector that is an estimate of the direction of the feature vector, but has a length that depends on the sum of the firing rates. If we are interested in the precise values of the stimulus features we have therefore to normalize the firing rates to the relative values and the sum in eqn 5.42 to the total firing rates,

$$\hat{r}_i = \frac{r_i - r_i^{\min}}{r_i^{\max}} \tag{5.43}$$

$$\hat{s}_{\text{pop}} = \sum_i \frac{\hat{r}_i}{\sum_j \hat{r}_j} s_i^{\text{pref}}. \tag{5.44}$$

This is the *normalized population vector* that can be used as an estimate of the stimulus. The absolute error of decoding orientation stimuli with this scheme in our example of eight neurons is shown in the last row of Fig. 5.14. The decoding error is very large for small orientations because this part of the feature space is not well covered by the population of neurons. Reasonable estimates are, however, achieved for the areas of the feature space that are covered reasonably well by the receptive fields of the neurons. The error is not uniform and depends on the feature values. The average error is, however, much less for the larger receptive fields compared to the average errors of the population of neurons with smaller receptive fields.

We have only shown results for the noiseless case. The real test is the performance of estimates with noisy data, which we leave as an exercise. It is well known that other estimation methods can easily outperform the population vector decoding-procedure outlined here. However, this might not be a problem in brain processes as the accuracy of the estimate might be sufficient for many brain processes. The population vector method can easily be implemented in the brain as it consists mainly of a dot product between firing rates and preferred direction vectors. Dot products between firing rates and synaptic weight vectors are calculated by simple nodes as we have seen in the last chapter; we have thus only to find a way to identify or train the weight vectors to represent preferred direction vectors. This can be achieved easily with associative nodes, which will be discussed in Chapter 7. Note, however, that the simple population vector decoding also has problems. For example, the method breaks down for an inhomogeneous distribution of receptive field, which seems to be more realistic in the brain than the idealized situations often considered in the literature.

5.4.5 Decoding in the brain

Many other estimation techniques are known from statistics, and specific discussions of different estimating techniques can be found in the literature. Here we want only to note that different estimation techniques often have different characteristics. For example, the numerical cost of the estimate can vary considerably. The choice of an estimation method therefore depends on the application. It is also obvious to ask which decoding schemes are used in the brain. This is still an unsolved question, and further research is necessary to understand this important issue for brain processing. Some proposals have been made as to how sophisticated estimation methods with performances similar to those of maximum likelihood estimations could be implemented in the brain. This includes networks of the type discussed in Chapter 9, called *continuous attractor models*. Before discussing more specific information processing in networks we want to focus in the remainder of this chapter on the distributed representation in the brain, which will form the basis for the network models later in this book.

5.5 Distributed representation

5.5.1 Distributed versus localized coding

Before going on to find experimental evidence for the kind of information representation used in the brain it is useful to outline some alternative strategies that can potentially be employed by the brain. We consider here that a stimulus is represented by a vector with components that have values given by the responses (such as firing rates) of neurons in a brain area. We then want to know how many components are actively used to represent a stimulus in the brain. We call each component a node in the following. We can distinguish roughly three classes of representations.

1. **Local representation.** In a local representation only one node represents a stimulus in that only one node is active when a particular stimulus is presented. A single node (or neuron) would be sufficient to indicate that a particular stimulus was present. Neurons with such characteristics have been termed *cardinal cells*, *pontifical cells*, or *grandmother cells*. A vector of length N could represent N different features with such a local representation. The number of stimuli that can be encoded by such a vector therefore increases linearly with the number of components (nodes). Reading out which stimulus is represented is very easy with a local representation. In the case of representations with binary nodes that are either 'on' or 'off', which we will frequently consider later, only one node would be 'on' for each possible stimulus.

2. **Fully distributed representation.** The fully distributed representation can be seen as the other extreme compared to the local representation. A stimulus is encoded by the combination of the activities of all the components in the vector representing a stimulus. In the case of binary nodes we can think of the case where the probability of each node to be *on* or *off* is equal (that is, 50%), so that the number of active nodes is 50% on average. The number of different stimuli that can be encoded with such vectors scales exponentially with the number of nodes and is therefore larger than in the case of local representations. The information stored in a node vector is, however, the same as they always specify a stimulus uniquely. An advantage of such a representation is that we can define similarities of stimuli by counting how many components of the vector have similar values. This is important for building associations as we will discuss in Chapter 7.

3. **Sparsely distributed representation.** Somewhat of a compromise between the above two representations is a sparse distributed representation. In this scheme only a fraction of the components (nodes) of a vector is involved in representing a certain stimuli. The number of stimuli that can be represented by a vector of length N is then somewhere in between the cases above. Note that the information content has to be the same in all cases because the information cannot change with the internal representation as long as the representation is information-preserving; only the information content of each node will be different. In the case of binary vectors we would have a larger probability for a node to be 'off' compared to the probability for the nodes to be 'on'.

5.5.2 Sparseness

We want to specify more quantitatively how many (or what percentage of) neurons are involved in the neural processing of individual stimuli, and we therefore define here formally a measure of the *sparseness* of a representation. A definition is obvious in the case of binary nodes. If we consider neurons that either fire ($r = 1$) or not ($r = 0$), then the average number of neurons that are firing throughout the set of stimuli is

$$a = \frac{1}{S} \sum_s \frac{1}{N} \sum_i r_i^s, \tag{5.45}$$

where S is the number of stimuli over which the sparseness is evaluated, and N is the number of neurons in the considered population. In other words, if we consider relative firing rates r that are either 0 if the neuron is not responding or 1 if it is responding, then the sparseness of the representation with binary nodes is defined by the average relative firing rate,

$$a = \langle r_i^s \rangle_{i,s}. \tag{5.46}$$

where the average is taken over the number of neurons in the population and the number of stimuli in the test set. A fully distributed binary representation has in this definition a sparseness of $a = 0.5$, and the sparseness of sparsely distributed representations is less if we restrict ourselves to the case where the number of nodes that are *on* is less than $N/2$. Note that the other case of having more *on*-nodes than *off*-nodes can be mapped to the previous case by redefining the values for the representation of *on* and *off*.

The definition of sparseness in the case of vectors with continuous (real-valued) components is not that obvious. We have then to decide how much weight we put on the contribution of small versus large firing rates. What we would like to do is to take the information in the firing rate as a weighting factor, which means that we should take the firing rate relative to the variance of the firing rate into account. We thus define the sparseness of a representation, again defined as average over a set of stimuli, as

$$a = \frac{\langle r_i^s \rangle_{i,s}^2}{\langle (r_i^s)^2 \rangle_{i,s}}. \tag{5.47}$$

For example, we can measure again the firing rate of N neurons in response to a set of S stimuli and estimate the sparseness as

$$a = \frac{(\frac{1}{S} \sum_s \frac{1}{N} \sum_i r_i^s)^2}{\frac{1}{S} \sum_s \frac{1}{N} \sum_i (r_i^s)^2}. \tag{5.48}$$

The definitions 5.45 and 5.47 are equivalent in the case of binary vectors with components of value 0 and 1.

5.5.3 What is a feature?

We talked above about the response of a population of nodes to a stimulus. Such a stimulus could be, for example, a visual image such as a face. When probing neuronal responses to external stimuli it is obvious to try and decompose the stimulus into features to find out to what feature a neuron is responding. For example, we could probe a neuron that responds to faces with stimuli made out of sections of the face image such as only the eyes, a mouth, or more complicated combinations of such parts. This is indeed what is done in many neurophysiological experiments. If the neurons are more specific to some features, then it is clear that only a subset of neurons that respond to a face will respond to a subset of features. We can therefore study the number of nodes that respond to a stimulus only in the light of the nature of the stimulus.

An alternative definition of a feature would be to view the population representation of a stimulus as a *feature decomposition* of the stimulus. Each component of the population vector would then define a feature. We will, indeed, call the population vector a *feature vector* in the discussions of object recognition in the following chapter, as this terminology is common in the literature on this subject. In this view it might be possible to find a stimulus that elicits only a very small number of cells (one node in the extreme case). The representation would then be local by definition. However, it is likely that even a highly specialized stimulus will elicit spikes from a number of cells in the brain. For example, neurons typically have tuning curves (see Fig. 5.2B), which demonstrates that they respond in a gradual way to certain features, and different cells may have overlapping tuning curves. Hence, even a very specific feature is distributed among neurons, and representations of natural stimuli with many features are likely to be based on a distributed code in the brain.

5.5.4 Population coding

Many neurophysiological explorations of brain functions suggest that information is represented and processed in a distributed fashion in the brain. Our aim in the following is to get some estimates of how coarse the typical distributed representation in the brain is, that is, how many neurons are involved in particular processes such as representing a visual object in the inferior-temporal cortex. Such questions are not easily answered with direct electrophysiological measurements as it is difficult to record from many neurons simultaneously. Optical recordings with or without voltage-sensitive dyes have shown that cell activity can be localized in small areas. However, such areas still consist of many thousands of neurons that would have to be recorded simultaneously. Also, optical recordings might not show us the activity of some few neurons in other areas that might be important for the object representation. However, the study of distributed representation is another area where information theory can help. We envision recording only a few cells and ask how much information about a specific external stimulus that was presented can be gained from their activity. By varying the number of neurons taken into account we might be able to extrapolate to the order of magnitude of the number of neurons involved in the object representation.

5.5.5 Using decoding schemes

To study the nature of representations in the brain we want to estimate mutual information between populations of neurons or between a set of external stimuli and a population of neurons. However, a difficulty in this estimation is the explosion of the probability spaces as we now have to consider joint response probabilities between all neurons, that is, $P(\mathbf{y}) = P(r_1^s, r_2^s, r_2^s, ...)$. We would need good estimates of such probabilities in order to calculate the mutual information

$$I^{\mathrm{mutual}} = \int_S \int_Y \mathrm{d}s \mathrm{d}\mathbf{y} P(s, \mathbf{y}) \log_2 \frac{P(s, \mathbf{y})}{P(s)P(\mathbf{y})}. \tag{5.49}$$

Instead of evaluating this mutual information directly we adopt another strategy. We can try to reconstruct from the neural signal $\mathbf{y}$, which consists of the responses of the population of neurons, the original signal or, more precisely, an estimate of the stimulus $\hat{s}$ as discussed in the last section. The average information

$$I^{\mathrm{mutual}} = \int_S \int_{\hat{s}} \mathrm{d}s \mathrm{d}\hat{s} P(s, \hat{s}) \log_2 \frac{P(s, \hat{s})}{P(s)P(\hat{s})} \tag{5.50}$$

is equivalent to the mutual information of eqn 5.49 if the decoding is done perfectly, that is, if we don't lose information in the decoding procedure. The average information of eqn 5.50 is lower than the mutual information 5.49 if the decoding is not perfect, that is, if we lose some information in the decoding scheme.

5.5.6 Population information in the inferior-temporal cortex

To estimate the information content in a set of neurons one has to record from several cells in an area simultaneously if the responses of the population neurons are not independent of each other. If, on the other hand, we assume that each neuron responds with a certain independent probability distribution to each stimulus, then we can use neurons that have been independently recorded. *Edmund Rolls* and colleagues have used such independent recordings from neurons in the inferior-temporal (IT) cortex to estimate the information in a population of neurons. The authors used neurons that responded to a set of 20 faces and used a statistical estimate for the decoding by fitting a Gaussian to the fluctuations of the neuronal response of each neuron. The results of the mutual information between the 20 face stimuli and the response of C recorded neurons is shown in Fig. 5.15A with star symbols. The information quickly rises with the number of neurons taken into account in the analysis until the information gain levels off.

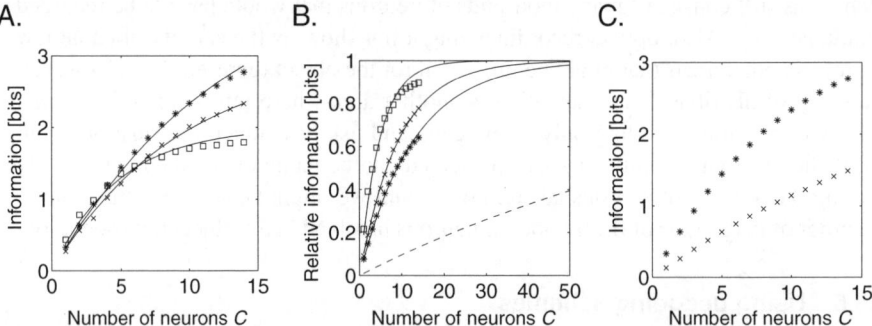

Fig. 5.15 (A) Estimate of mutual information between face stimuli and firing rate responses of C cells in the inferior-temporal cortex. The set of stimuli consisted of 20 faces (stars), 8 faces (crosses), and 4 faces (squares). (B) The information in the population of cells relative to the number of stimuli in the stimulus set. The solid lines are fits of the cell data to the ceiling model (eqn 5.51). The dashed line illustrates the values of the ceiling model for a stimulus set of 10,000 items and $y = 0.01$. (C) Estimate of mutual information with 20 faces when the neuronal response is derived from the spike count in 500 ms (stars) and 50 ms (crosses) [data from Rolls, Treves, and Tovee, *Exp. Brain Res.* 114: 149–62 (1997)].

The information for a stimulus set of 20 faces can be compared with the responses of the neurons to a reduced set of faces. The information, when analysed with a set of 8 faces (crosses) and a set of 4 faces (squares), is also shown in Fig. 5.15A. We see that the levelling off of the curves depends on the size of the stimulus set, and the linear regime of the scaling extends further with increasing size of the stimulus set. The deviation from the linear envelope for larger numbers of neurons used for decoding can be expected because the entropy of a set of N equally likely objects is given by $S_{\max} = \log_2 N$. The entropy for a fixed size of the stimulus set cannot therefore continue to grow. The deviation of the linear scaling is therefore a *ceiling effect*. The data points are indeed well fitted with a simple empirical model describing the ceiling effect by

$$I(C) = S_{\max}(1 - (1 - y)^C),$$ (5.51)

where the parameter y is the fraction of information carried by each neuron. The fits reveal a decreasing fraction y of information carried by each neuron with increasing size of the stimulus set. This can be expected as each new stimulus that has to be represented by a neuron diminishes the ability to discriminate among of the other stimuli represented by the neuron.

On the basis of the simple ceiling effect model we can also plot the data of Fig. 5.15A relative to the maximum entropy of the stimulus set. This is shown in Fig. 5.15B. From this it seems that only a small number of neurons are necessary to accurately represent a stimulus set. This conclusion is, however, obscured by the use of small stimuli sets in the experiments. If we extrapolate the data using the ceiling model we find that much larger numbers of neurons are necessary to gain sufficient information about a presented stimulus. An example of a set size of 10,000 items with $y = 0.01$ is shown as a dashed line in Fig. 5.15B. Larger set sizes with smaller y value would result in an even smaller increase of the information gain with the number of cells in the population. Also, the information gains in Fig. 5.15A and B are overestimates if we think that the amount of information in shorter time intervals is relevant. In the experiments a 500 ms time interval following the stimulus was used to estimate the information gain. The estimates for a smaller time interval (50 ms) are shown in Fig. 5.15C as crosses and compared to the information gain estimated with a 500 ms time interval (stars) from the previous analysis. The information gain therefore increases only slowly with the number of neurons in the population, which suggests a distributed representation of information.

5.5.7 Sparseness of object representations in the inferior-temporal cortex

We have still not answered the question of how sparse the representation in the IT cortex is. To do this we use the measured responses of individual neurons directly as input to eqn 5.47. The average firing rate of one particular IT neuron, averaged over many trials, in response to 70 individual stimuli is shown in Fig. 5.16A. The responses are thereby ordered so that the stimuli with the largest responses are given a small stimulus number. We indicate the spontaneous firing rate of this neuron, against which we have to judge the response of the neuron, as a horizontal dashed line. The individual average firing rate response of a neuron to some stimuli are markedly different from the spontaneous firing of the neuron, whereas most such differences are only minor. If all such differences are relevant than we see that the neuron responded to practically all stimuli so that we have to conclude that a very distributed code is used. However, if only large responses are relevant then we would conclude a more sparse representation because only a small number of stimuli drive the cell with large responses.

To take the relevance in the firing rate into account we include in the following a firing rate threshold and count only cell responses above this threshold as relevant. The relevant firing rate of the neuron for a given stimulus is therefore the mean firing rate for

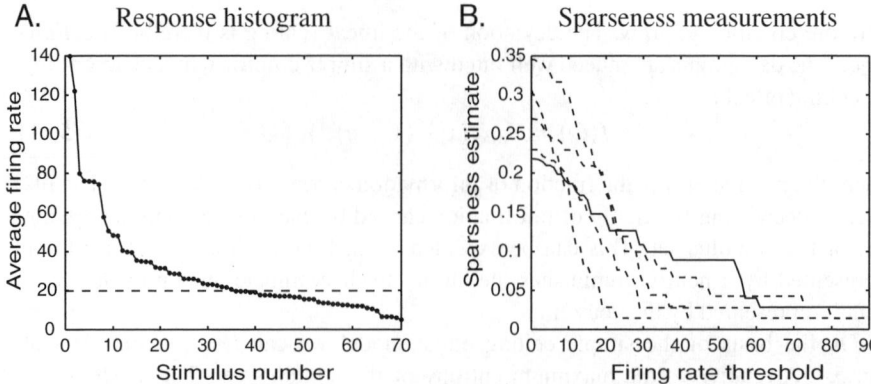

A.   Response histogram   **B.**   Sparseness measurements

Fig. 5.16 (A) Average firing rates of a neuron in the IT cortex in response to each individual stimulus in a set of 70 stimuli (faces and objects). The responses are sorted in descending order. The horizontal dashed line indicates the spontaneous firing rate of this neuron. (B) The sparseness derived from five IT neurons calculated from eqn 5.47 in which responses lower then the firing rate threshold above the spontaneous firing rate are ignored. The solid line corresponds to the data shown in (A) [data from Rolls and Tovee, *J. Physiol.* 73: 713–26 (1995)].

responses above the threshold minus the spontaneous firing rate, and is zero for firing rates below the threshold. These relevant firing rates are used in eqn 5.47 to estimate the sparseness. The sparseness estimate from data from five different neurons as a function of the firing rate threshold are shown in Fig. 5.16B. The solid line corresponds to the cell shown in Fig. 5.16A. All cells indicate fairly small values of sparseness. A value of around 0.3 is only reached if we also take small deviations of the spontaneous firing rate into account as relevant signal. The estimated sparseness, however, becomes much smaller if we take only high firing rates into account. It is also likely that this estimation of the sparseness in IT responses is an overestimation of the sparseness that we would get for much larger stimuli sets. The data set used in the experiments had several images that often drive responses in IT cell. The reported results indicate that sparsely distributed representations, roughly on the order of $a = 0.1$, may be employed in the brain.

Conclusion

We discussed in this chapter how information may be coded within the brain. We saw that there is currently little evidence of an elaborate code based on spike timings. However, this does not rule out the importance of specific spike timings. Indeed, we discussed experimental evidence that suggests that external events trigger a cascade of neuronal events that propagate rapidly through the brain within time intervals where at most only a few spikes can be relevant. It is this rapid transmission, rather than the elongated firing of neurons with specific patterns, that carries a lot of information.

Important for the models in the next chapters is the fact that information is represented and processed in the brain in a distributed fashion. A large number of neurons is likely to be responsible for representing a single event in the environment. Population coding has many advantages in information processing (for example, such codes are robust against loss of neurons). Population coding, however, makes it necessary to have decoding mechanisms in place so that other brain areas can utilize the information provided to them. We have discussed how this could be achieved in the brain. One of the simplest methods is to use a dot product between input stimuli and synaptic efficiencies, an operation for which neurons are well equipped.

Further reading

Edmund T. Rolls and Alessandro Treves
> *Neural networks and brain function*, Oxford University Press, 1998.
> Chapter 10 and Appendix 2 are relevant for this chapter.

Fred Rieke, David Warland, Rob de Ruyter van Steveninck, and William Bialek
> *Spikes: Exploring the neural code*, MIT Press, 1996.
> A very readable book that explores neural codes in nervous systems in much more detail. Very much recommended.

Allesandro Treves
> Information coding in higher sensory and memory areas, in *Handbook of biological physics IV*, Stan Gielen (ed.), Elsevier, 2001.
> This article gives a very good overview of the current state of the 'neural code'. You shouldn't miss this one, even if you have read the books above.

Stefano Panzeri, Simon R. Schultz, Alessandro Treves, and Edmund T. Rolls
> Correlation and the encoding of information in the nervous system, in *Proc. R. Soc.*, London B 266: 1001–12, 1999.
> This article includes some of the advanced techniques you need to know to extract information measures from neural data.

Edmund T. Rolls, Allesandro Treves, and Martin J. Tovee
> The representational capacity of the distributed encoding of information provided by populations of neurons in primate temporal visual cortex, in *Exp. Brain Res.* 114: 149–62, 1997.
> Several data shown in this chapter are taken from this original work.

Emilio Salinas and Laurence F. Abbott
> Vector reconstruction from firing fates, in *J. Comput. Neurosci.* 1: 89–107, 1994.
> This is a nice classical paper discussing several decoding schemes. There are also several more recent papers with advanced and biologically more plausible decoding techniques that should be read in preparation for serious research in this area.

6 Feed-forward mapping networks

In this chapter we explore the ability of networks of sigma nodes to represent mapping functions. Feed-forward mapping functions are essential in all kinds of brain processes and have dominated many models in cognitive science. The networks discussed in this chapter have the basic architecture of the multilayer perceptrons often considered in the literature. We start by exploring the effects of choosing appropriate values for the synaptic weights and explore the principal strategies of learning algorithms. We also outline some design issues involved in mapping networks, and mention some specific generalizations at the end of this chapter.

6.1 Perception, function representation, and look-up tables

6.1.1 Optical character recognition (OCR)

To illustrate the abilities of the networks that we are going to introduce in this chapter we will follow an example, that of optical character recognition. When we type a letter into a word processor on a computer we store a computer code in the computer memory that a program can interpret. The unique code is, for example, used to print an optical representation of each letter on the screen, and it is easy to change the fonts as the computer 'knows the meaning' of the character. [17] We can also use the digital representation of characters to form words that can be compared to those in a stored dictionary for the purpose of spell-checking. The problem starts when we want to reverse the process, that is, if we want to assign a symbol character to an image. For example, when we scan a handwritten page into the computer, we are not able to change the fonts of the words or to check the spelling of the words. The computer only stores a bitmap image and cannot attach meaning to the symbols. Transforming image sections of a bitmap to code that can be interpreted by other programs is a very difficult task. There are now some optical object recognition programs on the market, but if you have ever used one you know that the success rate its limited when characters are a little bit distorted, and there is nearly complete failure on handwritten texts.

Character recognition is a relatively easy task for us and it is interesting to study the human perceptual system in more detail. It is useful to distinguish two major components in the *perception* of the letter, the 'seeing' of an image of a letter and attaching a meaning to such an image. This is similar to the computer example, which includes a scanning phase that converts the printed image into a binary representation and an OCR program that transforms a binary image into a meaningful representation. When we read a letter, each letter has to be transformed from the representation on the paper to a brain representation that can be interpreted by the brain.

---

[17] More precisely, application programs have implemented an interpretation of specific code.

6.1.2   Scanning with a simple model retina

We will discuss this process in more detail by following a simple example, that of recognizing the letter 'A'. A printed representation of this letter is shown on the left side of Fig. 6.1. The first step towards perception of this letter is to get the signal into the brain. This is achieved by an eye with photoreceptors able to transduce light falling on to the retina into signals that are transmitted to the brain. We approximate this process with a simplified digitizing model retina of only $10 \cdot 10 = 100$ photoreceptors. Each of these photoreceptors transmits a signal with value 1 if the area covered by the receptor (its *receptive field*) is covered at least partially by the image. In this way we end up with the digitized version of the image shown on the right side of Fig. 6.1.

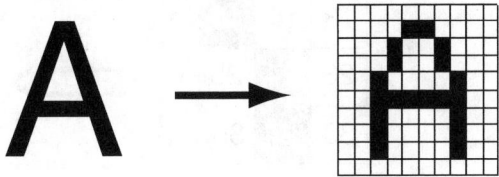

Fig. 6.1 (Left) A printed version of the capital letter A and (right) a binary version of the same letter using a 10×10 grid.

This model is certainly a crude approximation of a human eye. The resolution in the human retina is, of course, much higher; a typical retina has around 125 million photoreceptors. Also, the receptors in a human retina are not homogeneously distributed, as in our example, but have a distribution that is highest in the centre of the visual field, in the so-called *fovea*. In addition, it is well known that much more sophisticated signal processing goes on in a human eye before a signal leaves the eye through the axons of ganglion cells. However, this crude model is simply intended to illustrate a general scheme, and there is no need to complicate the model with more realistic but irrelevant details for the argument.

6.1.3   Sensory feature vectors

The output of the photoreceptors of our simple model retina is a collection of values that we can write in the form of a vector. That is, we give each model neuron an individual number and write the value of this neuron into a large column at a position corresponding to this number of the node (see Fig. 6.2). We call this vector the *sensory feature vector* as it is the set of feature values that are given to the system. Here we will represent the sensory feature vectors by the symbol $\mathbf{x}$, where the bold face indicates that it is a vector. Individual components are written with an index such as x_i, where the index i numbers the different feature values. Feature values can be binary, as in our example, real-valued, or any other symbolic representation. The number of feature values defines the *dimensionality* of the feature space, which is often very large (100 in our example; over a million for the visual feature vector generated by one eye if we take the number of axons of ganglion cells that enter the optic nerve in humans as basis). It is important to keep in mind that the precise form of sensory feature vectors depends on the specific functionality of the feature sensors (the retina in our example) and the related encoding procedures. We have either to model the biological details of the sensory system or to make valid assumptions that reflect the biological system if the form of the feature vector is crucial. However, the precise form of the sensory feature vector is often not needed to illustrate general information processing principles, in particular for the ideas presented in this chapter.

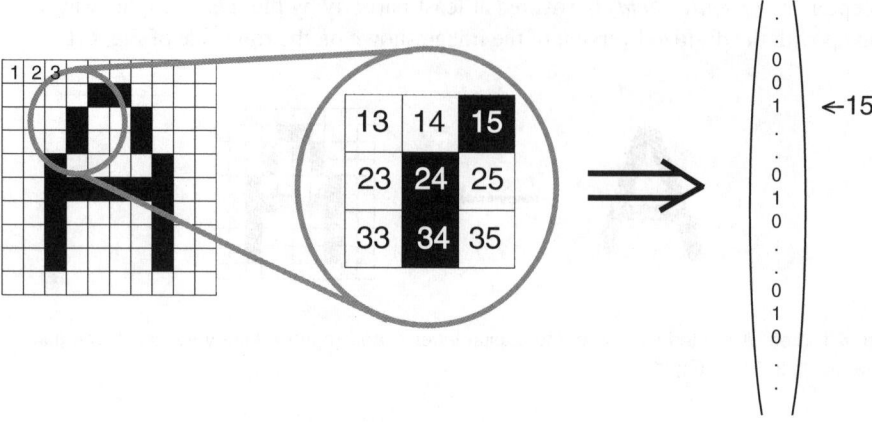

Fig. 6.2 Generation of a sensory feature vector. Each field of the model retina, which corresponds to the receptive field of a model neuron, is sequentially numbered. The firing value of each retinal node, either 0 or 1 depending on the image, represents the value of the component in the feature value corresponding to the number of the retinal node.

6.1.4 Mapping functions

A sensory feature vector is the necessary input to any object recognition system. Given such a sensory feature vector we can formulate the character recognition process as the problem of mapping this sensory feature vector on to a set of symbols representing the *internal representation* of the object. This internal representation corresponds, for example, to the ASCII code of the letter in the computer example, or to a largely unknown representation in the language area in our brain. In our example, where we want to be able to recognize a letter, we will simply represent this internal object vector with a single variable (one-dimensional vector), whose value represents the *meaning* of the letter. We choose therefore discrete values running from 1 to 26, corresponding to the position of the letter in the alphabet.

At this point we have reduced the recognition process to a vector function, which we will call *mapping* for short in the following. The example is a mapping from a binary vector (the feature vector) to an integer value (the object value). Generally, we can define a mapping as a vector function f from a vector $\mathbf{x}$ to another vector $\mathbf{y}$ as

Table 6.1 (A) Look-up table for a specific binary mapping function (Boolean AND function) in a two-dimensional feature space. (B) Partial look-up table for a sample function in a two-dimensional feature space with discrete but not binary feature values.

A. Boolean AND function

x_1	x_2	y
0	0	0
0	1	0
1	0	0
1	1	1

B. Non-Boolean function

x_1	x_2	y
1	2	-1
2	1	1
3	-2	5
-1	-1	7
⋮	⋮	⋮

$$f : \mathbf{x} \in \mathbf{S}_1^n \to \mathbf{y} \in \mathbf{S}_2^m, \tag{6.1}$$

where n is the dimensionality of the sensory feature space ($n = 100$ in our example), and m is the dimensionality of the internal object representation space ($m = 1$ in our example). $\mathbf{S}_1$ and $\mathbf{S}_2$ are the sets of possible values for each individual component of the vectors. For example, the components of the vector can consist of binary values ($\mathbf{S} = [0, 1]$), discrete values ($\mathbf{S} = \mathbf{N}$), or real values ($\mathbf{S} = \mathbf{R}$) such as the firing rates of neurons. In our example we have used a subset of the natural numbers, $\mathbf{S}_2 = [1, ..., 26]$, to represent the meaning of the 26 letters in the alphabet.

6.1.5 Look-up tables

How can we realize a mapping function? There are several possibilities, which we will discuss in turn. Our aim is to understand how the brain achieves this mapping, but it is instructive to think about some of the possibilities. The simplest solution is to construct a *look-up table*. A look-up table is a large table that lists for all possible sensory input vectors the corresponding internal representations. To illustrate this we further reduce the size of the feature vector and consider a binary feature vector that has only two feature values. Hence, the dimensionality of the sensory feature space is only 2 (not 100 as before). Let's call the first feature value x_1 and the second feature value x_2. Each sensory feature value should only have two possible values, 0 and 1. We have then four possible combinations of feature values, which is the set $\{(0\ 0), (0\ 1), (1\ 0), (1\ 1)\}$. A look-up table lists for each combination of the feature values y the corresponding value of the internal representation of the object. An example is illustrated in Table 6.1A representing one possible mapping function (the Boolean AND function) within this sensory feature space.

With binary feature values we can have $n_c = 2^2 = 4$ different combinations of the feature values in the feature vector $\mathbf{x}$. As each combination of feature values can be given an independent number of 0 and 1, we have $2^4 = 16$ possible binary mapping functions in the two-dimensional binary vector space. In general, the possible combinations of feature values are given by $n_c = b^n$, where b is the number of possible feature values ($b = 2$ for binary feature values) and n is the dimensionality of the feature space. The number of possible mapping functions is given by $n_f = b^{n_c}$, which can be very large. Thus, the size of the table grows very fast with the number of feature components and the number of possible feature values. For example, we need a look-up table of size $2^{100} \approx 10^{30}$ for the recognition of letters binarized by our simple model retina. The size rapidly increases with an increasing number of feature values, for example, by increasing the resolution of the digitization process. The size is even infinitely large if the feature values are real-valued. There are some fixes to the problem of the large sizes needed for look-up tables. For example, it might not be necessary to list all possible combinations of the feature values in the look-up table. This would also make sense because not all combinations correspond to a letter in our example. We could simply ignore those combinations and have our system generate a 'not a letter' result if the combination is not found in the partial look-up table. However, the number of possible representations of the letters is still very large. Furthermore, we have to ask how to build such a look-up table. In our example we would have to generate all possible representations of letters, which seems unrealistic at least with respect to the development of our perceptive system.

6.1.6 Prototypes

Another possibility for realizing a mapping function, which has often been proposed as a model for human perception, is the utilization of *prototypes*. A prototype is a vector that encapsulates, on average, the features for each individual object. In our letter representation example we would need to store only 26 prototypes in a look-up table, one for each different letter. We have then to add another process that maps each sensory feature vector on to a specific prototype vector. One way to achieve this is to calculate the 'distance' of the sensory feature vector from each prototype and choose the prototype that is 'closest' to the sensory feature vector. Some measures of the distance between two vectors are summarized in Appendix A. Different measures can be employed in the schemes, and different measures can lead to different performances of the recognition system.

A remaining question in the prototype scheme is how to generate the prototype vectors. A possible scenario is to present a set of sample letters to the system and to use the average (or another statistical measure) as a prototype for each individual letter. The interesting idea behind this scenario is that the generation of the prototypes is driven by examples. This is an example of a learning system that does statistical learning in the environment defined by the training set. Learning systems are of great interest to computational neuroscientists, and we will explore such systems in much more detail later. The disadvantage of the prototype scheme discussed here is that we have to compare each example to be recognized to each individual prototype vector. Thus, the computational load of this scheme scales linearly with the number of prototypes. The time for recognition might therefore exceed reasonable times in problems with a large set of possible objects. This is typically what we are confronted with when we would like to understand how humans are able to recognize objects in real life. The simple prototype scheme outlined here can therefore not be the whole story when it comes to human cognition.

To summarize this section, we have divided the recognition process into two parts, the encoding of an object into a sensory feature vector and the mapping of this sensory feature vector on to an internal representation representing the 'meaning' of objects. The latter step was first realized with a look-up table that we discovered increased rapidly in size with increasing dimensionality of the feature space and possible numbers of feature values. Not only does the size of the look-up table place a huge demand on the storage capacity of our recognition system, but it also creates the problem of generating all possible feature combinations in order to store them in the table. A superior solution in several respects seems to be the use of prototype vectors. With the prototype scheme we are able to reduce the size of the look-up table tremendously at the expense of having to compare each sensory feature vector with all possible prototype vectors. We will now explore whether such processes can be realized with processing elements resembling neurons in order to gain some insight into how perception processes can be realized in the nervous system.

6.2 The sigma node as perceptron

We demonstrate in this section that a simple neuron, one as simple as the sigma node, can represent certain types of vector functions. This can be done by assigning each individual feature value of a sensory feature vector to an individual input channel of a sigma node, by setting the firing rates of the related input channels to (see Fig. 6.3)

$$r_i^{\text{in}} = x_i. \tag{6.2}$$

The node in this simple perception system has to have at least n input channels, where n is the dimensionality of the feature space. For example, a simple perceptron in a two-dimensional feature space would look like the one illustrated in Fig. 6.3. The firing rate of the output defines a function

$$\tilde{y} = r^{\text{out}}. \tag{6.3}$$

We will discuss here which functions can be represented by simple sigma nodes. To define the system completely we also have to choose an activation function g for the node. We will start with the linear activation function $g(x) = x$ and discuss others below. The output of such a *linear perceptron* is calculated from the formula

$$\tilde{y} = w_1 x_1 + w_2 x_2, \tag{6.4}$$

where w_1 and w_2 are the weight values assigned to each input channel.

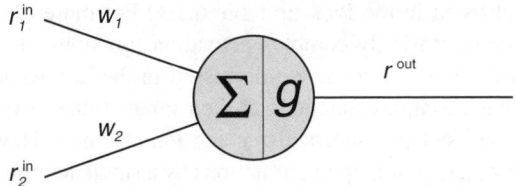

Fig. 6.3 Simple sigma node with two input channels as a model perceptron for a two-dimensional feature space.

6.2.1 An example of a mapping function

Let's see how this linear sigma node can represent a particular function, for example, the function listed partially in the look-up table in Table 6.1B. The first feature vector in this table is $x_1 = (1,2)'$, where the prime operator '$'$' stands for the transpose that simply transfers this row vector into a column vector (see Appendix A). We use the transpose only to be able to write this vector in a line in the text while still being precise with our notation. We also use commas in row vectors to separate the individual components for clarity. To calculate the value of the output, given the particular values of the input, we have to specify the weight values w_1 and w_2. We have the freedom to give them any value we want. Let's choose the values $w_1 = 1$ and $w_2 = -1$. The output of the node is then

$$\tilde{y}^1 = \tilde{y}(x^1) = \tilde{y}(x_1 = 1, x_2 = 2) = 1 \cdot 1 - 1 \cdot 2 = -1 = y^1. \tag{6.5}$$

It is not actually surprising that we get, $\tilde{y}^1 = y^1$, an exact match between the output of the node and the function value we want to represent. The reason for this is that we choose the weight values accordingly. Of course, we would have achieved the same result with other values, for example, $w_1 = -1$ and $w_2 = 0$. Indeed, there are an infinite number of solutions to represent the function value y_1 because we have only one constraining equation with two free parameters. The reason that we chose $w_1 = 1$ and $w_2 = -1$ is that we are then also able to represent the second value in the look-up table,

$$\tilde{y}^2 = 1 \cdot 2 - 1 \cdot 1 = 1 = y^2. \tag{6.6}$$

At this stage we have used up all our free parameters of the weight vector $\mathbf{w}$ in order to represent the first two entries in the look-up table, and all the other values of the function $\tilde{y}$ are uniquely defined. The third entry of the look-up table is also correctly represented by the perceptron, namely

$$\tilde{y}^3 = 1 \cdot 3 - 1 \cdot (-2) = 5 = y^3. \tag{6.7}$$

The reason for this match is that the third point lies on the two-dimensional *sheet* as illustrated in Fig. 6.4 that is defined by eqn 6.4. The fourth point in the look-up table, y^4, does not lie on this output sheet, and the specific linear perceptron can therefore not represent all the points of the look-up table in Table 6.1B.

What about other activation functions? If we choose the activation function to be $g(x) = sin(x)$ instead of a linear function we are able to represent all four points listed in the look-up table 6.1B (as can be verified easily). However, what about possible additional points not listed in the look-up table 6.1B? For more complex functions we have to introduce increasingly complex activation functions in order to be able to represent them with a sigma node. We discussed in the last section the fact that single neurons can have complex functions of their input–output mapping due to the interactions of different synaptic mechanisms and ion channels. However, an entire solution to the representation of mapping functions by a single neuron with a complex input–output relationship seems physiologically unrealistic.

A more realistic solution would be to employ many nodes in a network of neurons. This is what we are going to discuss further in this chapter and most of the remainder of the book. Indeed, we will follow this line in an extreme way by keeping the activation functions simple. But keep in mind that signal processing in the brain may well employ a combination of complex functions realized by a small number of neurons and the network properties of a collection of such neuron pools. A network solution does not rely on the capabilities of a single neuron and therefore has the additional advantage of stability against failure of a certain number of nodes (for example, cell deaths).

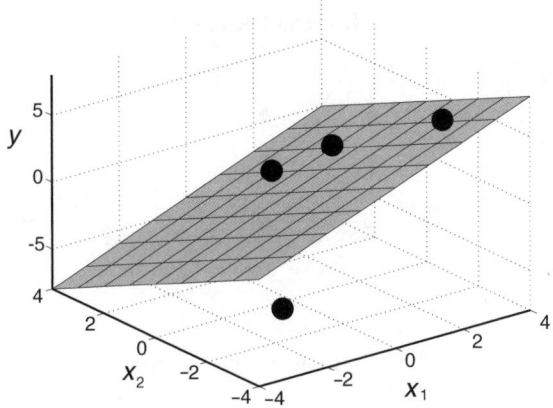

Fig. 6.4 Output manifold of sigma node with two input channels that is able to partially represent the mapping function listed in the look-up table 6.1B.

6.2.2 Boolean functions

Before leaving the single sigma node perceptron we briefly explore an important subset of the possible vector functions, the class of *binary functions* or *Boolean functions*. These are functions that have only two possible feature values for each feature component and only two possible output values, such as $\{true, false\}$, $\{-1, 1\}$, or $\{0, 1\}$. This important set of functions represents only a small subset of all the functions that we can represent in a look-up table. However, if we are able to represent these functions we can build other arbitrary functions from these basic functions. For binary functions it is natural to use a threshold function as the activation function of the node, because it limits the output values to the required binary values. A simple threshold perceptron is indeed able to represent a lot of Boolean functions.

An example, the *Boolean OR function*, is illustrated in Fig. 6.5A. As shown in the look-up table on the left, the OR function is equal to $y = 1$ if either one of the inputs is one, and $y = 0$ otherwise. The middle graph represents this function in a graphical way by representing the binary values of y with either a white or a black dot at the locations for the possible combinations of the input values x_i. This function can be represented by the single threshold sigma node shown on the right in Fig. 6.5A.

Indeed, all but two out of the 16 possible Boolean functions with two input features can be represented by the threshold sigma node with the appropriate choice of weight values. The activation of the sigma nodes with two weight values and a possible constant bias defines a linear function, and the threshold activation function sets the output to the appropriate value as long as the linear activation curve can separate the two classes of y values appropriately. We say that such classes of functions are *linear separable*. The only two functions that cannot be represented are the XOR (exclusive-OR) function and its corresponding reverse, non-XOR. The XOR function is defined by the look-up table in Fig. 6.5B. The graphical representation reveals the reason for the failure, it is not possible to divide the regions of the feature values with different y-values with a single line that can be implemented by the activation function of the sigma node. We say that this function is *not linear separable*.

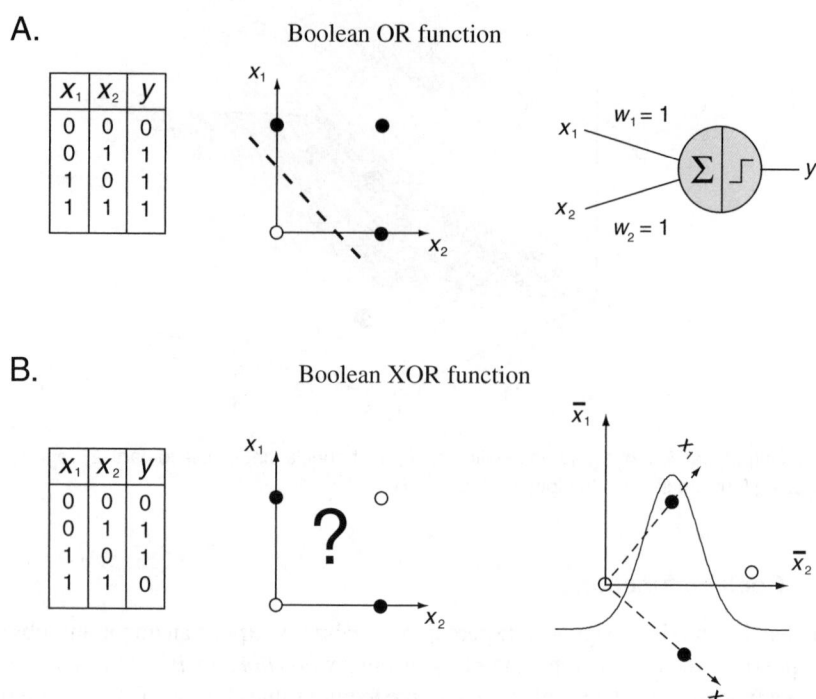

Fig. 6.5 (A) Look-up table, graphical representation, and single threshold sigma node for the Boolean OR function. (B) Look-up table and graphical representation of the Boolean XOR function, which cannot be represented by a single threshold sigma node because this function is not linear separable. A node that can rotate the input space and has a non-monotonic activation function can, however, represent this Boolean function.

It does not seem so bad that only two out of 16 functions cannot be represented. However, the number of nonlinear separable functions grows rapidly with the dimension of the feature space and soon outgrows the number of linear separable functions (see Table 6.2). The problem that only linear separable Boolean functions can be represented by a single threshold perceptron, often stated as the *XOR-problem*, greatly diminished the interest in single sigma nodes (also called *simple perceptrons*) for several years until multilayer mapping networks, which we will discuss in the next section, became tractable. However, it is worth realizing that we can represent even the XOR function with a single sigma node if we employ a nonlinear activation function, such as a rotation, combined with a non-monotonic function, such as a Gaussian function, as illustrated on the right in Fig. 6.5B. Such non-monotonic activation functions seem not very physiologically plausible when we have a single neuron in mind, as a stronger input to a neuron typically elicits stronger responses. However, as we stressed in Chapter 4, nodes are often intended to represent a collection of neurons. We will see in the next section that small networks of monotonic neurons can represent non-monotonic functions. Also, we have seen that single neurons have tuning functions that can often be described as Gaussian.

Table 6.2 Number of Boolean functions in an n-dimensional feature space and the number of linear non-separable functions.

n	Number of linear separable functions	Number of linear non-separable functions
2	14	2
3	104	151
4	1,882	63654
5	94,572	$\sim 4.310^9$
6	15,028,134	$\sim 1.810^{19}$

6.2.3 Single-layer mapping networks

Although we have discussed here examples with a single output node, it should be clear that the functionality of a single output node generalizes directly to networks with several output nodes to represent vector functions. It is therefore useful to introduce some notation that will greatly simplify the discussions of networks. In the case of a single node we were able to arrange the weight values into a vector. The generalization of this notation, arranging all weight values associating all the nodes of a feeding layer to all the nodes in a receiving layer, is called a *matrix* (see Appendix A). A matrix is similar to a large table. The weight values between the inputs and the output nodes can be written using the matrix convention (*weight matrix*) as

$$\mathbf{w} = \begin{pmatrix} w_{11} & w_{12} & w_{13} & \ldots & w_{1n^{\text{out}}} \\ w_{21} & w_{22} & w_{23} & \ldots & w_{2n^{\text{out}}} \\ . & . & . & \ldots & . \\ . & . & . & \ldots & . \\ . & . & . & \ldots & . \\ w_{n^{\text{in}}1} & w_{n^{\text{in}}2} & w_{n^{\text{in}}3} & \ldots & w_{n^{\text{in}}n^{\text{out}}} \end{pmatrix} \tag{6.8}$$

Each element in the matrix is labelled with two indices; the first is the index of the postsynaptic node; the second is the index of the presynaptic node (just remember 'to–from' when you read the indices from left to right). The numbers n^{in} in the last row is the number of presynaptic neurons, and n^{out} in the last column is the number of postsynaptic neurons. We represent matrices like vectors with bold symbols because vectors can be seen as special cases of matrices (an n-dimensional vector can be viewed as $(n \times 1)$-dimensional matrix), and the dimensionality of the matrix will be clear from the context. With this notation we can express the functionality of a *single-layer mapping network*, sometimes also called a *simple perceptron*, in a compact way as

$$\mathbf{r}^{\text{out}} = g(\mathbf{w}\mathbf{r}^{\text{in}}), \tag{6.9}$$

where g is an activation function. This equation is equivalent to writing the equations for all the components,

$$r_i^{\text{out}} = g\left(\sum_j w_{ij} r_j^{\text{in}}\right). \tag{6.10}$$

Each of the last two formulas summarizes the functionality of the single-layer mapping network. These equations are called the *update rule* of the single-layer mapping network.

6.3 Multilayer mapping networks

We have seen that the limited number of weight values in a single neuron limits the complexity of functions that we can represent using a single sigma node. An obvious solution therefore is to increase the number of nodes, and thereby the number of connections with corresponding independent weight values. However, by doing so we want to keep the number of input channels, representing the feature values, and the number of output channels, representing the dimension of the internal object representation, constant. The only solution therefore is to use *hidden* nodes. An example of such a *multilayer mapping network* is illustrated in Fig. 6.6. The middle layer is commonly called a *hidden layer* as the nodes in this layer have no direct input or output channel to the external world.

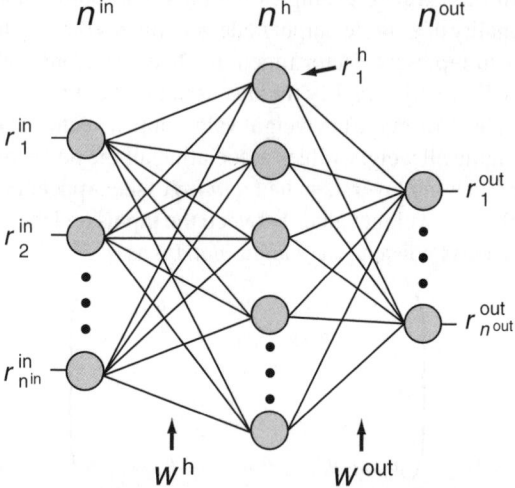

Fig. 6.6 The standard architecture of a feed-forward multilayer network with one hidden layer, in which input values are distributed to all hidden nodes with weighting factors summarized in the weight matrix $\mathbf{w}^h$. The output values of the nodes of the hidden layer are passed to the output layer, again scaled by the values of the connection strength as specified by the elements in the weight matrix $\mathbf{w}^{out}$. The parameters shown at the top, n^{in}, n^h, and n^{out}, specify the number of nodes in each layer, respectively.

The architecture is also frequently called a *multilayer perceptron*, in particular when trained using the back-propagation training algorithms described in Chapter 10. We discuss in this chapter the precautions that have to be taken when applying such networks in modelling brain functions, although feed-forward networks are certainly an important ingredient in brain processes. To try and distinguish the *problematic use of the networks* from the general *architecture*, which we think is important for brain processes, we will call these networks generally feed-forward multilayer mapping networks. The name multilayer perceptrons is only used when referring to the use of such networks in connectionist modelling.

The number of weight values, n^w, in networks with a hidden layer grows rapidly with the number of nodes n^h in the hidden layer and is given by

$$n^w = n^{\text{in}} n^{\text{h}} + n^{\text{h}} n^{\text{out}}, \qquad (6.11)$$

where n^{in} is the number of input nodes and n^{out} is the number of output nodes. We have neglected the weights of the input channels because we merely assigned to the input nodes the role of distributing the input value to all of the other nodes. These nodes are hence not directly computing nodes, and we adopt therefore the common nomenclature and call the architecture illustrated in Fig. 6.6 a *two-layer* network. It is straightforward to include more layers of hidden nodes as we will outline below.

6.3.1 The update rule for multilayer mapping networks

With the vector and matrix notation introduced at the end of the last section we can state the functionality of a multilayer mapping network in a compact form, directly generalizing the update rule for a single-layer mapping network discussed in the last section (eqn 6.9 or 6.10). An input vector is weighted by the weights to the hidden layer, $\mathbf{w}^{\text{h}}$, and all the inputs to one node are summed up. This corresponds to a matrix-vector multiplication

$$\mathbf{h}^{\text{h}} = \mathbf{w}^{\text{h}} \mathbf{r}^{\text{in}}, \qquad (6.12)$$

which is again just a compact form of writing all the equations for the individual components

$$h_i^{\text{h}} = \sum_j w_{ij}^{\text{h}} r_j^{\text{in}}. \qquad (6.13)$$

We call the vector $\mathbf{h}^{\text{h}}$ the *activation vector* of the hidden nodes.

The next step is to pass the activations of the hidden nodes through an activation function for each node, which results in a vector that represents the firing rates of the hidden layer

$$\mathbf{r}^{\text{h}} = g^{\text{h}}(\mathbf{h}^{\text{h}}). \qquad (6.14)$$

This firing rate vector becomes the input vector to the next layer, which can be the output layer or another hidden layer in the case of a multilayer network with more than one hidden layer. The final output vector is calculated from the outputs of the last hidden layer in a way similar similar to that used before,

$$\mathbf{r}^{\text{out}} = g^{\text{out}}(\mathbf{w}^{\text{out}} \mathbf{r}^{\text{h}}). \qquad (6.15)$$

We can summarize all the steps of the multilayer feed-forward network in one equation, namely

$$\mathbf{r}^{\text{out}} = g^{\text{out}}(\mathbf{w}^{\text{out}} g^{\text{h}}(\mathbf{w}^{\text{h}} \mathbf{r}^{\text{in}})). \qquad (6.16)$$

It is easy to include more hidden layers in this formula. For example, the operation rule for a four-layer network with three hidden layers and one output layer can be written as

$$\mathbf{r}^{\text{out}} = g^{\text{out}}(\mathbf{w}^{\text{out}} g^{\text{h3}}(\mathbf{w}^{\text{h3}} g^{\text{h2}}(\mathbf{w}^{\text{h2}} g^{\text{h1}}(\mathbf{w}^{\text{h1}} \mathbf{r}^{\text{in}})))). \qquad (6.17)$$

This formula looks lengthy, but it is a straightforward generalization of the two-layer network and can be easily implemented with a computer program.

Let us discuss a special case of a multilayer mapping network where all the nodes in all hidden layers have linear activation functions $(g(x) = x)$. Equation 6.17 then simplifies to

$$\mathbf{r}^{\text{out}} = g^{\text{out}}(\mathbf{w}^{\text{out}}\mathbf{w}^{\text{h3}}\mathbf{w}^{\text{h2}}\mathbf{w}^{\text{h1}}\mathbf{r}^{\text{in}})$$
$$= g^{\text{out}}(\tilde{\mathbf{w}}\mathbf{r}^{\text{in}}). \tag{6.18}$$

In the last step we have used the fact that the multiplication of a series of matrices simply yields another matrix, which we have called $\tilde{\mathbf{w}}$. Equation 6.18 represents a single-layer network as discussed before. It is therefore essential to include nonlinear activation functions, at least in the hidden layers, to make possible the advantages of hidden layers that we are about to discuss.

In general, we could also include connections between different hidden layers, not just between consecutive layers as shown in Fig. 6.6. It is possible to formulate the connectivity of such networks with a large matrix of size $N \times N$, where N is the total number of nodes in the network. In addition, sometimes there are nodes with constant values included to represent variables of the activation functions such as the threshold value. All such networks are still mapping networks in the sense described in this chapter as long as there are no recurrent connections, that is, connections feeding back to a node from which a node received some input. Recurrent networks will be described separately in Chapters 8 and 9.

6.3.2 Universal function approximation

Which functions can be approximated by multilayer mapping networks? The answer is, in principle, any. A multilayer feed-forward network is a *universal function approximator*. For example, a network that can represent the XOR function is shown in Fig. 6.7. Multilayer networks do not have the constraints of single-layer networks in two respects: (1) they are not limited to linear separable functions; (2) the number of free parameters is not restricted in principle. Let us clarify the last point in more detail. The number of free parameters in the single-layer network, which include the weight values and the firing thresholds of the nodes, are fixed by the number of input channels and the number of output nodes that have to agree with the respective dimensionalities of the function we want to approximate. However, the number of hidden nodes is not restricted in a multilayer network. We can hence increase this number arbitrarily, which in turn can increase the number of weight values sufficiently (see eqn 6.11). Indeed, it was proven that in this way multilayer networks can approximate any function arbitrarily well; hence the statement that these networks are universal function approximators. But this does not tell us how many hidden nodes we need, nor does it mean that we know if it is better to use more hidden layers or just to increase the number of nodes in one hidden layer. These are important concerns for practical engineering applications of those networks. Unfortunately, an answer to these questions cannot be given in general. It depends strongly on the function we want to approximate.

How well a particular feed-forward network can approximate a function depends on all network details including the activation functions of the nodes. The most commonly used activation function for such networks in engineering applications is the sigmoid function. This particular function has contributed greatly to the success of such

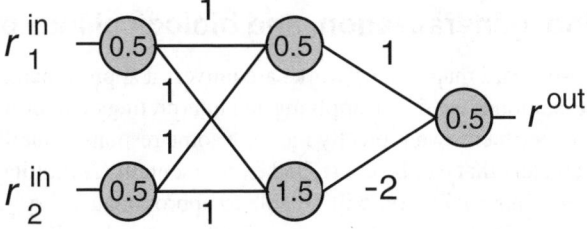

Fig. 6.7 One possible representation of the XOR function by a multilayer network with two hidden nodes. The numbers in the nodes specify the firing threshold of each node.

networks in technical applications. A partial reason for that is that the sigmoid function has the useful characteristic that it changes smoothly in a relatively confined area and is nearly constant outside this area. By combining several such sigmoid functions we can quickly generate complex functions. It is therefore very useful that we can adjust the approximation locally by adding a sigmoid function to a given approximation and subtracting a second sigmoid function with a slightly different offset. An example of approximating a sine function by the sum of a small number of sigmoid functions is illustrated in Fig. 6.8. Three dotted lines are drawn corresponding to sigmoid functions with the same slope and amplitude but different offsets. The sum of these functions, shown as a dashed line, does indeed approximate one period of the sine function (solid line) reasonably well. These characteristics of multilayer networks make them not only universal approximators in the mathematical sense, but also make them useful approximators to fit data sets in practice.

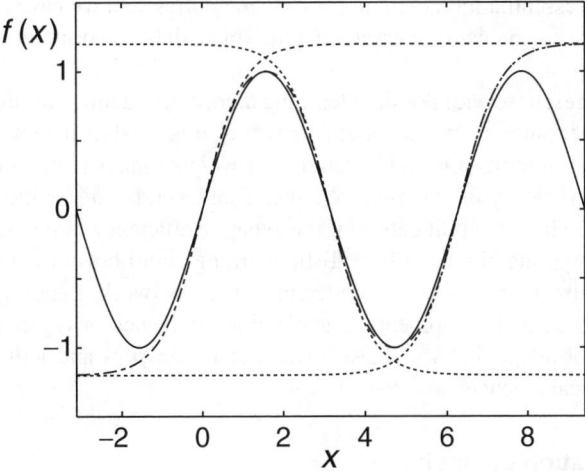

Fig. 6.8 Approximation (dashed line) of a sine function (solid line) by the sum of three sigmoid functions shown as dotted lines.

6.4   Learning, generalization, and biological interpretations

The fact that feed-forward mapping networks are universal approximators is very important, but some caution is needed in applying and interpreting them in computational neuroscience. In experiments we typically measure some response function as a function of some parameters that can be controlled experimentally. A multilayer mapping network, as universal approximator, will be able to approximate every such function arbitrary well with the right choice of parameters in the network. When such networks are used to fit experimental data one might be tempted to claim that these systems represent models of the brain on the basis that the processing nodes in the network resemble neurons. However, there are several problems with such 'models', and we hope to clarify some of the issues further below. For this it is important to understand the principles of learning and the generalization abilities of such networks, which we will now outline.

6.4.1   Adaptation and learning

We discussed the enormous potential of multilayer networks representing arbitrarily close approximations of any function by properly choosing values for the weights between nodes. How can we choose proper values? This seems an enormous task given the vast number of adjustable parameters and the exploding number of their combinations. The solution to this problem, one of the great achievements and attractions of neural network computing, lies in algorithms that can find appropriate values from examples. The process of changing the weight values to represent the examples is often called *adaptation*. Sometimes the examples are supplied by a teacher, and we can then think of the adaptive network as being the student. The adaptation algorithms are therefore also known as *learning* or *training algorithms*. Learning, or adaptation, is thought to be an essential ingredient in forming memories, and we have to understand related biological mechanisms in greater detail. This will be the principal theme of the next chapter.

Before we present several possible learning algorithms, defined as algorithms that adjust the weight values in abstract neural networks, it is good to understand what this entails in biological terms. Our interpretation of weight values is that they represent the strength or efficiency of synapses. We therefore have to explore the biochemical mechanisms that allow a modification of the synaptic efficiency more precisely when we want to incorporate biologically realistic learning algorithms in our models. The weight matrix also represents the architecture of the network. Learning algorithms can therefore represent developmental organizations of the nervous system. Of course, different types of adaptation will have distinct characteristics and will be based on different biological mechanisms.

6.4.2   Information geometry

A useful geometrical way of thinking about learning in mapping networks was proposed by *Shun-ichi Amari* and formalized in what he calls *information geometry*. Formalizing learning in such a way has proven to be very fruitful in describing and understanding statistical learning. We will only outline briefly the heuristic picture that can guide our intuition when studying learning algorithms in more detail later.

We have already argued that a network is specified by the set of all weight values between nodes. It is therefore useful to think about the space (or set) of all possible weight values of a network, which we call the *network-manifold*. The weight space of networks has, of course, very many dimensions, and we can only outline the idea schematically in a pseudo three-dimensional illustration as attempted in Fig. 6.9. Within this very high-dimensional space there is a *solution subspace* (given that the network is able to solve the problem) representing all the specific values of the weights, which allows the network to map all required input vectors to the required output vectors of the network. We will also call the solution subspace *solution manifold*. Each additional input–output pair is an additional constraint on the network. If we try to train each input–output pair correctly then the network becomes more and more constrained. Building memories with associative networks, which are the main focus of the next chapter, is an example of this type of learning. The solution subspace then depends on the number of training constraints.

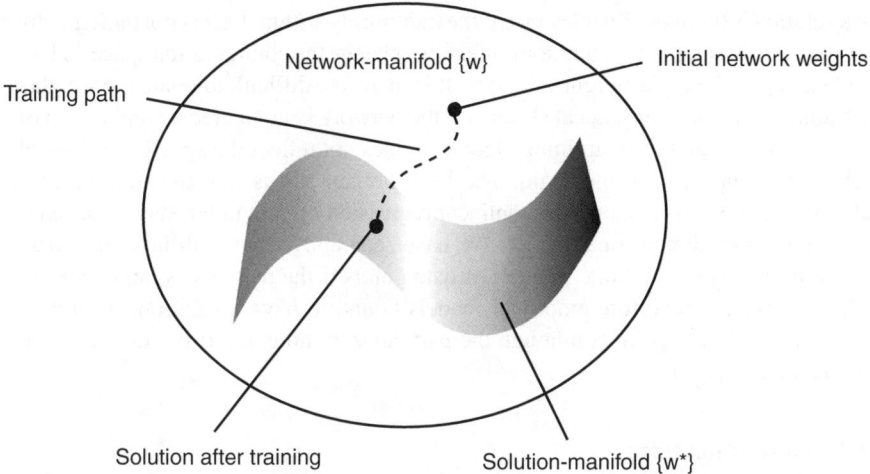

Fig. 6.9 Schematic illustration of the weight space and learning in multilayer mapping networks [adapted from Amari; see for example *Neural Networks* 8: 1379–408 (1995)].

Alternatively, we can think of the solution subspace as the realization of the 'true' mapping function, which we have to learn from examples. *Statistical learning rules*, which are designed to discover regularities in a vast amount of samples, are commonly used in applications of artificial neural networks. Such types of learning might also be employed by the brain, for example, in motor control as discussed in Chapter 10. We can think about such training in the graphical scheme illustrated in Fig. 6.9 in the following way. An initial condition of the network represents one point in the configuration space (network-manifold) of the network. A learning algorithm changes the state within the network-manifold of the network to a new point. This can be done in many different ways, for example, by randomly choosing a new point in the configuration space. Such a *random search* is a valid learning rule because we might get lucky and hit a point on the solution subspace. If we don't want to rely on our luck we might want to use other methods. *Gradient descent* methods, which we will discuss in Chapter 10, select new weight values along the gradient of an *error function* that measures the performance of the network. Such *supervised learning algorithms* minimize the *distance* between the actual state of the network and the solution subspace of the network. The proper definition of this distance is central to the design of the learning algorithms that includes a suitable *metric* that we impose on the weight space. Consecutive training steps in such learning schemes can be viewed as forming a path through the weight space toward the solution subspace as illustrated in Fig. 6.9.

6.4.3 The biological plausibility of learning algorithms

The intuitive picture of learning just outlined offers some insights into the necessity of biologically plausible learning algorithms. Learning algorithms are designed to find solutions to the training problem in the solution subspace, and different algorithms might find different solutions. The specific solution found by a particular learning algorithm depends on the algorithm employed. We thus cannot necessarily conclude that a solution is biologically relevance if the training algorithm itself is not biologically plausible. Furthermore, if we add some nodes we change the configuration space and we might end up with very different solutions. It is therefore difficult to relate the weights in a trained network to biological systems if the network is not a precise replication of the brain network and if the training algorithm does not reflect the specific biological mechanisms employed in the brain. The best we can say is that the training of a multilayer network corresponds to nonlinear regression of data in the statistical sense to a specific high-dimensional model. We have seen above that multilayer networks are able to represent any finite number of data points if the network is large enough. Such networks are therefore empirical models bound to have a solution in fitting a finite number of data points (although the particular training algorithm might not be able to find a solution).

6.4.4 Generalization

An important aspect of regression models is their *generalization ability*. *Generalization* refers here to the performance of the network on data that were not part of the training set. This can be illustrated with the help of Fig. 6.8, which illustrates the approximation of a sine function by three sigmoidal functions. We only need a few training examples to fix the parameters of the three sigmoidal functions to fit the sine function reasonably well, and the training points are correctly represented by the 'network'. Points in between these correctly represented points are also fairly well approximated by the network, and we would speak of good generalization. The generalization ability of sigmoidal networks is often quite good, in particular, when interpolating between points. The reason for this is that the systems we want to describe are often smooth. In contrast, the generalization ability of networks when extrapolating into domains not 'seen' by the network is often very poor. The reason for this is that sigmoidal networks basically assume a constant dependence for such data points that might not be appropriate, as in the example shown in Fig. 6.8.

Even interpolations of a network can be very poor as we demonstrate in Fig. 6.10. There we have chosen a training set of 6 data points (shown as stars) derived from the 'true' function $f(x) = 1 - \tanh(x - 1)$ (shown as a solid line). In addition we added some small random values to these points to represent, for example, noise in the measurements. A large network might actually come close to the 'true' solution during training, but at the end of the training, given that we train until the training data are precisely fitted, could look like the dashed line in Fig. 6.10. This is called *over-fitting*. Many methods have been proposed to prevent over-fitting, for example, to stop training after a while, to train until data points are fitted within a certain precision (called *regularization*), and to use only a small number of hidden nodes.

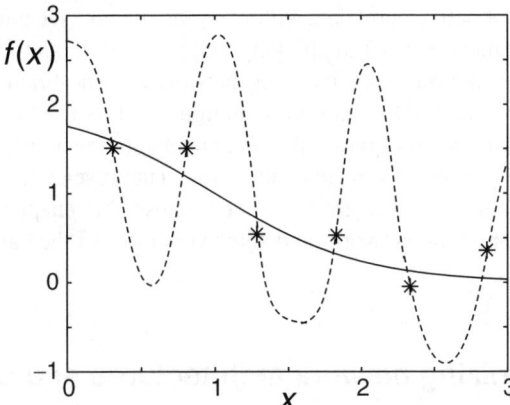

Fig. 6.10 Example of over-fitting of noisy data. Training points were generated from the 'true' function $f(x) = 1 - \tanh(x - 1)$, and noise was added to these training points. A small network can represent the 'true' function correctly. The function represented by a large network that fits all the training points is plotted with a dashed line.

6.4.5 Caution in applying mapping networks as brain models

The network approximation of a particular problem depends strongly on the number of hidden nodes and the number of layers. Adding more hidden nodes can drastically change the representation of learning examples in the network. An interpretation of hidden node activities in such networks is therefore questionable when the models are aimed at a neuronal level. In connectionist models it is common to drastically limit the number of hidden nodes to enable better generalization performances of such networks. Using a small number of hidden nodes results in a relatively smooth interpolation between training data, and a better generalization can be expected for smooth problems. However, the number of hidden nodes in biological systems is often very large, which makes this kind of analysis different from building actual brain models on a neuronal level. Another problem when interpreting the network models as models of the brain is the biological plausibility of learning algorithms. For example, the commonly used back-propagation algorithm, which we will introduce in Chapter 10, is biologically questionable.

For the above reasons we have to be very careful in interpreting the simplified neural networks as brain models when applying neural networks only generally to data analysis. Many applications have to be seen as fitting experimental data to a very general nonlinear 'model'. The nonlinear nature of the model together with the many free parameters and the universal approximator nature of such networks typically results in some reasonable fitting of experimental data sets. These heuristic models can be useful in their own right. For example, by fitting data to some general neural network model we might be able verify if some feature values are important for the prediction of a quantity that we suspect depends on these features. Those models are also useful in illustrating the difference of local versus distributed processing. However, the reverse conclusion, predicting the distributed nature of biological information on the grounds of the reasonable fit of behavioural data with mapping networks, cannot be drawn.

The caution that must be used in applying mapping networks to brain modelling does not negate the importance of mapping networks in the brain. Information is frequently mapped between different representations, and parts of cognitive abilities are supported by such mapping abilities. A better understanding of precisely how this is achieved in the brain is a major topic in computational neuroscience. Before outlining this further in the following chapters we want to close this chapter by mentioning some design issues of neural networks and some variations of the basic feed-forward mapping network.

6.5 Self-organizing network architectures and genetic algorithms

6.5.1 Design algorithms

In the previous discussions we have always started with a particular architecture of feed-forward networks by specifying the number of layers and the number of nodes per layer. The architecture is often crucial for the abilities of such networks. The universal-approximation theorem only tells us that mapping networks with enough hidden nodes can represent any mapping function. However, this does not give us any hint as to how many nodes we need in practice, nor how these nodes should be connected. Too few nodes might prevent the mapping network of being able to represent a mapping function, and too many nodes in the design can drastically diminish the generalization abilities of the network. Several algorithms have been proposed to help with the design of networks. We call these types of learning procedures *design algorithms*.

There is often a critical dimension in regard to the number of hidden nodes neces-sary for a mapping network to be able to represent a particular function to a desired accuracy. An example is shown in Fig. 6.11 where the average error (normalized sum over all training examples) and the worst error (largest error of one training example) are shown. The network was trained on the task of adding two binary 3-digit num-bers, and a new hidden node was created after some fixed amount of training with the specific number of hidden nodes of a mapping network as indicated by the numbers between the dashed lines in the figure. Although the average error decreased steadily (which can be expected as the number of parameters increases), some examples were always not well represented until seven hidden nodes were reached. This is an example of a procedure called a *node creation algorithm*, which uses this effect by starting with a small network and increasing the number of nodes systematically until the required performance of the network is reached. Other procedures, called *pruning algorithms*,

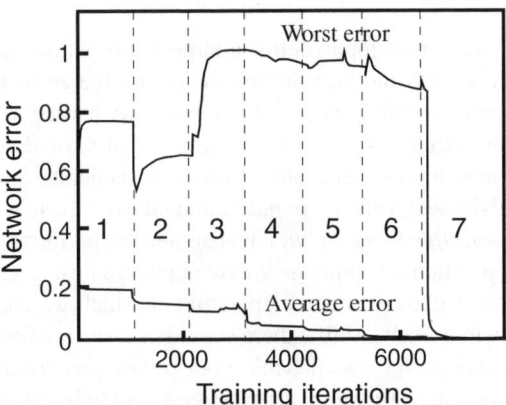

Fig. 6.11 Example of the performance of a mapping network with a hidden layer that was trained on the task of adding two 3-digit binary numbers. After a fixed amount of training a new hidden node was created (indicated by the horizontal dashed line) [redrawn from Ash, *Connection Science* 1: 365 (1989)].

start with a large number and decrease the number of nodes until a satisfactory solution is found. A particularly interesting version of a pruning algorithm is known as *weight decay*. Within this algorithm the values of the weights decay with time during learning, for example, as

$$w_{ij}(t+1) = w_{ij}(t) + \delta w_{ij}(t) - \epsilon^{\text{decay}} w_{ij}(t). \tag{6.19}$$

The term δw_{ij} stands for the change of weight values from those of standard learning algorithms, which we will discuss in the following chapters, and ϵ^{decay} is a small positive value. The decay term in this example is proportional to the values of the weights themselves, though there are many alternative ways of introducing weight decay terms, and we mention some other examples in later chapters (see, for example, Sections 7.5.2, 7.7.3, 9.5.2, 9.5.3, and 11.2.6). The principal effect of such procedures is that a decay term can eliminate connections that are not consistently reinforced by the training set. The weight decay also keeps the overall strength of the weight values bounded. This is important in biological systems as we will see in the next chapter when discussing biological mechanisms of synaptic plasticity.

6.5.2 Genetic algorithms

We want to end this section by mentioning a design algorithm that is very relevant in biological systems. This algorithm is an example algorithms, which are termed *genetic algorithms* because of their attempt to simulate evolutionary processes. Evolutionary computing and genetic algorithms are active research areas where further advances can be expected, in particular, in the area of *coevolution*. The genetic algorithms discussed here are certainly vastly abstracted from evolutionary processes in nature, but they are intended to illustrate some important basic ingredients.

Genetic algorithms work commonly on large vectors, typically called *genomes*, which describe structures to be optimized. To apply such algorithms to the design of neural networks we can, for example, define a large vector with components that are either one or zero. The ones indicate the presence, and the zeros the absence of a connection between corresponding nodes in a network. A particular vector therefore describes a particular architecture, which is analog to an *individual* in a biological system. The principle behind genetic algorithms is to generate a large set of such genome vectors (individuals), called a *population*, and to evaluate each single vector with the help of an *objective function*, which specifies the performance of each individual. From this population we generate a new population using *genetic operators*. There are many possible choices of such operators of which we only mention some that seem particularly important. At first there should be a *survival operator* that saves only the individuals that perform well while poor performers are discarded. A new population of new individuals (*offspring*) are generated only by the good performers. Offspring can, for example, be generated by combining parts of the genome of two good performers with an operation that is called *cross-over*. In addition, it is essential to allow some random modifications to the result. We can think of this operation as *random mutation*.

The generation of a new population with individuals that are likely to perform better than previous generations in terms of the objective function is essential for the optimization capability of a genetic algorithm. In this way genomes are optimized to perform a certain task. The application of these simple algorithms shows that many generations are necessary in order to generate satisfactory individuals. This fact often prevents the use of these algorithms for the design of networks in engineering applications. The slow convergence of such simplified algorithms also suggests that additional organizing principles that enable a more efficient optimization within evolutionary strategies might be important. It is, however, widely accepted that the principles behind evolutionary algorithms have helped to develop the major structure of the central nervous system. It is important to realize that the global structures of the brain are very similar in different individuals and are therefore likely to be coded genetically. The comparison of different species has shown that such structures have been evolving over time, with evolutionary forces driving new developments. In contrast to the evolutionary forces that allowed the evolution of important structures of the brain, the learning algorithms discussed throughout the book are mainly thought to fine-tune the brain within the genetically dedicated designs. Understanding the learning abilities of neural networks within the constraints of the brain organization is therefore a major concern of computational neuroscience and should not be forgotten in the discussions of neuronal networks.

6.6 Mapping networks with context units

6.6.1 Contextual processing

Feed-forward mapping networks are powerful function approximators (when used carefully) in the sense that they can map an input to any desired output. Many processes in nature can be modelled by such mappings, and feed-forward mapping networks have become a standard tool in cognitive science, particularly in the area of connectionist modelling. The study of cognitive abilities of humans, however, reveals that our behaviour, for example, the execution of particular motor actions, often depends on the context in which we encounter a certain situation. For example, we might encounter an equivalent situation on two consecutive days such as seeing a person in front of a house who is apparently studying the building in some detail. On the first day we might just think that this person is interested in architecture and we will probably continue on our way without acting further on this encounter. In the morning of the second day we might read in the newspaper about an increase in burglaries in the area, and seeing the person of the previous day again studying a house might very well prompt us to inquire about his or her intentions. This shows that the context of a sensory input can be important.

6.6.2 Recurrent mapping networks

A simple architecture that demonstrates some form of contextual processing was proposed by *Jeffrey Elman* and is often called a *simple recurrent network* or *Elman-net*. An example is outlined in Fig. 6.12, which illustrates a mapping network with 4 input nodes, 3 hidden nodes, and 4 output nodes. In addition to the standard feed-forward mapping component the networks also has *context nodes*. The 3 context nodes shown in the figure contain the activation of the hidden nodes of the previous time step. Furthermore, the activations of the context units are fed back into the system as internal inputs (rather than external inputs) for the next time step. The architecture therefore includes a new class of projections that feed into cells which in turn can influence the sending node at a later time. The network is therefore said to have *recurrences*. This type of physical back-projection should not be confused with the information flow that is used during training the networks such as in the error-back-propagation algorithm discussed in Chapter 10.

Output nodes

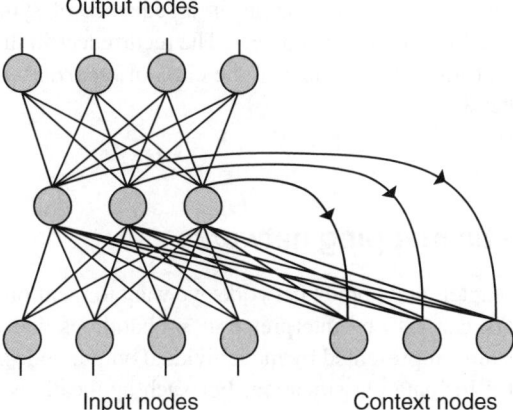

Input nodes   Context nodes

Fig. 6.12 Recurrent mapping networks as proposed by Jeffrey Elman consisting of a standard feed-forward mapping network with 4 input nodes, 3 hidden nodes, and 4 output nodes. However, the network also receives internal input from context nodes that are efferent copies of the hidden node activities. The efferent copy is achieved through fixed one-to-one projections from the hidden nodes to the context nodes that can be implemented by fixed weights with some time delay.

The network in Fig. 6.12 is only one example of this class of networks where we have included context nodes that receive inputs from the hidden nodes. We can also include context units that receive inputs directly from output or input nodes. The latter remembers the input of the previous time step, and such a mechanism is often termed *short-term memory*. We will discuss related but distinct forms of short-term memory as used in the physiology and psychology literature in Chapter 8.

6.6.3 The effect of recurrences

The context units in the example shown in Fig. 6.12 are designed to contain the activity (firing rate) of the hidden nodes at the previous time step. This can formally be achieved with linear units and setting the weight values to one while assuming some delay in the projections. The network functions as follows. For each input to the network, consisting of the external input from the input nodes and from the context nodes (which memorized the previous firing rates of the hidden nodes), we calculate the activation of the hidden nodes and then the activation of the output nodes. The activation of the hidden nodes is also copied to the context nodes. All this can be thought of as a basic time step of the network. Thus, we can treat the network at each time step as a standard feed-forward mapping network and can thus use all the algorithms as discussed before, including the training algorithm that can be applied to feed-forward mapping functions. However, the function of the network now has inherent time dynamics in that there is a new input to the hidden nodes at the next time step (even with the same external input from the input nodes as in the previous time step).

To take into account the context during training we have to train the network on whole sequences of inputs. The advantage is that the network can generate sequences of outputs, and we can thus use those models in a wider context. In particular, these networks can learn to predict the next output in a sequence of symbols, and these networks have been used mainly in this context. The recurrences in the network make this class of networks formally a member of the class of *recurrent networks* that we will discuss in Chapter 8.

6.7 Probabilistic mapping networks

Before closing this chapter on mapping networks we want to mention that the outputs of a mapping network can also be interpreted as probabilities that a specific input vector has certain features represented by the individual output nodes. Such networks are particularly useful for data classification. For such applications we can employ a feed-forward mapping network that has n^{out} output nodes, where n^{out} is equal to the number of possible classes to which an object represented by the input vector can belong. The activity of each output node can be interpreted as the probability of membership of the object to the class represented by the node. Such mapping networks are sometime called *probabilistic feed-forward networks*. Note that such probabilistic networks are different from *stochastic networks* in which the updating rule of the nodes has some probabilistic (stochastic) components (discussed further in Chapter 8).

To allow a probabilistic interpretation of the activities of the output nodes we have to normalize the sum of all output activities to one (see Appendix B),

$$\sum_i r_i^{\text{out}} = 1. \tag{6.20}$$

The reason for this is that the probability of the input vector belonging to any of the classes must be one if the nodes represent all possibilities of classes. Such a condition can be achieved, for example, with an output layer that competes for the output. This can be realized with collateral connections between the output nodes as shown in Fig. 6.13A. These collateral connections could, for example, be inhibitory so that the strong activity of one node inhibits the firing of other nodes. In the extreme case of very strong inhibition, so that only one node is active for each input vector, we speak of a *winner-take-all* architecture. Collateral connections introduce recurrences that we will discuss in more detail in Chapter 8, and competitive networks form the topic of Chapter 9. Here we only want to stress the probabilistic interpretation of the output of mapping networks.

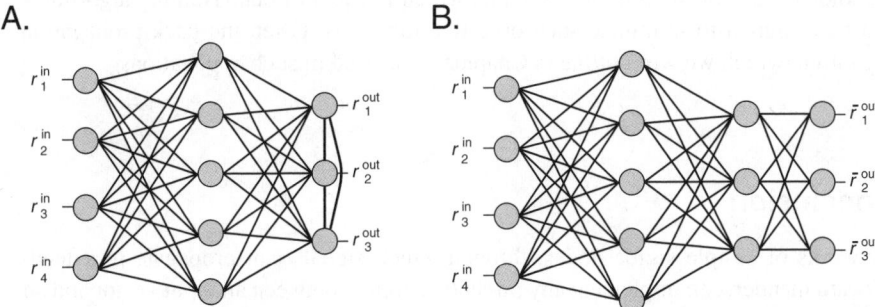

Fig. 6.13 (A) Mapping network with collateral connections in the output layer that can implement competition in the output nodes. (B) Normalization of the output nodes with the softmax activation function can be implemented with an additional layer.

6.7.1 Soft competition

A winner-take-all output layer in a mapping network can only indicate the class to which an object represented by an input vector is most likely to belong. More useful is a representation of class-membership probability so that we can also gain some information on the confidence with which the network has classified the input vector. This can be achieved with soft competition in the output layer. In practice we can simulate this with a normalization of the output activity using the *softmax* function

$$\bar{r}_i^{\text{out}} = \frac{e^{r_i^{\text{out}}}}{\sum_j e^{r_j^{\text{out}}}}, \tag{6.21}$$

where r_j^{out} are the firing rates of the output nodes before the normalization step, and $\bar{r}_i^{\text{out}}$ is the final output of the output nodes that can be interpreted as probabilities of class membership because $\sum_j \bar{r}_j^{\text{out}} = 1$. This can also be implemented with a strictly feed-forward network as illustrated in Fig. 6.13B, where the new output layer has a softmax activation function.

6.7.2 Cross-entropy as an objective function

Many learning algorithms are based on performance measures that represent the objective function. For the probabilistic interpretation it is appropriate to use cross-entropy

$$E = -\sum_\mu \sum_i t_i^\mu \bar{r}_i^{\text{out}}(r_i^{\text{in}}, W; \mu) \tag{6.22}$$

as an objective function that measures the distance in the probabilistic framework. The actual output of the network, $\bar{r}_i^{\text{out}}$, depends on the specific input vector, for example, μ, and the weights $\mathbf{w}$ of the network. The vector $\mathbf{t}^\mu$ is the desired target vector for the example μ, which consists of zeros for all but one node, the node that corresponds to the known class of the training example, which is equal to one. Training algorithms can be designed to minimize such objective functions. Often the back-propagation algorithm, which we will outline in Chapter 10, is used in such applications.

Conclusion

Networks of simple sigma nodes, through which signals can propagate in a feed-forward manner, can implement any possible mapping between an input vector and an output vector if synaptic weights are chosen appropriately. This is extremely useful for brain processes because an efficient representation of information is often the first step in solving an information-processing problem. There are many possible ways to choose the weight values, and such algorithms are generally known as teaching or learning algorithms. We will discuss several specific biologically motivated learning methods in the remainder of the book. Feed-forward networks are universal function approximators and as such useful in many technical applications. Information processing in the brain, however, uses many more mechanisms, which we think are necessary to understand how complex cognitive tasks can be solved with brain processes. We explore some of them in the following chapters.

Further reading

Simon Haykin

Neural networks: a comprehensive foundation, MacMillan (2nd edition) 1999. Many details of multilayer perceptrons and training algorithms important for technical applications can be found in this comprehensive artificial neural network book.

Peter McLeod, Kim Plunkett, and Edmund T. Rolls
> *Introduction to connectionist modelling of cognitive processes*, Oxford University Press, 1998.
> This is an example of a book that uses connectionist modelling to understand cognitive processes. It frequently utilizes multilayer perceptrons trained with error back-propagation. This book also includes a discussion of Elman-nets.

E. Bruce Goldstein
> *Sensation & perception*, Brooks/Cole Publishing Company (5th edition) 1999.
> We discussed only a few issues in perception and sensation. A thorough introduction to this area, including the psychophysics and the underlying physiology, can be found in this book.

7 Associators and synaptic plasticity

Association is a major cognitive ability, which reflects some basic operations of neural networks. In this chapter we start exploring the neural basis of associative mechanisms. These mechanisms depend strongly on synaptic plasticity. We review some of the biological mechanisms underlying synaptic plasticity in the nervous system and discuss some of their consequences on the operation of neurons and information processing in the nervous system.

7.1 Associative memory and Hebbian learning

The computational abilities of the perceptrons discussed in the last chapter relied heavily on the precise tuning of network parameters such as the general architecture and, in particular, the synaptic weights. There are many biological mechanisms that influence the precise realization of brain networks. The overall architecture is certainly strongly guided by genetic specifications, although the precise mechanisms are still to some extent unknown. The growth of neurites during brain development is not just random. Indeed, there are indications that the growth of neurites is guided by certain attractor substances. To find the general principles of brain development is one of the major scientific quests in neuroscience.

On the other hand, it is likely that not all characteristics of the brain can be specified by a genetic code. There are at least two reasons that support this hypothesis. First, it was recently estimated that the human genome has only around 30,000–40,000 genes, each carrying the information required for constructing particular proteins and thereby guiding the functionality of the organisms. If the genes work relatively directly as generally assumed, which means that high-dimensional multigene interactions are not dominant in the specification of information, then the number of genes would certainly be too small to specify all the details of the brain networks, which include neuritic morphologies and individual synaptic efficiencies. Secondly, it is also advantageous that not all the brain functions are specified genetically. This gives the opportunity to adapt to particular circumstances in the environment. Adaptive mechanisms are thus thought to be very important for brain functions, at least for the fine tuning of parameters in brain networks. We'll discuss in this chapter an important adaptation mechanism that is thought to form the basis of building associations. This is a specific form of adapting synaptic efficiencies (learning algorithm) as discussed more generally in the last chapter.

7.1.1 Synaptic plasticity

Synaptic plasticity is a major key to adaptive mechanisms in the brain, and we will explore some of the biological bases of such mechanisms in this chapter. Synaptic plasticity, in some abstract form, has been behind most of the excitement in neural networks. Learning rules used in applications of artificial neural networks are not always biologically realistic. The study of adaptation in abstract neural networks has sometimes given the impression that neural networks would be able to learn entirely from experience, so that genetic coding would be of only minimal importance in the brain development. This is certainly not the case. Though we have discussed the universal approximator characteristics of mapping networks in the last chapter, it is clear that we are facing a different type of problem in computational neuroscience in that we want to understand learning in a constrained system. So far, not much research has been done on this important issue. We will mention some advantages of learning in constrained systems in Chapter 11 when discussing the advantages of modular neural networks. For now it is sufficient to keep in mind that the mechanisms discussed in this chapter enable a neural network to self-organize, which provides it with associative abilities.

7.1.2 Hebbian learning

The principal ideas behind the learning rules discussed in this chapter were outlined by the famous Canadian psychologist *Donald O. Hebb* in his 1949 book entitled *The organization of behavior*. In this he stated that:

> When an axon of a cell A is near enough to excite cell B or repeatedly or persistently takes part in firing it, some growth or metabolic change takes place in both cells such that A's efficiency, as one of the cells firing B, is increased.

In 1949 the statement was just a wild guess and Hebb had no means in verifying this proposal. Nevertheless, his book had a great influence on many researchers as it was one of the first concrete proposals about brain mechanisms and how they can relate to behaviour. In particular Hebb proposed that cell assemblies, formed dynamically in response to external stimuli, carry out much of the information processing in the brain. The verification and more detailed specification of these mechanisms on a physiological level are still mostly beyond our experimental abilities. However, some recent experimental results have shed much light on the details of synaptic plasticity as proposed by Hebb over half a century ago. Some of these experimental results are reviewed below. They show that Hebb's qualitative description as quoted above is remarkably accurate and captures a great deal of the essence behind organization in the nervous system. We call the type of learning rules outlined below *Hebbian learning* in his honour.

7.1.3 Associations

Digital computers are now part of our lives and we often hear about computer memory. Memory in a digital computer consists of a device where information is stored in magnetic or other physical form. When recalling information from the computer memory we have to tell the computer internally where to find this memory using a memory address. Failing to specify the memory address precisely results in complete loss of this memory as it cannot then be recalled. Even an error in one bit in the memory address turns out to be fatal for memory recall. Natural systems cannot work with such demanding precision.

The human memory system is certainly different in many respects to conventional computer memory. We are often able to recall vivid memories of events from small details. For example, if you visit some places of your childhood it is likely that the images of the locations, although not exactly the same as years ago, will trigger vivid memories of your friends and certain incidents that occurred at these places; a picture with only part of the face of a friend can be sufficient for you to recognize him and to recall his name; or, if someone mentions a name of a friend, it is likely that you will be able to remember certain facial features of this person. In contrast to digital computers we are able to learn associations that can trigger memories based on related information. Only partial information can be sufficient to recall memories. It is this type of memory that we want to describe in this chapter. Essential for this type of memory, which we think is the basis for many cognitive functions realized by the brain, is the ability to form associations.

7.1.4 The associative node

Synaptic plasticity is the necessary ingredient behind associative abilities in the brain. This mechanism allows the brain to form (associative) memories of events as we will see. The principal mechanism that underlies associative learning in the brain can be

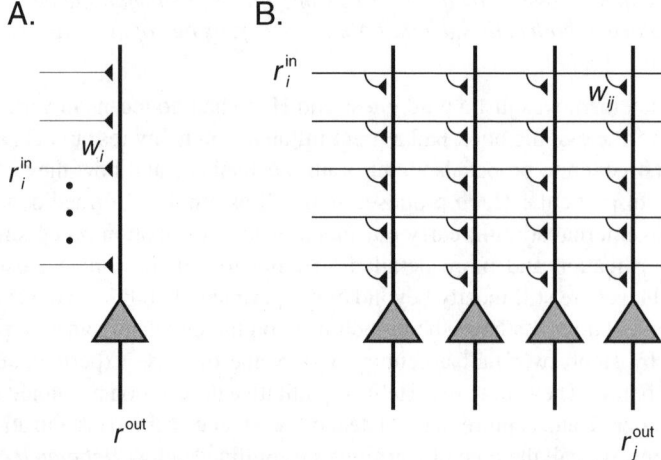

Fig. 7.1 Associative node and network architecture. (A) A simplified neuron that receives a large number of inputs r_i^{in}. The synaptic efficiency is denoted by w_i. The output of the neuron, r^{out} depends on the particular input stimulus. (B) A network of associative nodes. Each component of the input vector, r_i^{in}, is distributed to each neuron in the network. However, the effect of the input can be different for each neuron as each individual synapse can have different efficiency values w_{ij}, where j labels the neuron in the network.

outlined using a simplified neuron as illustrated in Fig. 7.1A. We have given this model neuron a triangular shape mimicking pyramidal neurons to distinguish the receiving end of the node, a dendrite extending from the top of the pyramid, from the output end, an axon extending from the base. However, we will commonly take this as a generic model of a neuron and could also interpret this as a node describing average firing rates of a neuron or a collection of neurons as outlined in Chapter 4.

We indicated only a few presynaptic axons that synapse on the model node in Fig. 7.1A. However, it is important to keep in mind in the following discussion that neurons receive input from a large number of sources as mentioned before. A cortical neuron receives on the order of 5000–10,000 or even more synapses from other cortical cells in the vicinity, as well as projections from other cortical and subcortical areas. However, only a small subset of active synaptic channels can be sufficient to elicit a presynaptic spike if the synaptic efficiencies are sufficient and the synaptic events fall within a certain temporal window. From the discussions in Chapter 4 we know that the number of sufficient synaptic inputs is much smaller than the total number of input channels of a neuron. For a cortical cell it is estimated that synaptic input from only a few synapses should be sufficient to elicit a postsynaptic spike. This is indeed likely if the times of the presynaptic spikes are within a few milliseconds of each other, enough to get sufficient additive effects from the different EPSPs. Of course, whether a certain input pattern of presynaptic spikes can elicit a postsynaptic spike depends on many other circumstances. These include the firing history, which determines the refractory time in which the neuron cannot fire, the presence of shunting inhibition, or even the availability of extracellular molecules necessary to generate the spike. We will neglect these circumstances here to simplify the discussion.

7.1.5 The associative network

The model neuron shown in Fig. 7.1A can easily be used to compose 'associative' networks as illustrated in Fig. 7.1B. For simplicity we have included only five presynaptic axons and four nodes in the illustration, but these numbers are not in principle restricted (other than biological plausibility and the physical restrictions in specific implementations). In most of the discussions in this chapter we concentrate on the mechanisms within one node (or neuron), although the mechanisms apply directly to networks of associative nodes.

7.2 An example of learning associations

Let us discuss the example where a certain event, such as the presence of a particular visual stimulus, is reflected in the firing of a subset of the input channels to the neuron in Fig. 7.2A. We have used only a subset of neurons to represent the input stimulus in accordance with distributed yet sparse coding as discussed in Chapter 5.

In Fig. 7.2A we represent the visual stimulus, say, the picture of a hamburger, with an input pattern that represents spikes in input channels 1, 3, and 5. The synaptic efficiencies of these input channels are assumed to be large enough to elicit a postsynaptic spikes. For example, we can take the spiking threshold to be $\theta = 1.5$ and assume initially that the synaptic weights of only these input channels have the values $w_i = 1$

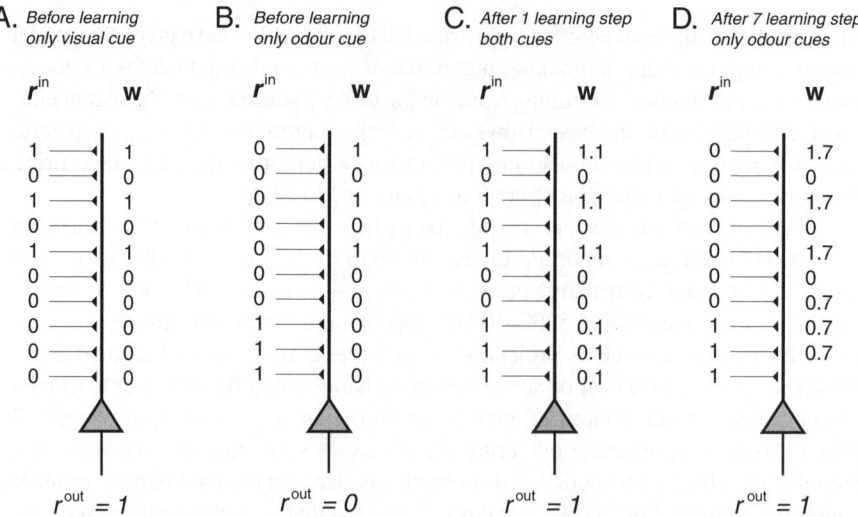

Fig. 7.2 Examples of an associative node that is trained on two feature vectors with a Hebbian-type learning algorithm that increases the synaptic strength by $\delta w = 0.1$ each time a presynaptic spike occurs in the same temporal window as a postsynaptic spike.

as shown on the right site of Fig. 7.2A. The input pattern is then sufficient to elicit a postsynaptic spike. This can easily be verified by comparing the internal activation of the neuron

$$h = \sum_i w_i r_i^{in} = 3 \tag{7.1}$$

with the firing threshold of the neuron ($\theta = 1.5$). These values of the synaptic weights are also sufficient to enable the neuron to fire in response to a partial input. This is very important for brain processes as the input pattern might not always be complete due to noisy processing or partial stimulation by partial sensory information. In this way the brain can achieve *pattern completion*, the ability to complete the representation of a partial stimulus.

Another stimulus, such as the smell of the hamburger represented as a presynaptic input to the last three efferents in Fig. 7.2B, is not able to elicit the response of this neuron with the current values of synaptic weights. In order to associate the visual cue of a hamburger with an odour signalling the smell of a hamburger, we have to modify the synaptic weights. For this we adopt the following strategy:

> *Increase the strength of the synapses by a value $\delta w = 0.1$ if a presynaptic firing is paired with a postsynaptic firing.*

This *learning rule* modifies the synaptic weights so that after one simultaneous presentation of both stimuli, the visual cue and the cue representing the odour, the weights have the values illustrated in Fig. 7.2C. After seven consecutive learning iterations we end up with the synaptic weights shown in Fig. 7.2D. How fast we achieve this point depends on a *learning rate* that we set to 0.1 in this example.

The odour of a hamburger alone is sufficient to elicit the response of the neuron after only 5 or 6 simultaneous presentations of both stimuli. The firing of this neuron is then associated with both, the presence of a visual image of a hamburger and the presence of a cue representing the smell of a hamburger. If this neuron (or this neuron in combination with other neurons in a network) elicits a mental image of a hamburger, then the presence of only one cue, the smell or the sight of the hamburger, is sufficient to elicit the mental image of a hamburger. Therefore, the odour and the visual image of a hamburger are associated with its mental image.

7.2.1 Hebbian learning in the conditioning framework

The mechanisms of an associative neuron outlined above relied on the fact that the first stimulus, the sight of the hamburger, was already effective in eliciting a response of the neuron before learning. This is a reasonable assumption as we could start with random networks in which some neurons would be responsive to this stimulus before learning. We have thus only selected such a neuron in our example from a large set of neurons that can also respond to other stimuli based on the random initial weight distribution. Although the synaptic strengths of these input channels changed with the learning rule outlined here, we call this stimulus for the moment the *unconditioned stimulus* (UCS) because the response of the neuron to this stimulus was maintained. The main reason for this labelling is that we want to compare this model with alternative models, and that we want to distinguish this input, which is already effective, to the second stimulus, the odour of the hamburger. For this second input the response of the neuron changes during learning, and we will therefore call the second stimulus the *conditioned stimulus* (CS). The input vector to the system is thus a mixture of UCS and CS as illustrated in Fig. 7.3A. This model encapsulates the principal idea behind the associative mechanisms that we think are fundamental for brain processes.

7.2.2 Alternative plasticity schemes

Several variations of this scheme are known to occur in biological nervous systems. One of these is outlined in Fig. 7.3B. We stressed that it is crucial for the UCS in the previous scheme to elicit a postsynaptic response of the neuron and we have incorporated this in our example above by sufficiently strong synapses in the input channels of the UCS. Alternatively, as illustrated in Fig. 7.3B, the fibres carrying the UCS can have many synapses on the postsynaptic dendrite that can ensure the firing of the node. For example, mossy fibres in the hippocampus are axons that generate a vast number of contacts with postsynaptic neuron are. Another example is climbing fibres in the cerebellum, which have many hundred or thousands of synapses with a single Purkinje cell. The cerebellum has been implicated in motor learning, and it was suggested that the UCS provides a motor error signal for such learning mechanisms. We will elaborate on this in Chapter 10.

Another model is illustrated in Fig. 7.3C. Here we have indicated a specific modulatory mechanism using a separate input to presynaptic terminals, which can induce presynaptic changes. Such mechanisms exist in invertebrates, for example, in the sensorimotor system of the aplysia. However, to achieve the same results as in the other two models, which depend globally on the firing of the postsynaptic neuron, we would have to supply the UCS signal to all synaptic inputs simultaneously. Such an architecture is much more elaborate to realize in larger nervous systems and seems absent

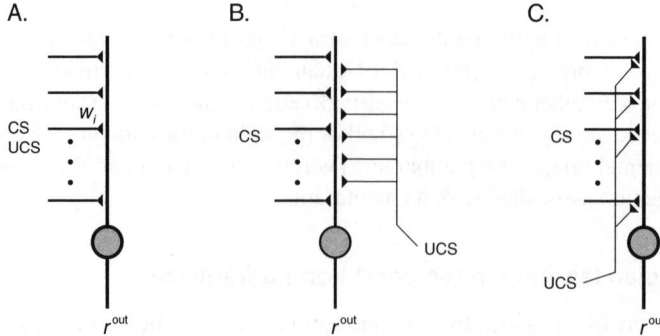

Fig. 7.3 Different models of associative nodes resembling the principal architecture found in biological nervous systems such as (A) cortical neurons in mammalian cortex and (B) Purkinje cells in the cerebellum, which have strong input from climbing fibres through many hundreds or thousands of synapses. In contrast, the model as shown in (C) that utilizes specific input to a presynaptic terminal as is known to exist in invertebrate systems, would have to supply the UCS to all synapses simultaneously in order to achieve the same kind of result as in the previous two models. Such architectures are unlikely to play an important role in cortical processing.

in vertebrate nervous systems. This demonstrates that different synaptic mechanisms and information-processing principles may be present in vertebrate and invertebrate nervous systems.

7.2.3 Issues around synaptic plasticity

The above example illustrates that it is possible to store information with associative learning; after imprinting an event–response pattern we can recall the response from partial information about the event. This is the primary reason that synaptic plasticity is thought to be the underlying principle behind associative memory and other important information-processing abilities in the brain. It is therefore important to study synaptic plasticity and the characteristics of associative networks in more detail to gain deeper insights into the working of the brain. In the following we will formulate the learning rules more precisely and relate them to biological mechanisms in neurons in order to demonstrate their biological plausibility.

We will also study several other interesting and important details and questions. For example, the learning rule outlined in the example increased all the relevant synaptic weights. This *synaptic potentiation* results in an increase of the complete weight vector after some time if the presentation of many different input vectors is coupled with some stochastic firing of the presynaptic node. The synaptic efficiencies would then become too large so that the response of the node is less specific to input patterns. Some form of synaptic weakening, called *synaptic depression*, will therefore be relevant as we will see in the following. Furthermore, the temporal structure of the firing pattern can be relevant in the learning rule as already suggested by Hebb, and we will review recent experimental findings that verify this hypothesis.

7.3   The biochemical basis of synaptic plasticity

We reviewed in Chapter 2 how a presynaptic spike elicits the release of neurotransmitters in chemical synapses through which it can influence the state of the postsynaptic neurons in various ways. A prominent example is an excitatory synapse through which a presynaptic spike can generate a postsynaptic EPSP as discussed before. The magnitude of the EPSP, that is, the strength with which the presynaptic spike alters the postsynaptic membrane potential, can vary under various conditions. For example, we can consider the presence or absence and the location of synapses during brain development as important factors. Furthermore, extracellular chemicals can influence the effectiveness of receptors (various drugs rely on such mechanisms). In contrast to those mechanisms we want to concentrate in the following on activity-dependent synaptic plasticity. We have already speculated that these changes in the synaptic efficiency are essential for learning.

7.3.1   Activity-dependent synaptic plasticity

An essential part of the learning rule in Section 5.1 is the dependence on the co-activation of pre- and postsynaptic neurons. Very active research is currently dedicated to potential mechanisms that can mark such co-activation in presynaptic or in postsynaptic terminals and induce some changes altering the efficiencies of individual synapses. For example, *backfiring* in dendritic trees, based on active spike generation in the dendrites, is currently being investigated. Such backfiring can convey information about the occurrence of a postsynaptic spike back to an ion channel at a synapse. However, most of the details regarding the relevance of such backfiring for neuronal and information-processing mechanisms are still unknown.

   Another identified biochemical mechanism that can provide the basis of signalling the postsynaptic state that is necessary for Hebbian learning (the Hebbian condition) is based on NMDA receptors that are known to be abundant in nervous systems that display synaptic plasticity. Recall that the release of glutamate transmitter effects several types of synapses, such as metabotropic receptors, activating secondary messenger pathways, and transmitter-gated ion channels such as AMPA receptors and NMDA receptors. The AMPA receptors allow the influx of sodium (Na^+) and are therefore most directly responsible for the EPSP generated by the transmitter release. In contrast, NMDA receptors can be blocked by magnesium ions (Mg^{2+}), which are only removed when the postsynaptic membrane becomes depolarized. A further distinguishing feature of the NMDA receptor is that it allows the influx of calcium ions (Ca^{2+}) in addition to sodium in its open condition. The excess of intracellular calcium can thus indicate the co-activation of pre- and postsynaptic activity as the opening of the NMDA channel depends on the presynaptic glutamate release and the postsynaptic depolarization. The hypothesis that such mechanisms are involved in synaptic plasticity is backed up by the demonstrations that the blockade of NMDA with antagonistic drugs or the prevention of the rise in postsynaptic calcium concentrations abolishes synaptic plasticity.

   What is still largely unknown is how this calcium signal is used to accomplish longlasting changes in synaptic efficiencies. Many possible mechanisms have been investigated experimentally, for example, that the effectiveness of AMPA receptors can depend on a cascade of biochemical reactions triggered by the intracellular calcium. The change in the synaptic efficiency would, in this scenario, only depend on postsynaptic mechanisms. However, there is also some evidence that presynaptic modifications resulting in a change of the amount of transmitters released by the presynaptic terminal can contribute to the change in synaptic efficiencies. This would demand the use of *retrograde messengers* that can signal the Hebbian conditions to the presynaptic terminal. Popular scenarios involve chemicals such as *nitric oxide*. Such retrograde messengers are not required in invertebrate systems as outlined in Fig. 7.3C, and presynaptic changes seem indeed to be dominant in such systems that rely on the temporarily linked transmitter release of the UCS input and the activation of a specific presynaptic terminal.

7.3.2 Longlasting synaptic changes

To study the influence of synaptic changes on information-processing abilities in the brain it is very important to know how long those synaptic changes can last. Some molecular mechanisms outlined in the literature, in particular those involving the *phosphorylation of proteins*, can last from several hours up to days and weeks. For longer lasting changes, supporting lifelong memories, separate mechanisms are necessary of which some candidates have been identified. These include mechanisms in which the phosphorylation becomes independent of the secondary messengers. Interestingly, there is also some evidence of structural changes such as the specific increase or decrease in the number of synapses dependent on learning circumstances. The biochemical bases of such changes are at present largely unknown.

7.4 The temporal structure of Hebbian plasticity: LTP and LTD

Activity-dependent synaptic plasticity has been investigated predominantly in slices of brain tissues and in cultivated brain cells that can be kept alive for at least several hours. In such conditions it is possible to record intracellularly from one neuron while stimulating another. The stimulation of a single presynaptic neuron is often not enough to elicit a spike in the postsynaptic neuron. This is actually advantageous when we want to measure the size of the EPSP, or the related strength of the current (EPSC). The mean level of an EPSC can be compared to the size of EPSCs before and after conditions that induce synaptic plasticity. With such experiments it is possible to measure changes in synaptic efficiencies as long as we have reasonable evidence that the changes in the EPSC amplitudes are indeed produced by synaptic plasticity. As in the last section, we are particularly interested in activity-dependent synaptic plasticity. It is possible to stimulate the presynaptic and postsynaptic neurons independently and thereby force the neurons to spike in various temporal relations to each other. With such experiments it has been shown that neither the exclusive spiking of the presynaptic neuron nor the exclusive spiking of the postsynaptic neuron induces synaptic plasticity. Only when the postsynaptic neuron is triggered within a short time window with a specific temporal relation to a presynaptic spike, can synaptic plasticity be observed.

7.4.1 Experimental example of Hebbian plasticity

Several experiments on neocortical and hippocampal slices, in cultured neurons, and in developing retinotectal synapses *in vivo* have recently explored the temporal domain of synaptic plasticity in more detail. We outline some results of experiments by *Guo-chiang Bi* and *Mu-ming Poo* in hippocampal cultures, which have been confirmed by other studies. Results of experiments with varying pre- and postsynaptic conditions as outlined above between glutamatergic neurons are shown in Fig. 7.4. In Fig. 7.4A the change in magnitude of EPSC amplitudes over time is shown relative to the measured amplitudes before a synaptic efficiency changing condition was introduced. EPSC amplitudes are relatively stable as long as there is no postsynaptic activity. However, the situation changes after the postsynaptic neuron is stimulated in addition to the presynaptic neuron at the time defined as $t = 0$ and marked by the horizontal dashed line in Fig. 7.4.

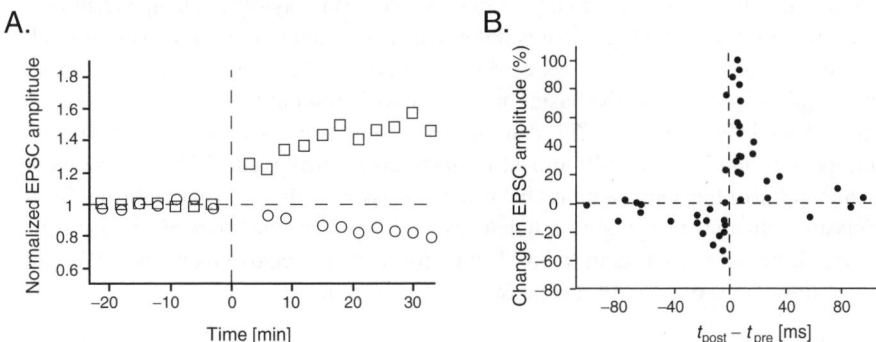

Fig. 7.4 (A) Relative EPSC amplitudes between glutamatergic neurons in hippocampal slices. A strong postsynaptic stimulation was introduced at time $t = 0$ for 1 minute that induced spiking of the postsynaptic neuron. The postsynaptic firing was induced in relation to the onset of an EPSC that resulted from the stimulation of a presynaptic neuron at 1 Hz. The squares mark the results when the postsynaptic firing times followed the onset of EPSCs within a short time window of 5 ms. The enhancement of synaptic efficiencies demonstrates LTP. The circles mark the results when the postsynaptic neuron was fired 5 ms before the onset of the EPSC. The reduction of synaptic efficiencies demonstrates LTD. (B) The relative changes in EPSC amplitudes are shown for various time windows between the onset of an EPSC induced by presynaptic firing and the time of induction of spikes in the postsynaptic neuron [data re-drawn from Bi and Poo, *J. Neurosci.* 18: 10464–72 (1998)].

We combined in Fig. 7.4A the results of two separate experiments with different postsynaptic activity patterns, which started at $t = 0$ and lasted for only 1 minute. In one experiment, illustrated by squares, a postsynaptic depolarization was induced shortly after the onset of a subthreshold EPSP that resulted from a 1 Hz activation of a presynaptic neuron. This stimulus pattern induced an increase of the magnitude of EPSCs relative to the test before the synaptic change-inducing stimulus pattern was applied. This synaptic enhancement was persistent for more than 30 minutes. In contrast to the first experiment, the postsynaptic spike was induced before the onset of an EPSP in the second experiment (illustrated by circles). This resulted in a reduction of synaptic efficiencies that was also persistent over the monitored time window. Similar experiments with NMDA antagonists that blocked NMDA receptors (not shown in Fig. 7.4A) indicated no changes in the synaptic efficiencies. This suggests the relevance of NMDA receptors for this kind of synaptic plasticity.

7.4.2 LTP and LTD

The synaptic changes in the experiments outlined above were persistent in these experiments for at least 30 minutes. Synaptic plasticity of this form was first demonstrated experimentally by *Timothy Bliss* and *Terje Lømo* in 1973, and other experiments have shown that such activity-dependent changes in synaptic plasticity can last much longer, for days, or even for weeks. It is therefore common to call the amplifications in the synaptic efficiency *long-term potentiation* (LTP) and the reductions in the synaptic efficiency *long-term depression* (LTD). It is usually difficult to keep slices of brain tissue or neuronal cultures alive for much longer. Whether such synaptic changes can persist for the lifetime of an organism is unknown. This could only be achieved through specialized biochemical processes as discussed above. Nevertheless, such forms of synaptic plasticity support the basic model of association outlined in the first section of this chapter. In particular, LTP can enforce associative response to a presynaptic firing pattern that is temporally linked to postsynaptic firing, and LTD can facilitate the unlearning of presynaptic input that is not consistent with postsynaptic firing. Such longlasting changes of synaptic efficiencies can also form the basis of mechanisms that enable the nervous system to build some forms of associative memories. We will explore these lines of thought further in the next chapter.

7.4.3 Time window of Hebbian plasticity

As mentioned above, LTP and LTD were demonstrated long before the experiments of Bi and Poo. However, Bi and Poo have further investigated the crucial temporal relation between pre- and postsynaptic spikes by varying the time between pre- and postsynaptic spikes. This temporal domain in synaptic plasticity should not be confused with the time-course of synaptic changes itself, which is a separate topic to be discussed further below. The time difference between pre- and postsynaptic spikes is defined here as the difference between onset of the EPSC, induced by the stimulation of the presynaptic neuron, and the peak of the postsynaptic action potential, induced by stimulating the postsynaptic neuron beyond its firing threshold. Results for different values of these time differences are shown in Fig. 7.4B. They indicate a critical time window of around $|\Delta t| \approx 40$ ms. No synaptic plasticity could be detected for differences with absolute values much larger than this time window. Changes in EPSC amplitudes are largest for small positive differences (LTP) or small negative differences (LTD) in the pre- and postsynaptic spike times. The decrease of synaptic plasticity can be fitted reasonably by an exponential decay with the absolute time of the difference between spiking. The resolution of the experimental data is, however, not sufficient enough to establish if the forms of LTD and LTP are markedly different. The consequences of such possible differences are discussed below.

7.4.4 Variations of temporal Hebbian plasticity

The *asymmetric form of Hebbian plasticity* found by Bi and Poo is illustrated again schematically in Fig. 7.5A. Besides the example of hippocampal cells illustrated, similar relations have been found in neocortical slices. In these examples, the magnitude and the sizes of the critical difference windows for LTP and LTD are similar. However, this form is not the only possible dependence of synaptic plasticity on the relative spike times. Some additional examples are schematically illustrated in Fig. 7.5. Examples are known in which the critical difference window for LTD is much larger compared to that for LTP as illustrated schematically in Fig. 7.5B. LTP and LTD can also be reversed relative to the difference in pre- and postsynaptic spiking. For example positive differences in spike timings induce LTD when synaptic plasticity is triggered in Purkinje cells in the cerebellum. Purkinje cells are known to be inhibitory, and such changes can therefore reduce inhibition in the targets of Purkinje cells. There are also some examples of *symmetric Hebbian plasticity* illustrated in Fig. 7.5D and 7.5E. It is an important task for computational neuroscientists to explore the functional roles of the different forms of Hebbian plasticity. Some suggestions are discussed in the next section.

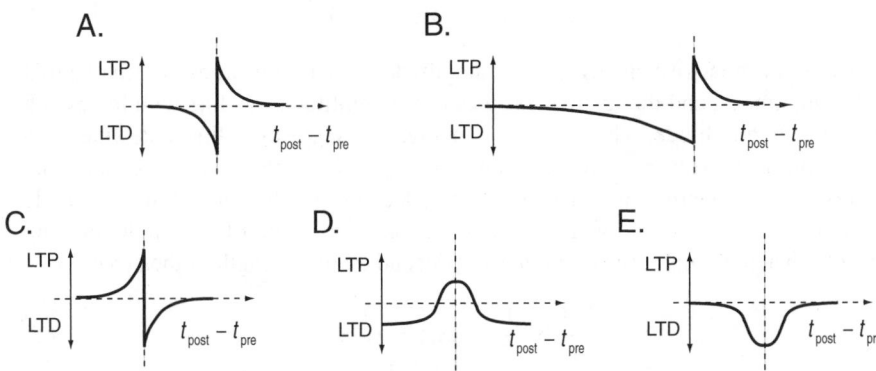

Fig. 7.5 Several examples of the schematic dependence of synaptic efficiencies on the temporal relations between pre- and postsynaptic spikes [adapted from L. F. Abbott and S. B. Nelson, *Nature Neurosci.* 3: 1178–83 (2000)].

7.4.5 Dependence of synaptic changes on initial strength

Before leaving the study of Bi and Poo we have to mention another crucial result of their study, which has important consequences for the information-processing abilities of the nervous systems discussed in the next section. So far we have not discussed whether the size of synaptic changes depends on the strength of a synapse. Bi and Poo investigated this important question, and their results are plotted in Fig. 7.6.

These data indicate that the relative change for depression $\Delta A/A$ (decrease of EPSC amplitude ΔA relative to the amplitude A before the synaptic-changing condition was introduced) does not vary within the observed fluctuations when results for different initial EPSCs are compared. Note that this ratio is plotted as a percentage

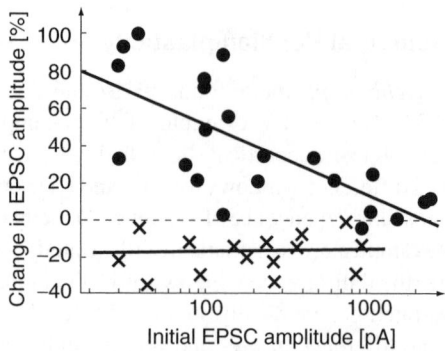

Fig. 7.6 Dependence of LTP and LTD on the magnitude of the EPSCs before synaptic plasticity is induced [re-drawn from Bi and Poo, *J. Neurosci.* 18: 10464–72 (1998)].

in Fig. 7.6. This result therefore suggests that the absolute strength of the synaptic efficiencies in LTD is proportional to (depends in a multiplicative way on) the initial synaptic efficiency

$$\frac{\Delta A}{A} \propto \text{const} \rightarrow \Delta A \propto A. \tag{7.2}$$

In contrast, the relative changes of EPSC amplitudes for LTP are largest for small initial EPSC amplitudes, and the relative changes in the amplitudes decline with increasing initial EPSC amplitude. The fitted line in Fig. 7.6 suggests a linear decline on a logarithmic scale. However, the data can similarly well be fitted with en exponential function on the logarithmic scale, so that the decrease in absolute EPSC amplitude would be inversely proportional to the initial magnitude of the EPSC amplitude. The absolute change in this case would thus not depend on the strength of the synapses,

$$\frac{\Delta A}{A} \propto \frac{1}{A} \rightarrow \Delta A = \text{const.} \tag{7.3}$$

This corresponds to an additive rule of synaptic changes. Only synapses with relatively weak initial amplitudes have been included in the data shown in Fig. 7.6.

7.5 Mathematical formulation of Hebbian plasticity

We can describe synaptic plasticity by a change of weight values **w** between nodes in neural network models. The weight values are hence not static but can change over time. We do not know much about the precise time course of synaptic changes, and most experiments can only explore the modification after certain training intervals. It is hence appropriate to express the variation of weight values only after time steps Δt in a discrete fashion as

$$w_{ij}(t + \Delta t) = w_{ij}(t) + \delta w_{ij}(t_i^{\text{f}}, t_j^{\text{f}}, \Delta t; w_{ij}). \tag{7.4}$$

We have here indicated the dependence of the weight changes on various factors. For activity-dependent synaptic plasticity it is clear that the weight change should depend on the firing times of the pre- and postsynaptic neuron. We also included the possibility of a dependence of this change on the value of the synaptic strength itself as we have seen in the experiments by Bi and Poo. Even without the knowledge of these experiments it is clear that we have to include such a term as the strength of biological synapses can certainly only vary within some interval (for example, the strength is bounded from below at zero and cannot increase infinitely).

7.5.1 Hebbian learning with spiking neurons

In networks of spiking neurons we can start to explore the computational consequences of the experimental findings in more detail. We can describe the dependence of the synaptic changes on the relative timing of postsynaptic spiking t^{post} and presynaptic spiking t^{pre} with the following equation

$$\delta^{\pm}w_{ij} = f^{\pm}K^{\pm}(t^{\text{post}} - t^{\text{pre}}) - f^{\text{decay}}, \tag{7.5}$$

where we have allowed different functional forms of LTP (labelled with superscript '+') and LTD (labelled with superscript '−'). For completeness we have also included a general decay term labelled f^{decay}. We will see below that decay terms have important computational consequences.

The precise form of the functional dependence of synaptic plasticity on various factors is not yet known from experiments. It is therefore useful to study several alternatives and to understand the computational consequences of different models. For the so called *kernel* function $K^{\pm}(t^{\text{post}} - t^{\text{pre}})$ we will assume here an exponential form as indicated by the experimental data above, for example,

$$K^{\pm}(t^{\text{post}} - t^{\text{pre}}) = e^{\mp \frac{t^{\text{post}} - t^{\text{pre}}}{\tau^{\pm}}} \ \Theta(\pm[t^{\text{post}} - t^{\text{pre}}]), \tag{7.6}$$

where we have allowed a different decay scale for potentiation and depression given by the constant $\tau^{\pm}$. $\Theta(x)$ is a threshold function that simply restricts LTP and LTD to the correct domains of positive and negative differences between pre- and postsynaptic spiking, respectively. The amplitude factor $f^{\pm}$ in eqn 7.5 can also depend on various circumstances and has therefore to be considered as a function of some variables such as the absolute value of the weights. These functions have to include some form of nonlinearity to restrict the synaptic efficiencies to certain intervals. For example, we can choose them to be constant in a certain interval of synaptic efficiencies in what we will call the *additive rule with absorbing boundaries*,

$$f^{\pm} = \begin{cases} a^{\pm} & \text{for } w_{ij}^{\text{min}} \le w_{ij} \le w_{ij}^{\text{max}} \\ 0 & \text{otherwise} \end{cases}. \tag{7.7}$$

Our alternative rule is a *multiplicative rule* with more graded nonlinearity when approaching the boundaries,

$$f^{+} = a^{+}(w^{\text{max}} - w_{ij}) \tag{7.8}$$

$$f^{-} = a^{-}(w^{\text{min}} - w_{ij}). \tag{7.9}$$

These are only examples that demonstrate the consequences of different forms of synaptic changes on neuronal properties. For example, we will study the distributions of synaptic weights in a single neuron as a result of these plasticity rules in the next section where we will see that they critically depend on these functional forms of the rules. The experiments of Bi and Poo indicated roughly an additive rule for potentiation and a multiplicative rule for depression (see Fig. 7.6 and note that there the relative change in percentage is plotted). However, other neurons might have different forms.

7.5.2 Hebbian learning in rate models

In models that describe only the average behaviour of neurons or cell assemblies (*rate models*) we cannot incorporate the spike timings. We have therefore to fall back on a description of plasticity that depends only on the activity of pre- and postsynaptic nodes. Some of the essence of the Hebbian learning rules is then captured by the observation that the plasticity depends on the average correlation of pre- and postsynaptic firing. We will lose thereby some details of the synaptic mechanisms, and we have to verify that these are not essential for the conclusions that we draw from these models. Many studies in computational neuroscience have been based on rate models, although many findings have been confirmed in models with spiking neurons.

Most Hebbian plasticity rules used in rate models have a functional form that can be expressed as

$$\delta w_{ij} = f_1[(r_i - f_2)(r_j - f_3) - f_4] \tag{7.10}$$

between a postsynaptic node i with a firing rate r_i and a presynaptic node j with firing rate r_j. We included several parameters (constants or functions) in order to incorporate the variety of rules within this general form. The first term f_1 represents the overall strength of changes and is often called the *learning rate*. The learning rate can depend on various factors such as the weight values themselves, and can also be explicitly time-dependent to describe, for example, external modulation of synaptic plasticity. For the moment it is sufficient to give it a constant value, $f_1 = k$. The values f_2 and f_3 are *plasticity thresholds* on which the direction of synaptic change depends. These values are therefore most commonly set equal to the average firing rates as only the deviation of the average firing rate is relevant. The final term, f_4, is a general *weight decay* term of which we will study some examples later. Hebbian learning rules without this decay term are commonly written in the form

$$\delta w_{ij} = k(r_i - \langle r_i \rangle)(r_j - \langle r_j \rangle), \tag{7.11}$$

where the angular brackets $\langle . \rangle$ are used to denote the average of the quantity enclosed.

To explore the connection of this rule with the general principles of synaptic changes in biological systems we calculate the average change of the weight values within this model. These are given by

$$\langle \delta w_{ij} \rangle = k \langle (r_i - \langle r_i \rangle)(r_j - \langle r_j \rangle) \rangle \tag{7.12}$$
$$= k(\langle r_i r_j \rangle - \langle r_i \rangle \langle r_j \rangle).$$

The average change of synaptic weights is thus proportional to the covariance of the pre- and postsynaptic firing. This is also called a cross-correlation function when k is a proper normalization constant. The essence captured by this learning rule is that LTP is induced when the presynaptic firing rate and the postsynaptic firing rate are both systematically below or above their average firing rate. In contrast, when one of the firing rates is below average while the other remains above, then LTD is induced.

We can distinguish the two conditions of LTD further. When the firing rate of a presynaptic node is above average while the firing rate of the postsynaptic node is below average, then the weight of this specific synapse should be reduced as it cannot be relevant for the functional relevant postsynaptic firing. This only concerns a specific synapse, and this form of plasticity is therefore termed *homosynaptic LTD* from the Greek prefix *homo=same*. In contrast, when the postsynaptic node is active above the average firing rate, then all synapses of presynaptic nodes that are firing below average should be reduced to make the neuron respond more specifically to relevant input. We call this form of plasticity *heterosynaptic LTD* with the Greek prefix *hetero=other* to stress that this type of plasticity concerns all the synapses on to a postsynaptic neuron.

7.6 Weight distributions

Synaptic efficiencies are continuously changing as long as learning rules are applied. The changes of weight have to be relatively rapid if the nervous system relies on such mechanisms to adapt to single events (*one-shot learning*). However, rapid changes of weight can lead to instabilities in the system, even with restricted weight values. This seems to pose a conflict. On the one hand, the neuron should adapt to rapid changes, while, on the other hand, it seems that a neuron should roughly maintain its main firing characteristics such as its firing rates. A solution to this conflict can be achieved if the weight distribution becomes stationary, that is, if the overall weight distribution stays relatively constant while the individual synapses can change. The weight distributions of most Hebbian learning rules can approach characteristic distributions that we will analyse in more detail.

We will see later in this book that some characteristics of Hebbian models depend on the form of the weight distribution. Also, weight distributions of models can be compared to distributions of synaptic efficiencies in biological neurons. Measurements of EPSCs for various synapses, as well as experiments marking synapses with fluorescent antibodies, have found evidence of unimodal and positively skewed distributions. An example of results with the latter method is shown in Fig. 7.7. In these experiments synapses on spinal neurons were labelled with antibodies that emit fluorescent light with intensities indicating the efficiencies of the synapses. Additional experiments have studied the variation of such distributions under various experimental conditions. We are here concerned with outlining the dependence of the weight distributions with respect to some details of the learning rule and with exploring the functional consequences of different learning rules on the signal-processing abilities of the brain.

7.6.1 Example of weight distribution in a rate model

It is straightforward to produce weight distributions in numerical simulations; we just have to run a single-neuron simulator on many training examples (inputs) and apply the Hebbian rules until a stable weight distribution is achieved. There are also

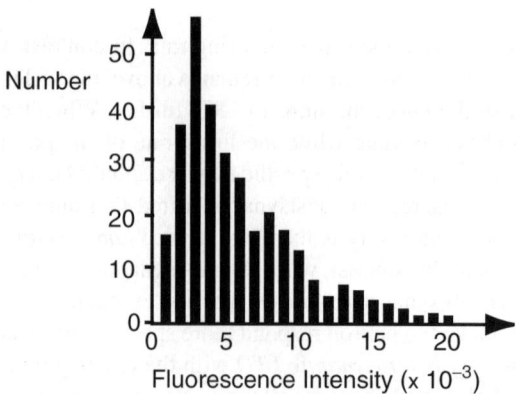

Fig. 7.7 Distribution of fluorescence intensities of synapses from a spinal neuron that were labelled with fluorescence antibodies, which can be regarded as an estimate of the synaptic efficiencies [redrawn from O'Brian *et al.*, *Neuron* 21: 1067–78 (1998)].

useful analytical methods available that can deepen our understanding of the weight distributions of learning rules further, some of which we will mention below.

Before showing some simulation results we should think about what we expect for synaptic weight distributions, for example, for the weight values produced with the important Hebbian rule 7.11 that is dominant in models with nodes describing firing rates. In most studies based on this rule random patterns are used to represent the input patterns of the presynaptic nodes. The random numbers for r_i and r_j, resulting from a specific stimulus–response pairing, are assumed to be independent of each other, at least during the learning phase. This might sound like a conflict as we want to have a specific input pattern resulting in a specific output response of a neuron. However, if this is already the case then learning should cease. We will therefore concentrate on the more interesting case where independent patterns, represented by independent random input vectors, are associated with a specific firing of a neuron. We have thus to consider independent firing rates for the pre- and postsynaptic nodes. After subtracting the mean from each random number for the pre- and postsynaptic firing rates, respectively, as demanded by the Hebbian rule 7.11, we end up multiplying random variables with zero mean. Those new random variables, which can have a different distribution than the distributions used to generate the firing pattern (see Appendix B), are then added together for each pairing of input and output firing rates. The central limit theorem then tells us that the sum is (under the conditions of the theorem) a random variable that is Gaussian-distributed (see Appendix B).

This is indeed what we typically find in numerical simulations of this rule. To demonstrate this we used exponentially distributed firing rates for the input and output patterns in the numerical experiment. The exponential distribution was chosen to mimic experimental findings as we have shown in Chapter 5 (see Fig. 5.16), but different distributions, such as the frequently studied pattern made out of equally distributed binary numbers, produce similar results. We normalized the weight values in the following to the total number of training examples, N^{pat}, which correspond to a learning rate of $k = 1/N^{\text{pat}}$ in eqn 7.11. This normalization is commonly used in this rule in artificial neural networks to keep the weight values within reasonable ranges but is not essential for the following demonstration. We will discuss weight normalization procedures in more detail below. A normalized histogram of all the weights after many learning steps of different input–output patterns is shown in Fig. 7.8. The details can be found in the file `hebb_dis.m` and are discussed further in Chapter 12. We have included a fit of the simulated weight distribution to the Gaussian function that shows that the data are indeed well approximated by a Gaussian distribution as we predicted above.

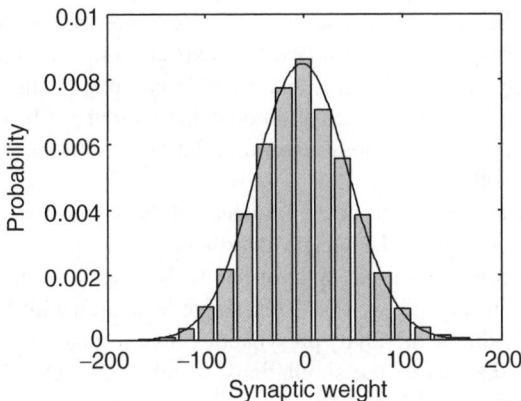

Fig. 7.8 Normalized histograms of weight values from simulations of a simplified neuron (sigma node) simulating average firing rates after training with the basic Hebbian learning rules 7.11 on exponentially distributed random patterns. A fit of a Gaussian distribution to the data is shown as a solid line.

We have shown that rate models of recurrent networks trained with the Hebbian training rule on random patterns have Gaussian distribution weight components. We will use this fact in later analysis. The distribution is centred around zero if we use the Hebbian covariance rule. Negative weight values have to be considered as inhibitory synapses, while positive weight values should be considered as excitatory synapses.

7.6.2 Change of synaptic characteristics

Biological neurons make either excitatory or inhibitory synapses (*Dale's principle*), and the synapses from a specific presynaptic neuron cannot change its specific characteristics. In contrast, in the simulations above we did not restrict the synapses to be either inhibitory or excitatory. During the learning phase it is indeed possible that some weight values cross the boundaries between positive and negative values, which is, of course, physiologically unrealistic. However, it is fairly easy to include such a constraint in the simulation. For example, we can initially set the weight values randomly to be either excitatory or inhibitory. We can then maintain the sign of the weight values by clipping all weight changes that would cross the sign boundary. Simulations with such constraints produce similar results for the distribution of the weight matrix components, and it is therefore common to relax this biological detail in the simulation.

7.6.3 Examples with spiking neurons

Let us now turn to the asymmetric Hebbian rules for spiking neurons. The dynamic process of learning an input pattern consisting of random spike trains will make this an ongoing stochastic process. It is clear that the strength of synapses that consistently drive the spiking of the postsynaptic neuron is increased under the rules outlined above. In contrast, spikes of presynaptic neurons that are uncorrelated with the postsynaptic spiking are equally likely to occur before or after the postsynaptic spike. However, only a few patterns are assumed to elicit a postsynaptic spike. It is therefore much more common that an uncorrelated spike occurs in conjunction with a postsynaptic spike, and the synaptic weight will thus decrease over time for patterns that are uncorrelated to the firing of the postsynaptic neuron. We thus expect a bimodal distribution with some synapses tending towards the maximal possible synaptic strength while others should decay to zero. In particular, such a bimodal distribution can be expected if we force the postsynaptic spiking in close temporal relation to some of the presynaptic spike trains, but not to others.

Instead of demonstrating this basic consequence of the asymmetric Hebbian rules we will study in the following the weight distribution of a spiking neuron that is entirely driven by uncorrelated input spike trains without exciting the postsynaptic neuron externally. As an example we simulate a single IF-neuron with 1000 excitatory synapses that are individually driven by presynaptic Poisson spike trains with average firing rates of 20 Hz. Details of the simulations, which follow closely the work of *Sen Song*, *Kenneth Miller*, and *Larry Abbott*, are given in Chapter 12. When we start the simulations we set all the synaptic weights to large values. As a consequence the postsynaptic neuron fires with large frequencies in a very regular manner because the average synaptic input exceeds the firing threshold of the neurons as discussed in Chapter 3. We then apply an additive Hebbian rule of the form of eqn 7.7 with marginally stronger LTD than LTP. The average synaptic weights decrease under this synaptic rule for reasons outlined in the next section. The result is that the neuron responds with lower firing rates and an increased coefficient of variation as illustrated in Fig. 7.9A. We are here first interested in studying the resulting weight distributions after the neurons have reached the state where the average firing rates do not decrease any longer.

The resulting weight distribution after 5 minutes of simulated time with the additive Hebbian rule is shown in Fig. 7.9B. There are many synaptic weights that cluster around very small values while there are a few synapses that have large synaptic weights approaching the upper limit imposed in the simulations. Analytical results also indicate that weight distributions with the additive Hebbian rule have bimodal distributions with peaks approaching the limiting values.

The weight distribution, however, depends on the specific Hebbian rule utilized for synaptic changes. The resulting weight distribution of similar simulations with the multiplicative Hebbian rule for LTD, as indicated by the experiments of Bi and Poo, was analysed by *Mark van Rossum* and collaborators. They found a unimodal and positively skewed weight distribution resembling the experimental data shown in Fig. 7.7. They have confirmed this result (as well as the bimodal nature of the weight distribution with additive Hebbian rules) with analytical techniques based on the Fokker–Planck equation. The Fokker–Planck equation is a stochastic differential

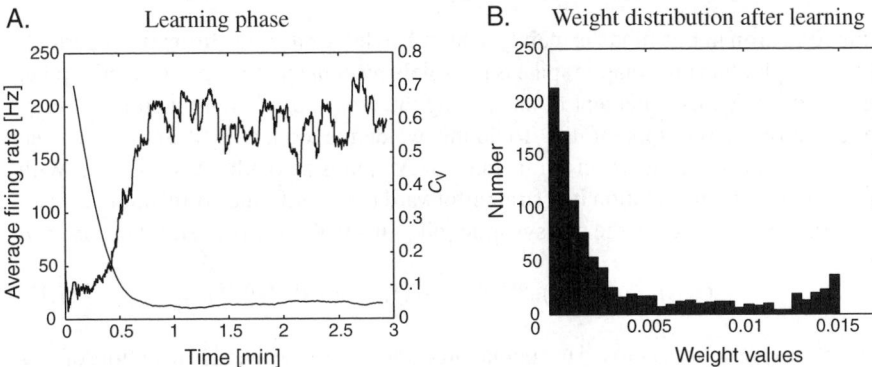

Fig. 7.9 (A) Average firing rate (decreasing curve) and C_V, the coefficient of variation (increasing and fluctuating curve), of an IF-neuron that is driven by 1000 excitatory Poisson spike trains while the synaptic efficiencies are changed according to an additive Hebbian rule with asymmetric Gaussian plasticity windows. (B) Distribution of weight values after 5 minutes of simulated training time (which is similar to the distribution after 3 minutes). The weights were limited to be in the range of 0–0.015. The distribution has two maxima, one at each boundary of the allowed interval.

equation that can be used to describe the drift of the weight values due to the stochastic changing of the synaptic efficiencies by defining change operators in small time steps that depend on the form of different learning rules. The authors also showed that no hard limits on the individual weight values have to be introduced. The resulting unimodal weight distribution effectively restricts the weight values to values near the peak of the distribution, and a hard limit on the weights is then not essential. However, the overall strength of the synaptic weight values has also to be adapted on a longer time scale to ensure that the postsynaptic firing rate is within the expected firing rate. This is a question of synaptic scaling that it is important to discuss in some more detail.

7.7 Neuronal response variability, gain control, and scaling

Spike-time-dependent Hebbian learning is a positive feedback mechanism in the sense that it makes strong synapses stronger and weak synapses weaker. This would make neurons highly specific to certain stimuli, for which they would respond vividly, while other stimuli would basically have no effect. This does limit the associative abilities of neurons, which we hypothesize are based on Hebbian mechanisms, because only a few already effective synapses get reinforced while others have only a small chance of becoming effective. In this section we will discuss some more computational consequences of spike-time-dependent synaptic plasticity as well as other mechanisms, such as synaptic competition, that can control neuronal spike characteristics.

7.7.1 Variability and gain control

We discussed in Chapter 3 the fact that IF-neurons respond with a regular firing even to random spike trains if the synaptic efficiencies are very strong so that the average current is larger than the firing threshold of the neuron (see Fig. 3.8). The firing time of the IF-neuron in this mode is mainly determined by the average firing input current, and each individual presynaptic spike is not solely responsible for a postsynaptic spike. We can quantify this statement by measuring the cross-correlation function between pre- and postsynaptic spike trains. To do this we define a quantity $s(\Delta t)$ that has the value $s = 1$ if a spike occurs in a time interval Δt and is zero otherwise ($s = 0$). With this spike train representation it is straightforward to quantify the correlation between the presynaptic spikes and the postsynaptic spikes using the *cross-correlation function*

$$C(n) = \langle s^{\mathrm{pre}}(t)s^{\mathrm{post}}(t + n\Delta t)\rangle - \langle s^{\mathrm{pre}}\rangle\langle s^{\mathrm{post}}\rangle, \qquad (7.13)$$

where the expressions enclosed in angular brackets denote the average over time of that quantity. The correlation is zero if the pre- and postsynaptic spike trains are independent (that is, $\langle s^{\mathrm{pre}}s^{\mathrm{post}}\rangle = \langle s^{\mathrm{pre}}\rangle\langle s^{\mathrm{post}}\rangle$). The correlation is larger than zero when there is, on average, more incidence of close relations between pre- and postsynaptic spikes. The quantity is negative if presynaptic spikes would result in a consistent reduction of postsynaptic spikes (anti-correlation).

Measurements of such cross-correlations confirm that a single input spike train has little correlation with the output spike train when the IF-neuron is in the regular firing regime. An example is shown in Fig. 7.10, which displays the average cross-correlations of an IF-neuron driven by 1000 presynaptic Poisson spike trains with average firing rates of 20 Hz. The average cross-correlations, indicated with star symbols and interpolated with a dashed line, are calculated from simulations where the synaptic weight values are fixed to a value of 0.015 which corresponds to the initial conditions in the simulations shown in Fig. 7.9. This resulted in a regular firing of the postsynaptic neuron with a frequency of around 270 Hz. The average cross-correlation in these simulations is consistent with zero if we consider the results within their variance indicated by error-bars. The similar values of the average cross-correlations for positive and negative time shifts Δt indicate that the occurrence of a presynaptic spike before a postsynaptic spike is equally as likely as the occurrence of a presynaptic spike after a postsynaptic spike.

The equal average occurrence of presynaptic spikes before and after a postsynaptic spike means that statistically LTP occurs as much as LTD. The average behaviour of the neuron would thus not change if the effects of LTP and LTD were equal. However, if the effect of LTD is a little bit stronger than that of LTD, that is, if the area under the LTD branch in Fig. 7.5 is larger than the area under the LTP branch, then we get some reliable decrease of the weight values over time. Such conditions were therefore included in the simulations of *Song et al.* as mentioned above. Such conditions were also observed in some cortical neurons (Fig. 7.5B) where the time window of LTD is larger then LTP. Another possibility, not shown in Fig. 7.5, is that the amplitude of LTD is larger than that of LTP while the time windows are the same.

The average cross-correlations between all presynaptic spikes and the postsynaptic spikes with the weight matrix after learning (shown in Fig. 7.9) are indicated in Fig. 7.10 by squares. This weight matrix caused a much lower and more realistic frequency of the postsynaptic spike train around 18 HZ. The peak for small negative values of Δt in the cross-correlation curve indicates that some presynaptic spike trains were responsible for eliciting a postsynaptic spike. This increased correlation of some spike trains supports the LTP necessary to stabilize the corresponding synapses. *Song*

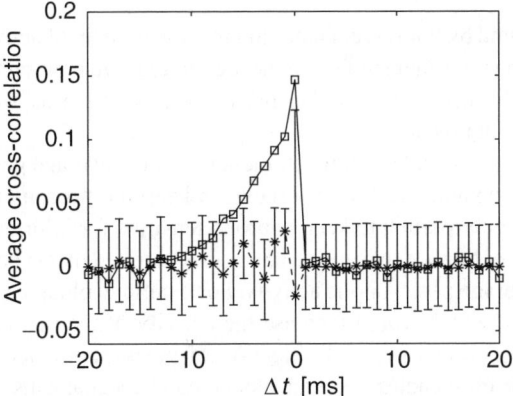

Fig. 7.10 Average cross-correlation function between pre-synaptic Poisson spike trains and the postsynaptic spike train (averaged over all presynaptic spike trains) in simulation of an IF-neuron with 1000 input channels. The spike trains that lead to the results shown by stars were generated with each weight value fixed to a value 0.015. The cross-correlations are consistent with zero when considered within the variance indicated by the error bars. The squares represent the simulation results from simulations of the IF-neuron driven by the same presynaptic spike trains as before, but with the weight matrix after Hebbian learning shown in Fig. 7.9. Some presynaptic spike trains caused postsynaptic spiking with a positive peak in the average cross-correlation functions when the presynaptic spikes precede the postsynaptic spike (see also S. Song and L. F. Abbott, *Neurocomputing* 32 & 33: 523–8 (2000). No error bars are shown for this curve for clarity.

and *Abbott*, who first elaborated the arguments outlined here, also demonstrated that a steady state with a reasonably large coefficient of variation can be reached for different firing frequencies in the input spike trains. Of course, the time it takes to reach the steady state after changing the input frequency depends on the time it takes to modify the synaptic weights. Such biological details are not yet known experimentally. In the case of a very fast transition to a steady state it would mean that the neuron could also adapt to a change in the firing frequencies of inputs, something that has been termed *gain control*. However, a too rapid permanent change of synaptic weights has other disadvantages. It is also possible that other neuronal mechanisms, some of which we will mention below, can contribute to (if not dominate) the gain control of biological neurons.

7.7.2 Synaptic scaling

We have already seen an example of *Song* and *Abbott* of how asymmetric Hebbian plasticity can adjust the response of a cell so that postsynaptic spike trains become correlated with presynaptic events that include the scaling of the overall strength of the weights so that the postsynaptic neuron fires in the desired high-variability domain. In this example the investigators used an additive Hebbian rule with fixed synaptic boundaries that results in bimodal weight distributions with many weak synapses and the appropriate number of strong synapses that drive the postsynaptic neuron with an average firing rate and coefficient of variation of the required order of magnitude. The number of strong channels is reduced for higher frequency inputs. This is biologically realistic as it was found by fluorescent labelling that the number of active channels does vary in such situations. Further studies are needed to explore how such functionalities could be achieved by different Hebbian plasticity rules such as weight-dependent (multiplicative) weight updates.

The dependence of overall synaptic efficiencies on the average postsynaptic firing rate, which we call synaptic scaling, is crucial to keep the neurons in the regime of high variability that is important to keep neurons sensitive for information processing in the nervous system. Many experiments have demonstrated that synaptic efficiencies are scaled by the average postsynaptic activity. For example, blocking some inhibitory channels of neurons, which would increase the activity of the neurons, scales down synaptic efficiencies, whereas forced reduction of postsynaptic activity leads to an increase of synaptic efficiencies. Several biochemical mechanisms in the brain can contribute to such effects including the direct activity-dependent modulation of ion channels such as DL-alpha-amino-3-hydroxy-5-methyl-isoxazole proprionate (AMPA) and NMDA that, in turn, affect the long-term plasticity in the neuron.

Bienenstock, Cooper, and *Munro* have suggested that the threshold where LTP is induced can depend on the time-averaged recent activity of the neuron. Only postsynaptic activity that exceeds this threshold can induce LTP, whereas otherwise LTD is induced. We have already incorporated this in the mathematical description of Hebbian plasticity in rate models in eqns 7.10 and 7.11. In the latter equation the postsynaptic term $r_i - \langle r_i \rangle$ is positive (corresponding to LTP) only if the rate for the postsynaptic node is larger than the average activity of this node. In general it is possible to study different variations of this model by including different threshold terms f_2 in eqn 7.10.

Other mechanisms are possible and under active investigation. Many such proposals introduce some form of competition between the synapses. For example, we can imagine a pool of common resources that can demand a constant sum of synaptic efficiencies. Such normalizations are often used in models for simplicity, although the biological mechanisms are not yet completely verified. There are, however, several possible cellular mechanisms that can effectively result in a weight normalization. These include the previously mentioned weight decay, for which we now give an example.

7.7.3 Oja's rule and principal components

A particularly interesting case of weight normalization through heterosynaptic depression was proposed and studied by the Finnish scientist *Erkki Oja*. He considered the Hebbian rule

$$\delta w_{ij} = r_i^{\text{post}} r_j^{\text{pre}} - (r_i^{\text{post}})^2 w_{ij}, \tag{7.14}$$

where a standard Hebbian correlation term is augmented with a multiplicative (dependent on w_{ij}) decay term that is proportional to the square of the postsynaptic firing rate. Let us study the consequence of this rule on a linear node with two input channels (Fig. 7.11A) that is driven by random input taken from a two-dimensional probability distribution with zero mean. The details of this computer experiment are included in the examples in Chapter 12. The learning examples of the pairs $(r_1^{\text{pre}}, r_2^{\text{pre}})$ define points in the r_1^{pre}–r_2^{pre}-plane that are indicated as dots in Fig. 7.11B. Before applying Oja's learning rule we initialized the weight vector with a random values that is shown as a cross in the figure. The trajectory of the weight vector during learning when applying Oja's rule is shown as the line extending from the cross that marks the initial weight values.

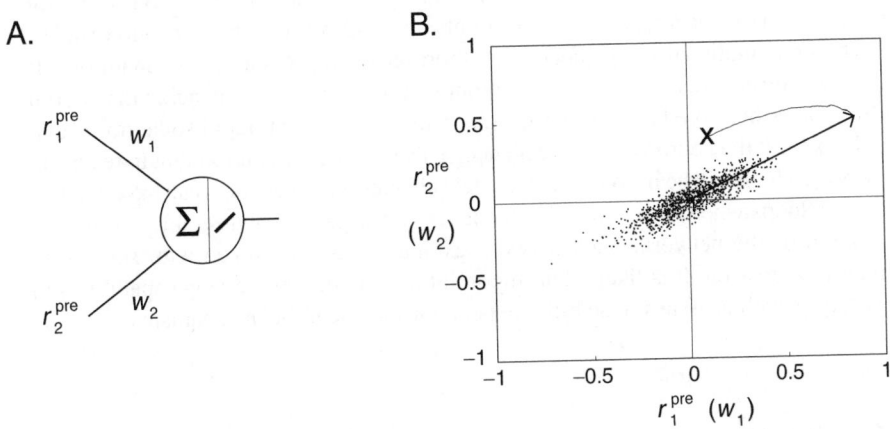

A.

B.

Fig. 7.11 Simulations of a linear node trained with Oja's rule on training examples (indicated by the dots) drawn from a two-dimensional probability distribution with mean zero. The weight vector with initial conditions indicated by the cross converges to the weight vector (thick arrow), which has length $|w| = 1$ and points in the direction of the first principal component.

At the end of the training session, after the network 'saw' a large number of sample points from the distribution, the weight vector converged to the one indicated in the figure as an arrow. This weight vector not only has a length of $|w| = 1$, but it can also be shown that it points in the direction of the maximum variance of the data set. This is the first *principal component* of the distribution. A single node can, of course, only express a single quantity describing the input data. The first principal component is such a quantity, which describes the data set along the direction of maximum variance and therefore incorporates as much of the data as can be done with a single variable. Using similar methods we can design networks in which other nodes represent further principal components on the order of maximum variations perpendicular to the previous principal components. The ability of neural networks to implement principal component analysis efficiently is exciting as it enables brain processes to utilize methods that have been shown to be very useful in analysing stochastic data.

7.7.4 Short-term synaptic plasticity and neuronal gain control

Long-term synaptic plasticity does not have to be the sole mechanism behind gain control and the information-processing abilities of single neurons. We discussed in Chapter 2 that cortical neurons typically have a transient response with a decreasing firing rate to a constant input current. This can be modelled with time- and usage-dependent weights that are weakened for a short time after the arrival of a presynaptic spike. This is an example of *short-term synaptic plasticity*, which is often called *short-*

term depression (STD). The effect does not even have to rely on the depression of an excitatory ion channel. For example, we saw in Chapter 2 that a slow calcium channel can produce such fatigue effects (see Fig. 2.12B). Furthermore, postsynaptic mechanisms are not the only source of STD. Presynaptic mechanisms of STD can, for example, be based on a reduction in the number of readily releasable synaptic vesicles.

The computational consequences of short-term depression can be manifold. It allows a neuron to respond strongly to input that has not been influencing the neuron recently and therefore has a strong novelty value. In contrast, rapid spike trains that would exhaust the neuron can be weakened. This is particularly important in recurrent networks, which will be introduced in the next chapter. Such networks can have positive feedback loops where strong positive feedback can trigger an undesirable continuous excitation of the network. Such network excitation can be broken with short-term synaptic depression. It is likely that important information-processing capabilities in the brain as well as issues of stability depend on such neuronal mechanisms.

7.8 Features of associators and Hebbian learning

We introduced in this chapter the associative node, which consists of a neuron-like integrator with synaptic weights that are trained with training rules that basically utilize only the Hebbian conjecture of synaptic plasticity based on the correlation of pre- and postsynaptic spike trains. Such an associative architecture has important characteristics that seem to incorporate some fundamental characteristics of brain-like information processing. We will end this chapter by summarizing some characteristics that will be particularly important for the models discussed in the following chapters. Note that the abilities of associative networks summarized here do rely on some form of distributed representation. We argued in Chapter 5 that this seems to be a reasonable conjecture for information processing in the brain.

7.8.1 Pattern completion and generalization

We have seen that a stimulus that captures only part of a pattern associated with an object can trigger a memory of an object. This means that the network supports the recall of details of an object that were not part of the stimulus. We thus talk about *pattern completion* or the ability to *recall from partial input*. Important for this ability is a distributed representation of an object stimulus so that some missing components of a feature vector can be added. The ability is based on the calculation of the overlap or similarity of an input vector with a weight vector that represents the pattern on which the node was trained. An example of such a similarity measure is the dot product between the pattern input vector and the vector of the synaptic efficiencies of a neuron which is the main functionality implemented by a sigma node. We also need a nonlinearity in the output function, such as a firing threshold, which is also characteristic of neurons. Then, as long as the overlap between input pattern and trained pattern is large enough, an output is generated that is equivalent to the output of a trained pattern. It follows that the output node responds to all patterns with a certain similarity to the trained pattern, an ability that is called *generalization*.

7.8.2 Prototypes and extraction of central tendencies

A closely related ability of associative nodes or networks is the ability to extract central tendencies. This occurs in the case when we train the nodes on many similar but not equivalent examples. The weight vector then represents an average over these examples, something that we could interpret as a *prototype*. Such training sets are typical in natural environments; for example, each person has an individual face while there are still many common features. Another possible reason for training sets with many slightly different patterns is that the training set is derived from the same object, which is represented over time with fluctuating patterns due to some noisy processes in the encoding procedure. The prototype extraction ability of associators can then be used to achieve *noise reduction*, an ability that is important in the case of noisy processing systems such as the brain and also has major technological applications.

7.8.3 Graceful degradation

The loss of some components of the system, such as some synaptic connections or even whole neurons, should not make the system fail completely. The loss or inaccuracy of only a few synaptic connections does not affect the system markedly. The system often degrades gracefully in that a large amount of synapses has to be removed before the system produces a large amount of error. Even the loss of whole nodes can be tolerated if the information feeds into an upstream processing layer with associative abilities that can archive pattern completion from partial information. Such a *fault tolerance* is essential in biological systems.

7.8.4 Biologically faithful learning rules

Machine learning has been studied widely over the last decades, and many useful algorithms have been introduced. Biologically motivated learning rules have contributed to many advancements in technical applications, and further advancements can be expected with increasing knowledge of how nature has solved problems that are still hard to mimic in artificial systems. The learning algorithms in artificial neural networks do not have to be biologically faithful, but complying to such restrictions can help us to find superior solutions. For example, many learning algorithms for artificial neural networks need to store the training data set in order to be able to cycle through it frequently. This might be reasonable for small data sets but is impractical when thinking about large data sets, for example, large images from a video stream. Complying to general restrictions of biological systems, such as demanding true *online learning* rather than using *batch learning*, can help us to find superior solutions in the long run. Such general restrictions can also guide brain models where specific details of the learning rules realized in the brain are not known.

The associative Hebbian learning rules introduced in this chapter have the following characteristics that we think are essential for most biologically faithful models:

1. **unsupervised:** No specific learning signal from the external world to a specific neuron is provided. The learning rule will make any input that is consistent with the firing of the output node being effective in eliciting the response of the output neuron. The firing of the output node can be initially random (random initial weights). The network then self-organizes under the learning rule, only

as a function of the firing history of pre- and postsynaptic neurons. The lack of a separate teaching signal motivates the term 'unsupervised' for this learning rule. Some forms of supervision without detailed firing instructions for the network are sometimes also labelled as unsupervised. These include reinforcement learning in which only reward and/or punishment are provided as teaching signals. We will discuss such learning rules in Chapter 10.

2. **local:** The basic associative learning rule outlined in this chapter is *local* in the sense that only presynaptic and postsynaptic observables are required to change the synaptic weight values. This is in marked difference to other learning rules such as the error-back-propagation algorithm mentioned in the last chapter and discussed further in Chapter 10. Locality in the learning rule seems to be an important requirement in order to be able to train large systems that can benefit from true parallel distributed processing. Local learning rules do not require an elaborated specialized message-passing network and can be implemented with minimal architectural requirements.

3. **online:** The learning rule does not require storage of firing patterns or network parameters. Many learning rules in artificial neural networks require the storage of large training sets or a history of the training process. It is unlikely that such requirements can be fulfilled in biological systems where we try to build systems that can handle realistic input patterns such as natural high-resolution images.

Note that the classifications are not always clear. For example, the Hebbian rule is commonly classified as unsupervised, whereas the delta-rule, which we will discuss in Chapter 10 is regarded as supervised despite the fact that the rules are closely related. [18] Also, online learning can incorporate some storage of information. For example, the Hebbian rule discussed in this chapter used average firing rates that can, however, be calculated online without storage of the training pattern. This can be contrasted to batch learning algorithms in which the training patterns are stored in the training phase so that the training patterns can be repeatedly used during the learning process.

Conclusion

The basic ability of neurons (or networks of neurons) to form associations between co-occurrences of sensory stimuli is thought to be essential for many information-processing mechanisms in the brain. The implementation of such abilities is possible in the brain through synaptic plasticity mechanisms that depend mainly on the firing of the presynaptic and postsynaptic neuron. Neurons are indeed able to implement such local learning mechanisms efficiently. Donald Hebb speculated over 50 years ago that such mechanisms might be essential in implementing brain functions in neuronal networks, and recent studies have explored and verified many details of synaptic plasticity. These studies have confirmed Hebb's basic postulates, and have also succeeded in measuring specific dependences of such synaptic plasticities. The experimental findings have to be explored computationally to comprehend the specific consequences of such plasticities for information-processing abilities in the brain. Most of the models in later chapters utilize basic Hebbian plasticity mechanisms.

---

[18] The delta rule is equivalent to the Hebbian rule with weight decay as shown in Chapter 10.

Further reading

Edmund T. Rolls and Alessandro Treves
    Neural networks and brain function, Oxford University Press, 1998.
    See Chapter 2.

Daniel J. Amit
    Models of realistic cortical Hebbian modules, in *Handbook of brain theory and neural networks*, Michael A. Arbib (ed.), MIT Press, 1998.
    The handbook is an important collection of papers about many neural network issues, some of which are relevant for brain functions. The article by Daniel Amit summarizes nicely some of the issues of recurrent networks.

Guo-chiang Bi and Mu-ming Poo
    Activity-induced synaptic modifications in hippocampal culture, dependence on spike timing, synaptic strength and cell type, in *J. Neurosci.* 18: 10464–72, 1998.
    Already a classic on which we based most of the discussions of asymmetric Hebbian learning in this chapter.

Mark C. W. van Rossum, Guo-chiang Bi, and Gina G. Turrigiano
    Stable Hebbian learning from spike timing-dependent plasticity, in *J. Neurosci.* 20(23): 8812–21, 2000.
    Useful additional reading as mentioned in the text. This article also includes the use of the Fokker–Planck equation to calculate weight distribution and a practical method (though not necessarily biologically plausible) to normalize the weights to achieve a desired postsynaptic firing rate.

Laurence F. Abbott and Sacha B. Nelson
    Synaptic plasticity: taming the beast, in *Nature Neurosci. (suppl.)*, 3: 1178–83, 2000.
    Very readable overview of asymmetric Hebbian learning and synaptic scaling. This supplementary issue of *Nature Neuroscience* is dedicated to computational neuroscience, a very much recommended reading.

Sen Song, Kenneth D. Miller, and Laurence F. Abbott
    Competitive Hebbian learning through spike-time-dependent synaptic plasticity, in *Nature Neurosci.* 3: 919–26, 2000.
    Shows how synaptic plasticity could regulate the firing rate of neurons.

8 Auto-associative memory and network dynamics

Recurrent auto-associative networks use the associative nodes introduced in the last chapter but are different from the basic associative networks and mapping networks in that they have, in addition, collateral or recurrent connections between the nodes in the neural layer. These types of connections are abundant in the brain and we will see that such systems have interesting properties. Systems with recurrent connectivity are formally dynamic systems, and we will explore the dynamics of such systems in this chapter.

8.1 Short-term memory and reverberating network activity

8.1.1 Short-term memory

Many mental abilities rely, at least partly, on the ability to hold information temporarily in an accessible form. We generally tend to call such memory *short-term memory*, although this term is very general and is often used in slightly different circumstances. For example, some form of human short-term memory has been characterized on a psychophysical level by a *recency effect*, which labels our tendency to remember recent objects in an object-stream better than objects further in the past. An obvious proposal to implement such short-term memory on a physiological level would be to hold corresponding neural activity over a certain duration. This type of physiological short-term memory could be related to short-term memory as measured by the recency effect. It could also support a type of short-term memory necessary for working memory, which we will discuss in Chapter 11.

8.1.2 Maintenance of neural activity

The maintenance of neural activity over a delay period has been observed experimentally. An illustrative example (though not the first observation of delay activity) of such an experiment by *Funahashi*, *Bruce*, and *Goldman-Rakic* is shown in Fig. 8.1. In their experiment a monkey was trained to maintain its eyes on a central fixed spot until a 'go' signal, such as a tone, indicated that it should move the eyes and to focus on one of several possible targets peripheral to the fixed spot. The choice of the target to which the eye should be moved in each trial was indicated by a short flash, but the subject was not allowed to move its eyes that lasted from the disappearance of the target cue until the 'go' signal indicating that the subject was to move its eyes. Thus, the target location for each trial had to be remembered during the delay period. The experimenters recorded from neurons in the dorsolateral prefrontal cortex (area 46) and found neurons that were active during the delay period. These neurons were sensitive to the particular target direction (the neuron shown only responded when the target was at 270 deg in the example of Fig. 8.1), and this neuron could therefore indicate by its delayed activity to which target the eye should be directed after the delay period.

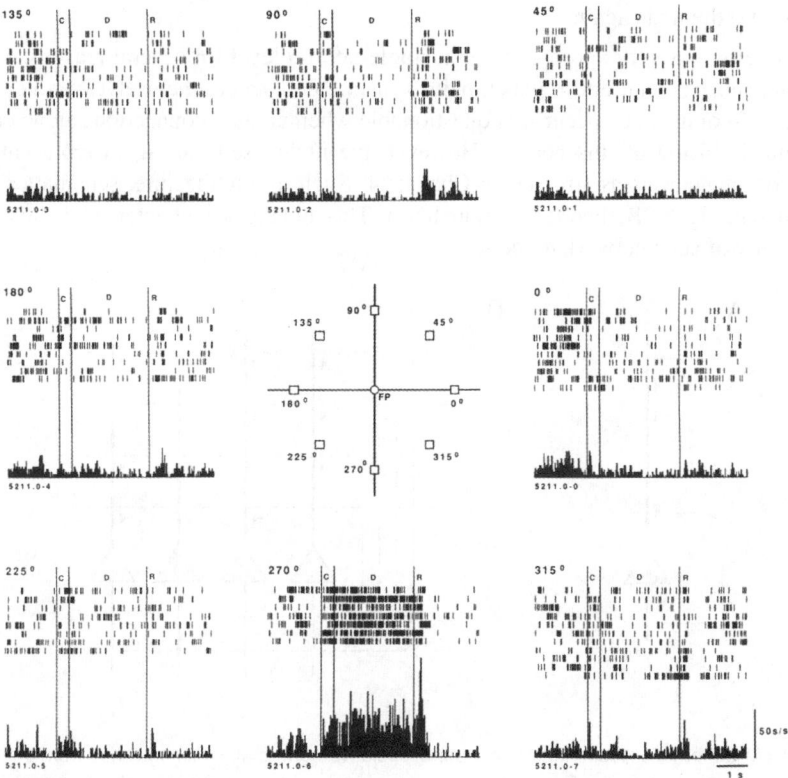

Fig. 8.1 Maintenance of delay activity in physiological experiments (reprinted from Funahashi, Bruce, and Goldman-Rakic, *J. Neurophysiol.* 61: 331–49 (1989)].

8.1.3 Recurrences

The experiment by *Funahashi et al.*, as well as other experiments, demonstrates that delay activity can be used to store activity over some period of time. Our question in the following is how such activity can be maintained by a neural network. We have seen that the membrane potential of a single neuron is reset after each spike. A single neuron is therefore not able to maintain activity on its own. The only solution seems to be that the neuron gets reliably activated again and again once it is active in the first place. A possible implementation of such a proposal is a model in which the neuron feeds its output back to itself. Such feedbacks are also called *recurrences*. A model of a recurrent node is illustrated in Fig. 8.2A. This node is able to maintain its firing, inflicted on it from an external input, through the recurrent connections as long as:

1. the recurrent pathway is strong enough;
2. there is some delay in the feedback so that the *re-entry* signal does not fall within a refractory time of the node; and
3. a possible leakage in the recurrent pathway is small enough so that it is possible to fire the node again.

This model is not very appropriate on a single-neuron level for several reasons. For example, neurons tend not to make major synaptic contacts to themselves, and even if such self-connections occur it is questionable whether such connections alone can maintain the firing of this neuron. However, the node itself can again represent a collection of neurons as stressed in Chapter 4. Such an architecture, schematically illustrated in Fig. 8.2B, then seems more likely. This and the next chapter are dedicated to the study of such network models.

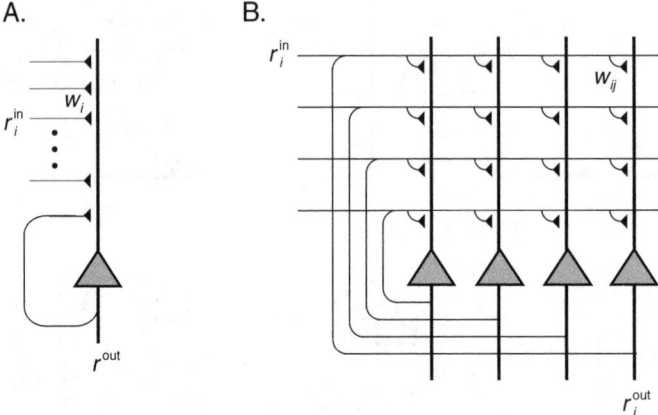

Fig. 8.2 (A) Schematic illustration of an auto-associative node that is distinguished from the associative node as illustrated in Fig. 7.1A in that it has, in addition, a recurrent feedback connection. (B) An auto-associative network that consists of associative nodes that not only receive external input from other neural layers but, in addition, have many recurrent collateral connections between the nodes in the neural layer.

8.2 Long-term memory and auto-associators

The network model illustrated in Fig. 8.2B is similar to the model network of associative memory (Fig. 7.1B), except that the input of each node is fed back to all of the other nodes in the network. This type of model is therefore commonly referred to as *auto-associative memory*.[19] In such a *recurrent network* model the activity of each node gets supported by all the nodes in the network. This creates the problem that each

---

[19] Note that we can also realize an auto-associative memory as feed-forward architecture, commonly also called an *auto-encoder*. However, in this chapter we are particularly interested in recurrent networks, and we will use the auto-associative network synonymously with the specific form of a recurrent network.

node could become active once a specific subset of nodes representing a meaningful pattern has become active. To solve this problem we have to carefully tune the recurrent connections. A biologically plausible mechanism to change synaptic efficiencies was introduced in the last chapter. If we include associative learning in this network it is indeed possible to maintain network activity that is pattern-selective.

The back-projections in this associative network introduce attractive features into the model that we will now discuss. We have already seen that associators are able to perform some form of pattern completion. After an external input pattern is presented to the network, the network will respond with an output pattern more similar to a trained pattern when Hebbian learning is applied. This response is fed back as input to the same network, so that the network will respond with a pattern that is again made closer to a learned pattern. We therefore expect that the cycling in a recurrent network can enhance the pattern completion ability of simple one-step associative nodes. The dynamic process of changing responses in the recurrent network will stop when a learned output pattern is reached. The network then keeps this pattern active until another pattern is applied to the network. The recurrent network model is therefore a prime candidate for *short-term memory* of the kind that supports sustained activity. At the same time it is also a model for *long-term memory* based on persistent synaptic changes.

The recurrent model is to some extent anatomically faithful in the sense that neurons within brain structures that have been associated with some form of memory functions frequently have massive collateral and back-projections. Collateral connections and intercortical back-projections are indeed abundant in the neocortex. The neocortical anatomy therefore supports the possibility that many recurrent networks of the type discussed in this chapter could be responsible for the measured delay activity in the prefrontal cortex. However, delay activity is not prominent in some other brain areas, for example, in the motor cortex. We therefore think that the types of connections in such areas are specialized in supporting long-term memories, such as the learning of motor skills.

8.2.1 The hippocampus and episodic memory

An area in the medial temporal lobe is called the *hippocampus* for its shape resembling a seahorse. This is the same structure as that in which *Bliss* and *Lomø* discovered LTP. The hippocampus has been particularly associated with the form of long-term memory called *episodic memory*. This type of memory refers to the storage of events, where each single event can be a vast collection of associations such as the encounter with an particular object at a certain location at a certain time. Within the hippocampus there is an area called CA3 that has well developed *axon collaterals* connecting to neurons within the same structure. This region has been assumed by many modellers to act as a recurrent auto-associative network and to therefore perform recall and pattern completion task.

The hippocampus has been implicated in the acquisition of episodic *long-term memory* because it has been shown that a patient in whom it was necessary to remove this structure suffered subsequently from a form of amnesia marked by the inability to form new long-term memories of episodic events. In contrast, long-term memory that was acquired before the removal of this structure, as well as the ability to learn new semantic memories and motor skills, was not impaired. The precise involvement of the hippocampus in memory acquisition is still under intense debate, and the neural mechanisms of human episodic memory abilities are mostly unclear. So far, it seems that the hippocampus can store some forms of memory for a considerable length of time until the content is stored permanently in other cortical areas. This type of memory is therefore also called *intermediate-term memory*. We will discuss some forms of spatial memory in which the hippocampus is known to be involved in the next chapter while we will concentrate here on some more general features of recurrent networks.

8.2.2 Learning and retrieval phase

Before we outline the important consequences of recurrent connections on the information-processing abilities of the brain we must mention a difficulty that occurs when combining associative Hebbian mechanisms with recurrences in the networks. Associative learning crucially depends on relating presynaptic activity to postsynaptic activity that is imposed by an unconditioned stimulus. The recurrent network will, however, drive this postsynaptic activity rapidly away from the activity pattern we want to imprint if the dynamic of the recurrent network is dominant. A solution to the problem, which we will follow in the discussions below, is to divide the operation of the networks into two phases, a *training phase* and a *retrieval phase*. These phases can, of course, be interleaved. There is no inherent difficulty in assuming such different phases as we rarely want to store and retrieve information at the same time.

There are several proposals as to how the switching between the learning and retrieval phase could be accomplished in the brain. For example, in the hippocampus there are, in addition to the massive collateral connections and the direct feed-forward connections from higher cortical areas (via the *entorhinal cortex*) on to the CA3 neurons in the hippocampus, *mossy fibres* from the dentate granule cells in the hippocampus on to CA3 neurons. These fibres could provide the necessary signal to indicate when learning should take place. Interestingly, mossy fibres have the largest synapses found in the mammalian brain, which the eminent British scientist *David Marr*, who developed many important models of brain functions, termed *detonator synapses*. Thus, it is possible that signals along mossy fibres command the firing of specific CA3 neurons at the right time to enable the formation of new associations through Hebbian learning.

Another proposal, investigated by *Michael Hasselmo*, is that chemical agents, such as *acetylcholine* (ACh) and *noradrenaline* (also called *norepinephrine*), could modulate learning and thereby enable the switching between a retrieval and learning phase. It has been shown that such agents, also called *neuromodulators*, facilitate synaptic plasticity and that their presence enhances the firing of the neurons. At the same time it suppresses excitatory synaptic transmission, which can suppress the effects of the recurrent collaterals. The neurons are therefore mainly responsive to external input to the system, which mirrors the proposal of the learning phase that we will generally apply in the following models. Such neuromodulatory control of the responsiveness of a system to external events can be important for an organism that has to respond to novel and potentially dangerous situations. The switching between learning and retrieval phase might also be necessary for the transfer of intermediate-term memories stored in the hippocampus to long-term memory stores in other cortical areas. This might happen during sleep, and fluctuations of ACh during sleep have indeed been detected.

There are further interesting questions, such as the control of the neuromodulation and the time scale of the switching between learning and retrieval. Further studies of such neuromodulatory mechanisms might reveal many more important brain mechanisms.

8.3 Point-attractor networks: the Grossberg–Hopfield model

Many researchers have speculated about the importance of feedback connections within brain networks, for example, *Gerald Edelman* who termed the feedback *re-entry*. The independent work of *Stephen Grossberg* and *John Hopfield*, together with that of their colleagues, has particularly contributed to the understanding and popularization of recurrent auto-associative memory of the form outlined in Fig. 8.2B. Their influential work has motivated and guided much of the further research in this area, and it is appropriate to call such models *Grossberg–Hopfield models* in their honour. [20] The models discussed in this chapter are designed to shed light on some consequences of recurrences in associative networks. The rigorous abstractions are therefore intended to make it possible to study the dynamic properties of such systems in more depth by making the systems to some extent analytically tractable.

The networks we are going to analyse are formally dynamic systems. We will concentrate here on so-called *attractor states* because such states have considerably advanced our understanding of dynamical systems. Attractor states are states that a dynamic system reaches asymptotically. Indeed, it may take a considerably long time for the system to come close to such states, and the relevance of such states for brain processes has therefore been questioned. It is indeed likely that the brain does not rely entirely on the necessity of settling into an attractor state. Rather, the rapid path towards an attractor state, or even a distinguishable behaviour of the response of a recurrent network to different initial states, may be sufficient to use such networks as associative memory devices. This would much better suit the real-time processing demands of brain functions. Interesting research into this issues is now emerging. We will, nevertheless, mainly discuss attractor states in the context of shedding light on the properties of recurrent neural networks.

8.3.1 Network dynamics

When studying recurrent networks we have to realize that the time domain becomes important because we need to understand not only the response of the system to a particular input but also the further development of the output states due to their recurrent influence on the input of the system itself. We will outline the features of such models with a network of sigma nodes governed by the leaky integrator dynamic introduced in Chapter 4. The change of the internal state of node i in the network is then given as

---

[20] The physics and artificial neural network literature often refers to such models as *Hopfield* models because Hopfield motivated many physicists to analyse such systems. However, Grossberg and his colleagues were certainly instrumental in analysing and popularizing such networks in connection with brain processes even before the important work by Hopfield.

$$\frac{1}{\tau}\frac{\mathrm{d}h_i(t)}{\mathrm{d}t} = -h_i(t) + \sum_j w_{ij}r_j(t) + I_i^{\text{ext}}(t), \tag{8.1}$$

where τ is a time constant setting the time scale of the leaky integrator dynamic, r_j is the firing rate of the node that is related to the internal state of the system through the gain or activation function $r_j = g(h_j)$, and $I^{\text{ext}}(t)$ is the current representing external input to the system. We can also study a discrete version of the system that is equivalent to the continuous system 8.1 for infinitesimal time step Δt, namely

$$h_i(t+\Delta t) = (1 - \frac{\Delta t}{\tau})h_i(t) + \frac{\Delta t}{\tau}\left[\sum_j w_{ij}r_j(t) + I_i^{\text{ext}}(t)\right]. \tag{8.2}$$

We also want to discuss this discrete model for time steps equal to the time scale of the dynamics by setting $\Delta t = \tau = 1$. Equation 8.2 simplifies in this case to

$$h_i(t+1) = \sum_j w_{ij}r_j(t) + I_i^{\text{ext}}(t). \tag{8.3}$$

Note that the *stationary states* of the continuous system (eqn 8.1), which are the states that do not change any more under the dynamics of the system and which are hence marked by $\mathrm{d}h_i/\mathrm{d}t = 0$, are also the *fixpoints* of the discrete system (eqn 8.3). Both models can thus illustrate the asymptotic features of the model. It is only when we are interested in the *transient response dynamics* of the model, for example when modelling the reaction times of the system, that we have to specify the dynamics of the system more carefully.

8.3.2 Hebbian auto-correlation learning

It is important for the recurrent networks discussed in this chapter that the weights w_{ij} are not random (like those we studied in Chapter 4) but are specifically self-organized by Hebbian learning. We want to study here the behaviour of such networks when trained with a Hebbian learning rule; in particular, we want to find out how many distinguishable memory states can be imprinted into these networks (we call this *capacity analysis*) and what the recall abilities of such networks are. These are important quantities to know when arguing about the employment of such mechanisms in brain processes. For simplicity, we illustrate this for a binary system in which the firing rates are either zero or one, that is, $r_i \in \{0, 1\}$. Equivalently, we can use the variables

$$s_i = 2r_i - 1 \tag{8.4}$$

in the equations, which will simplify some of the further discussions. We train the system on a set of binary random patterns $\xi_i^\mu \in \{-1, 1\}$ where the index μ labels the pattern and can run from 1 to the number of trained patterns P. We here chose the values of the pattern components to $+1$ and -1 with equal likelihood so that the mean of each component of the pattern is zero. The corresponding firing rates would have values of either 0 or 1 with a mean value of 0.5. The transformation (eqn 8.4) simplifies the Hebb rule (eqn 7.11) by using variables that have zero mean. In particular, when studying the network patterns in the $\{-1, 1\}$ representation it is appropriate to utilize the Hebbian rule

$$w_{ij} = \frac{1}{N} \sum_{\mu} \xi_i^{\mu} \xi_j^{\mu},$$ (8.5)

where we have normalized the weight matrix with N, the number of connections per neuron, which is appropriate as we will see within the following analysis of the network dynamics. We will here consider the discrete dynamic given by eqn 8.3. To specify the network completely we also have to define the gain function. In our specific example it is obvious to consider the threshold gain function of the net inputs h_i (membrane potentials) in the form of a step function

$$r_i = \text{sign}(h_i)$$ (8.6)

that keeps the state values (firing rates r_i) in the binary states of either -1 or $+1$.

8.3.3 Signal-to-noise analysis

We want to see how the network behaves on its own and we therefore set the external input to zero ($I^{\text{ext}} = 0$). A network with states $s_i(t)$ at time t has the following state values at the next time step

$$s_i(t+1) = \text{sign}[\sum_j w_{ij} s_j(t)]$$ (8.7)

and by inserting the Hebbian-trained weight matrix (eqn 8.5) we get

$$s_i(t+1) = \text{sign}[\frac{1}{N} \sum_j \sum_{\mu} \xi_i^{\mu} \xi_j^{\mu} s_j(t)].$$ (8.8)

Let us test the network when initialized with one of the trained patterns. Without loss of generality we can choose $\mu = 1$ for the demonstration. It is then useful to split the terms in the sum over μ for a term for the first training pattern and a second term with the rest of the training pattern, for example,

$$s_i(t+1) = \text{sign}[\frac{1}{N} \xi_i^1 \sum_j \xi_j^1 s_j(t) + \frac{1}{N} \sum_j \sum_{\mu=2}^{P} \xi_i^{\mu} \xi_j^{\mu} s_j(t)].$$ (8.9)

The expression $\xi_j^1 s_j(t)$ in the first term simplifies for the initial condition $s_j(t) = \xi_j^1$. This product is always one with the choice of the training pattern (either 1^2 or $(-1)^2$), and the sum of these ones just cancels the normalization factor N. We therefore end up with the expression

$$s_i(t+1) = \text{sign}[s_i(t) + \frac{1}{N} \sum_j \sum_{\mu=2}^{P} \xi_i^{\mu} \xi_j^{\mu} s_j(t)].$$ (8.10)

We want the network to be stationary for the trained pattern, and the first term does indeed point in the right direction. We call this term the 'signal' part as it is this that we want to recover after the updates of the network. The term $\frac{1}{N} \sum_j \sum_{\mu=2}^{P} \xi_i^\mu \xi_j^\mu \xi_j^1$ describes the influence of the other stored pattern on the state of the network and is called the *cross-talk term*. This cross-talk term is thought to be analogous to interference between similar memories in a biological memory system. The cross-talk in our formal analysis is a random variable because we used independent random variables ξ_i^μ as training patterns (see appendix B). It can therefore be considered as 'noise'. The activity of the node i remains unchanged in the next time step as long as the cross-talk term is larger than -1. The probability of the node changing sign depends on the relative strength of signal and noise, hence the name of the analysis. In this signal-to-noise analysis we can distinguish two cases.

8.3.4 One pattern

The special case of a network with only one imprinted pattern ($N^{\text{pat}} = 1$) is particularly easy to analyse. In this case we do not have a cross-talk term so the network stays in the initial state when started with the imprinted pattern. The imprinted pattern is thus a fixpoint of the dynamics of this network,

$$s_i(t+1) = \text{sign}[\xi_i] = s_i(t). \tag{8.11}$$

What happens if we do not start the network with the trained pattern, but instead start the network simulation with a noisy version of this pattern where some of the components are randomly flipped? In this case we have to go back to eqn 8.9 and to notice that the sum $\sum_j \xi_j s_j(t)$ is always positive as long as we change fewer then half of the signs of the initial pattern. We therefore retrieve the learned pattern even when we initialize the network with a moderately noisy version of the trained pattern. The retrieval is also very fast as it takes only one time step. The pattern will remain stable for all following time steps. The trained pattern is therefore a *point attractor* of the network dynamics because initial states close to the trained pattern are attracted by this point in the state space of the network. As an interesting side note, if the number of flipped states is more than half of the number of nodes in the network, then the inverse pattern $s_i = -\xi_i$ is retrieved. The inverse state is therefore also an attractor of the network trained on one pattern in this specific network.

8.3.5 Many patterns

Now let's turn to the situation when we train the network on more than one pattern. The states of the network initialized with the first pattern are preserved only if the contribution from the cross-talk (noise) term is smaller than -1, and we must therefore study this term in more detail. The cross-talk term is a random variable because we used random variables ξ_i^μ as training pattern as mentioned before. The mean of the random term is zero because the individual components are independent and the mean of the individual components is zero. We can therefore expect some cases in which some of the many trained patterns are stable. However, the probability of the cross-talk term reversing the state of the node depends on the variance of the noise term. We therefore have to estimate the variance of the cross-talk term in more detail.

In our example we used a training pattern with components that are equally distributed between values -1 and 1. Each part of the sum in the cross-talk term is therefore an equally distributed number with values -1 or 1. The sum of such binary random numbers over the training patterns (except the pattern that we used as initial condition for the network) is a binomially distributed random number with mean zero and variance $\sqrt{P-1}$. The large sum of such random numbers over the number of nodes in the network is then well approximated by a Gaussian distribution with mean zero and variance $\sqrt{(P-1)N}$ (see Appendix B). We finally have to take the normalization factor N in the cross-talk term into account, and the variance of this 'noise' term is therefore,

$$\sigma = \sqrt{\frac{(P-1)}{N}} \approx \sqrt{\frac{P}{N}} = \sqrt{\alpha}. \tag{8.12}$$

In this equation we have introduced the *load parameter* $\alpha = P/N$, which specifies the number of trained patterns relative to the number of nodes in the network.[21]

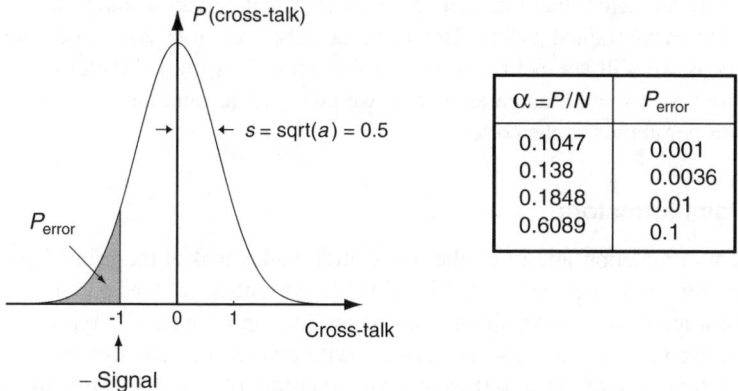

$\alpha = P/N$	P_{error}
0.1047	0.001
0.138	0.0036
0.1848	0.01
0.6089	0.1

Fig. 8.3 The probability distribution of the cross-talk term is well approximated by a Gaussian with mean zero and variance $\sigma = \sqrt{\alpha}$. The value of the shaded area marked P_{ERROR} is the probability that the cross-talk term changes the state of the node. The table lists examples of this probability for different values of the load parameter α.

Knowing of the probability distribution of the cross-talk term we can specify the probabilities of the cross-talk term changing the activity value of the node. The probability of the cross-talk term producing values less than -1 is illustrated in Fig. 8.3 and can be calculated with the error function summarized in Appendix B. This is given by

$$P_{error} = \frac{1}{2}[1 - \mathrm{erf}(\frac{S}{\sqrt{2\sigma}})], \tag{8.13}$$

where S is the strength of the signal (which is equal to one in our case), and σ is the variance of the cross-talk term as estimated above. To ensure that the probability that the cross-talk term does change the sign given by the signal term is less than a certain value, that is, $P_{error} < P_{bound}$, we need a load parameter less than

---

[21] The number of nodes in the network is equivalent to the number of connections per node in the fully connected networks considered here. We will see later that the relevant number is indeed the number of connections per node.

$$\alpha < \frac{1}{2[\mathrm{erf}^{-1}(1 - 2P_{\mathrm{bound}})]^2}. \tag{8.14}$$

The expression for the cross-talk term distribution tells us that the probability of the component flipping the sign is small when the load parameter of the network is small ($\alpha \ll 1$), that is, if the number of patterns in the training set is much smaller than the number of connections per node in the network ($P \ll N$). In particular, if we train the network on a number of patterns that is 10% of the number of nodes in the network, that is, $P = 0.1 * N$, then the probability that the component will flip signs is less than 0.1. Some other examples are listed in the table in Fig. 8.3. The signal-to-noise analysis shows that it is likely that trained patterns are still fixpoints of the network dynamics for a moderate number of trained patterns. On the other hand, with an increasing number of training patterns we will increase the probability of flipping signs in the activities of the nodes. The flipped states can cause further nodes to flip signs in the next time step. We will see below that this can cause an avalanche effect leading the network state away from the trained pattern. However, the network is robust for moderate load parameters as we will see below. Note that the number of trained patterns can still be large even with small load parameters if we take into account the large number of connections per neuron in the cortex.

8.3.6 Point attractors

The pattern completion ability of the associative nodes makes the trained patterns *point attractors* of the network dynamics in networks with small load parameters. We can demonstrate this easily in simulations as outlined in Chapter 12. Typical results are summarized in Fig. 8.4A for the network with quasicontinuous dynamics of eqn 8.2 using a time step of $\Delta t = 0.01$ and a time constant of $\tau = 10$. The network has $N = 1000$ nodes that are trained on $P = 100$ random binary patterns. We allowed the nodes of the network to have real-valued activities, and used the sigmoidal activation function $g(h) = \tanh(h)$, which restricts the firing rates to the interval $r_i \in [0, 1]$ corresponding to state components in the interval $s_i \in [-1, 1]$. We monitored the state of the network by calculating the distance defined by

$$d(\mathbf{a}, \mathbf{b}) = \frac{1}{2}\left(1 - \frac{\mathbf{a}'\mathbf{b}}{||\mathbf{a}|| \, ||\mathbf{b}||}\right) \tag{8.15}$$

between the training vector $\mathbf{a} = \xi^1$ and the state of the system $\mathbf{b} = \mathbf{s}$. Some alternative distance measures are listed in Appendix A. This distance measure is proportional to the dot product between two vectors. We only normalized the dot product so that the value would be equal to the value of the percentage of changed signs in the training vector that was used to initialize the network. The results shown in Fig. 8.4A demonstrate that the network converges to the trained pattern if the initial distance is less than a certain value around $d_0 \approx 0.3$. The trained pattern is therefore a point attractor under the dynamics of the network with a *basin of attraction* of size d_0.

It is interesting to realize that the time of convergence is very rapid compared to the time constant of the nodes. This can only be understood in terms of the cooperation of all the nodes in the network. Note also that the transition can be even faster in networks of spiking neurons as we argued in Section 4.3.4. Furthermore, it is possible that only

A.
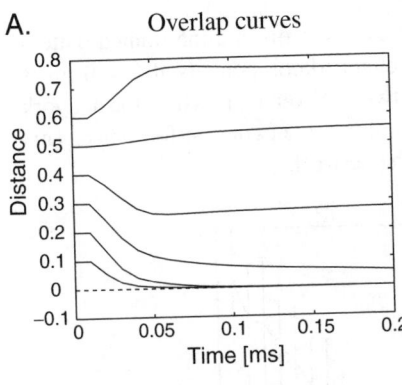

B.
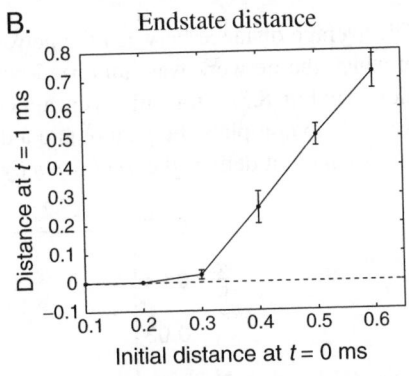

Fig. 8.4 Simulation results for an auto-associative network with quasicontinuous leaky-integrator dynamics, $N = 1000$ nodes, and a time constant of $\tau = 10$ ms. The time step was set to $\Delta t = 0.01$ ms. (A) Distance between a trained pattern and the state of the network after the network was initialized with a noisy version of one of the 100 trained patterns. (B) Dependence of the distance at $t = 1$ ms on the initial distance at time $t = 0$ ms.

the initial response of the system is sufficient to use such networks as associative memory devices.

In Fig. 8.4A we can see that the network did not converge to the trained pattern vector ξ^1 when the initial distance was too large, and the distance even increased over time. The average distance at $t = 1$ ms, which is larger than the convergence time, is plotted for 10 independent runs against the initial distance at $t = 0$ ms in Fig. 8.4B. The relatively sharp transition between the domain in which the network can restore a noisy version of a training pattern and the domain where the network is not able to retrieve the pattern has the signature of a *phase transition*, a transition between domains of a system with different properties.

8.4 The phase diagram and the Grossberg–Hopfield model

8.4.1 The load capacity α_c

The experiment shown in Fig. 8.4 was conducted with a fixed number of trained patterns relative to the number of connections per node in the network. Detailed analysis of networks with sparse connectivity shows that the relative load parameter is the number of trained patterns relative to the number of connections per node, which we will denote by C. We therefore define the load parameter as

$$\alpha = \frac{N^{\text{pat}}}{C}. \tag{8.16}$$

The number of connections per node is, of course, equal to the number of nodes in a fully connected network. We have seen previously that an increasing load parameter will increase the probability that the cross-talk term can drive the network away from a trained pattern. Consecutive iterations can make things worse and we have to expect that the network performance decreases with an increasing load of trained patterns.

The average distance between the network state at $t = 1$ ms and the trained pattern in which the network was initialized with 10% changed components at $t = 0$ ms is plotted in Fig. 8.5. A transition is again visible between a domain in which the network was able to complete the pattern and a domain with pure network performance. This transition point defines the *load capacity* α_c of the network.

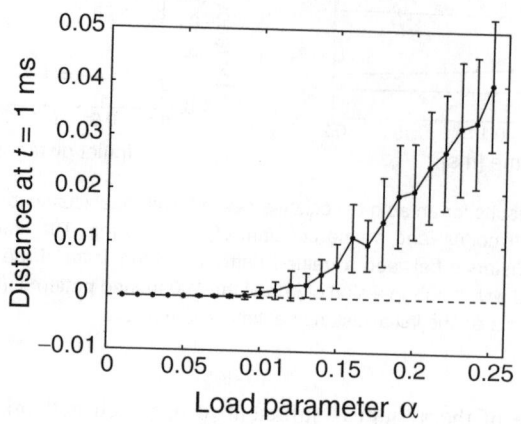

Fig. 8.5 Simulation results for an auto-associative network equivalent to the network used in Fig. 8.4 but with different numbers of patterns in the training set N^{pat}. The distance of the network state from the first training pattern at time $t = 1$ ms is shown. The network was initialized with a version of the first trained pattern with 1% reversed components.

8.4.2 The spin model analogy

The load capacity of auto-associative networks can be analysed in much more depth by realizing a useful correspondence of the recurrent network models to the so-called *spin models* developed in statistical physics. This correspondence enables us to apply sophisticated methods developed for the analysis of spin models directly to neural network models. We will summarize the principal idea behind the correspondence and the methods and outline some of the results of such analyses because they shed light on some important characteristics of recurrent neural networks.

Central to the correspondence of spin models and the recurrent network models discussed here is the realization that the binary states of the nodes can be interpreted as spins, or little magnets, that can have two orientations, up and down. These 'magnets' interact with all the other magnets in the network. The other magnets would try to align a particular node in the dominant direction of the other magnets if the influence of the other magnets were positive, which corresponds to positive weight values. In magnetic material there is, however, another force that tends to randomize the direction of the spins of the nodes. This force is *thermal noise*, which increases with increasing temperature T. The competition between the magnetic force, which tends to align the magnets, and the thermal force, which tends to randomize the directions, results in a sharp transition between a *paramagnetic* phase, in which there is no dominant direction of the magnets, and the *ferromagnetic* phase, in which there is a dominating direction of the elementary magnets. These phases have very different physical properties, and the transition is therefore called a *phase transition* similar to the transition between the liquid and vapour phase of water.

8.4.3 Frustration and spin glasses

Powerful analytical methods have been developed to describe systems of interacting spins such as ferromagnets. The situation in auto-associative networks is, however, further complicated by the fact that the force between the nodes is not consistently positive. Indeed, we saw in Chapter 7 that the Hebbian rule employed in these networks has a Gaussian distribution with positive a negative weights. Those conflicting forces can result in complicated spin states of the system. Such systems are known to physicists as *frustrated systems* or *spin glasses*; the latter alludes to the correspondence with glasses that have similar properties. Spin glasses are complicated systems and only partially tractable with standard methods of statistical mechanics, such as *mean field theory*. However, a breakthrough mathematical trick called the *replica method* was able to generalize the methods to be applied to spin glasses. Physicists such as *Daniel Amit* have considerably advanced our understanding of attractor neural networks using such methods. With these analytical methods we are not only able to calculate transition points, but can also specify to some extent the different natures of the different phases.

8.4.4 The phase diagram

The phase diagram of the Grossberg–Hopfield model is summarized schematically in Fig. 8.6. This phase diagram outlines the phase boundaries as a function of two parameters, the load parameter α on the abscissa and the temperature T (specifying the noise in the network) on the ordinate. Noise in the network can be simulated with probabilistic updating rules for the network dynamics. For example, we can replace the activation function

$$r_i(t) = \text{sign}(h_i(t)) \tag{8.17}$$

with a probabilistic version

$$P(r_i(t) = \pm 1) = \frac{1}{1 + \exp(\mp 2h_1(t)/T)} \tag{8.18}$$

that depends on the noise parameter T. We recover the noiseless update activation function in the limit of $T = 0$.

A detailed analysis of such noisy network models shows that the shaded region in the phase diagram shown in Fig. 8.6 is where point attractors exist that correspond to the trained patterns. The network in this phase is therefore useful as associative memory. This phase corresponds to a ferromagnetic phase in the magnet analogy. For vanishing noise $(T = 0)$ a transition point to another phase occurs at around $\alpha_c(T = 0) \approx 0.138$. For load parameters larger than this value the network is in a *frustrated phase* (spin glass phase), in which point attractors of trained memories become unstable. This frustrated phase is reached for smaller values of the load parameter if we include noise in the network. For strong noise the behaviour of the system is mainly random (*random phase*), corresponding to a *paramagnetic phase* in the magnet analogy. Simulations have confirmed the validity of the analytical results.

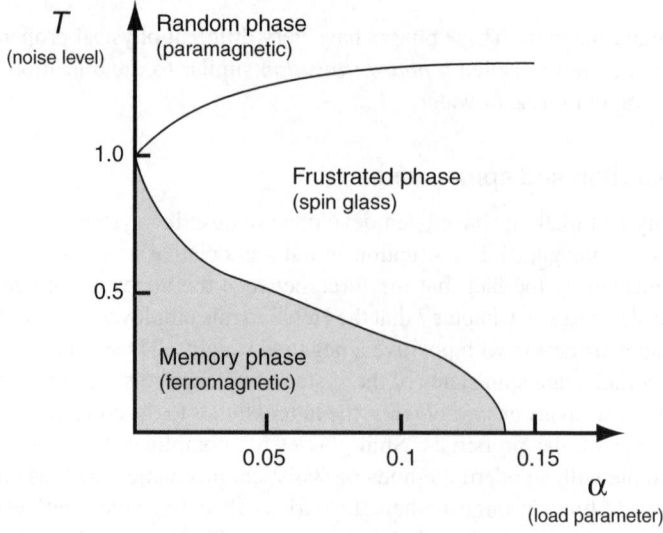

Fig. 8.6 Phase diagram of the attractor network trained on a binary pattern with Hebbian imprinting. The abscissa represents the values of the load parameter $\alpha = N^{\mathrm{pat}}/C$, where N^{pat} is the number of trained patterns and C is the number of connections per node. The ordinate represents the amount of noise in the system. The shaded region is where point attractors proportional to the trained pattern exist. The network in this region can therefore function as associative memory.

8.4.5 Spurious states

It seems that noise is only destructive for the memory abilities of the associative network. However, this is not entirely the case. To see how noise can help the memory performance, let us study how a mixture of some trained patterns behaves under the dynamics of the network. For example, let us start the network with a pattern that has the sign of the majority of the first three patterns,

$$s_i(t = 0) = \xi_i^{mix} = \mathrm{sign}(\xi_i^1 + \xi_i^2 + \xi_i^3). \tag{8.19}$$

The state of the node after one update of this node with the discrete dynamics (eqn 8.2) is then

$$
\begin{aligned}
s_i(t = 1) &= \mathrm{sign}(\frac{1}{N} \sum_j \sum_{\mu=1}^{N^{\mathrm{pat}}} \xi_i^\mu \xi_j^\mu \mathrm{sign}(\xi_j^1 + \xi_j^2 + \xi_j^3)) \\
&= \mathrm{sign}(\frac{1}{N} \sum_j (\xi_i^1 \xi_j^1 + \xi_i^2 \xi_j^2 + \xi_i^3 \xi_j^3) \mathrm{sign}(\xi_j^1 + \xi_j^2 + \xi_j^3) \\
&\quad + \frac{1}{N} \sum_j \sum_{\mu=4}^{N^{\mathrm{pat}}} \xi_i^\mu \xi_j^\mu \mathrm{sign}(\xi_j^1 + \xi_j^2 + \xi_j^3))
\end{aligned} \tag{8.20}
$$

The last term is again a cross-talk term, which can be shown to be of the same order as the cross-talk term in the case of a trained pattern. The magnitude of the signal term itself can have different values. For example, if the components ξ_i^1, ξ_i^2, and ξ_i^3 all have

Table 8.1 The possible states of three binary nodes and their summed value.

ξ^1	ξ^2	ξ^3	$\xi^1 + \xi^2 + \xi^3$
1	1	1	3
1	1	-1	1
1	-1	1	1
1	-1	-1	-1
-1	1	1	1
-1	1	-1	-1
-1	-1	1	-1
-1	-1	-1	-3

the same value, which happens with the probability of 1/4, then we can pull out this value from the sum in the signal term,

$$\frac{1}{N} \sum_j (\xi_i^1 \xi_j^1 + \xi_i^2 \xi_j^2 + \xi_i^3 \xi_j^3) \mathrm{sign}(\xi_j^1 + \xi_j^2 + \xi_j^3) = \xi_i^1 \frac{1}{N} \sum_j (\xi_j^1 + \xi_j^2 + \xi_j^3) \mathrm{sign}(\xi_j^1 + \xi_j^2 + \xi_j^3).$$

(8.21)

The components in the remaining sum of the signal term can have various values as can be seen from the possible combinations of components as listed in Table 8.1. We get a contribution of 3 if the signs of all three nodes ξ_j^1, ξ_j^2, and ξ_j^3 are the same. This happens again with probability 1/4. The remaining 3/4 of the time we get a contribution of 1. On average we therefore get a signal that is 3/2 of the signal of a trained pattern. However, the signal term is different when one sign of the nodes ξ_i^1, ξ_i^2, or ξ_i^3 is different from that of the other ones, which happens with probability 3/4. For example, if ξ_i^3 has a different sign from ξ_i^1 and ξ_i^2, we can write the signal term as

$$\frac{1}{N} \sum_j (\xi_i^1 \xi_j^1 + \xi_i^2 \xi_j^2 + \xi_i^3 \xi_j^3) \mathrm{sign}(\xi_j^1 + \xi_j^2 + \xi_j^3) = \xi_i^1 \frac{1}{N} \sum_j (\xi_j^1 + \xi_j^2 - \xi_j^3) \mathrm{sign}(\xi_j^1 + \xi_j^2 + \xi_j^3).$$

(8.22)

The average values of the remaining sum can again be evaluated as before, showing that this is equal to 1/2. We conclude that we have on average a signal that has the strength of

$$\frac{1}{4} * \frac{3}{2} + \frac{3}{4} * \frac{1}{2} = \frac{6}{8}$$

times the signal when updating a trained pattern. This indicates that there can be mixture states of the trained patterns that are also attractors in the system. These are an example of what is called a *spurious state* in the network, attractors that are different from the pattern used to train the network. Those strange memory states have sometimes been termed *schizophrenic* states to allude to apparent memory recalls with strange contents, although a direct relation to mental disorders has not been proven. But it shows that it is possible that under certain conditions some memory recalls are contaminated by other memory states, even more than through the usual cross-talk term analysed above.

8.4.6 The advantage of noise

We found that spurious states exist that are attractors under the network dynamics, in addition to the attractors corresponding to the trained patterns. However, we have also seen that the average strength of the signal for the spurious states is less than the signal we found for a trained pattern. This makes the spurious states under normal conditions less stable than attractors related to trained patterns. We can use this to our advantage by forcing the system out of the spurious states should it be attracted by an initial pattern close to them. We can do this by introducing noise into the system. This can be done in various ways as we discussed in Chapter 3. With an appropriate level of noise we can kick the system out of the basin of attraction of some spurious states and into the basin of attraction of another attractor. It is likely that the system will then end up in a basin of attraction belonging to a trained pattern as these basins are often larger for moderate load capacities of the network. Noise can therefore help to destabilize undesired memory states.

The analysis of the network in this section can only give us a picture of the average behaviour of the networks. The derivation of the phase diagram demands large, formally infinitely large, networks. This is a good assumption in the sense that networks in the nervous system are comprised of many thousands of highly interconnected neurons. The transitions are less sharp in finite systems of networks with a finite number of nodes. Also, in a signal-to-noise analysis we have analyse only the expected magnitude of the signals driving the dynamics. It should be clear that the behaviour of a particular network depends strongly on the specific realization of the training pattern. The phase diagram is therefore specific to the choice of training pattern, which we chose to be random binary numbers. The example nevertheless gives us the general picture that attractor states can be expected in recurrent networks and that phase transitions, a breakdown of the memory system sometimes called *amnesia*, are possible under various circumstances. The breakdown occurs only after the network is brought to it's limits. However, even a load capacity of $\alpha_c \approx 0.138$ means that over 1000 memories can be stored in a system with nodes receiving 10,000 inputs as is typical for neurons in the brain. These are plenty of states when considering situations for which we need short-term memories.

8.5 Sparse attractor neural networks

The load capacity for the noiseless Grossberg–Hopfield model with standard Hebbian learning on random binary patterns is $\alpha_c \approx 0.138$. These training patterns are, on average, uncorrelated. The sensory signals driving the learning in our brain are, on the contrary, often correlated in some way. For example, the visual patterns on the retina are correlated across space, and specific images (for example, a fish) is commonly seen in conjunction with other images (for example, water). Correlations between the training patterns worsen the performance of the network as the cross-talk term can yield high values in this case. A solution to this problem is to use a preprocessing step in which the representations of the training pattern get modified to yield orthogonal patterns. Orthogonal patterns have the property that the dot product between them is zero, that is

$$\xi^{\mu\prime}\xi^{\nu} = \delta^{\mu\nu}, \tag{8.23}$$

in which the *Kronecker symbol* $\delta^{\mu\nu}$ is defined as

$$\delta^{\mu\nu} = \begin{cases} 1 & \text{if } \mu = \nu \\ 0 & \text{elsewhere} \end{cases}. \tag{8.24}$$

The cross-talk term for such patterns is exactly zero so that the network can store up to C patterns, that is, $\alpha_c=1$.

8.5.1 Expansion coding

Such strategies seems indeed to be realized in the brain at least in an approximate sense as orthogonalization can be realized by expansion coding. An example illustrating the principle of expansion coding using a single-layer perceptron is shown in Fig. 8.7. The system has two input channels that can have four different combinations of input values if we restrict ourselves again to binary patterns, namely $(0,0), (0,1), (1,0)$, and $(1,1)$. The perceptron has four output nodes, and we have included a variable bias in the activation function (firing threshold) implemented by an additional input node with constant activation value. In the figure we specified an example of the weight values for which a network with threshold output nodes transforms the initial pattern representation into an orthogonal representation. All four patterns can therefore be stored in a subsequent auto-associative memory with only four nodes. Expansion coding can also be realized with competitive networks, which are the main subject of the next chapter. Common to the expansion schemes is that the number of nodes representing a pattern is expanded while at the same time the representation is made more sparse.

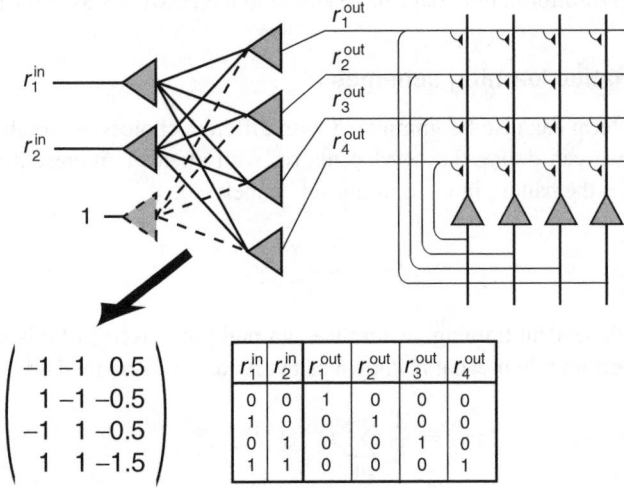

$$\begin{pmatrix} -1 & -1 & 0.5 \\ 1 & -1 & -0.5 \\ -1 & 1 & -0.5 \\ 1 & 1 & -1.5 \end{pmatrix}$$

r_1^{in}	r_2^{in}	r_1^{out}	r_2^{out}	r_3^{out}	r_4^{out}
0	0	1	0	0	0
1	0	0	1	0	0
0	1	0	0	1	0
1	1	0	0	0	1

Fig. 8.7 Example of expansion coding that can orthogonalize a pattern representation with a single-layer perceptron. The nodes in the perceptron are threshold units, and we have included a bias with a separate node with constant input. The orthogonal output can be fed into a recurrent attractor network where all these inputs are fixpoints of the attractor dynamics.

8.5.2 Sparse pattern

The expansion coding indicates that the load capacities of attractor networks can be larger for patterns with sparse representations. We have already discussed that the representation in the brain is most probably distributed but sparsely, not fully distributed as we have assumed so far in this chapter. We can define the sparseness a of a representation by

$$a = \frac{\langle r \rangle^2}{\langle r^2 \rangle} \tag{8.25}$$

as outlined in Section 5.5.2. *Alessandro Treves* and *Edmund Rolls* analysed the storage capacity of attractor networks with such sparsely distributed random patterns and found in the limit of small a and large networks a storage capacity of

$$\alpha_c \approx \frac{k}{a \ln(1/a)}, \tag{8.26}$$

where k is a constant that depends weakly on some details of the network and the pattern representations, but is roughly on the order of 0.2–0.3. The number of patterns in a pattern set with sparseness $a = 0.1$ that can be stored in a attractor network with nodes that have 10,000 synapses can therefore exceed 20,000, more than enough for many short-term memory requirements in the brain. Note, however, that the information content does not change. The enhanced storage capacity of the network has to be compared with the reduction of the amount of information that can be stored in a sparse representation compared to that in a representation with more active components. The information is proportional to $a \ln(1/a)$, the denominator in eqn 8.26. We conclude that the amount of information that can be stored in the network stays constant.

8.5.3 Alternative learning schemes

The cross-talk term can also be minimized with different choices of weight matrices. For example, we can define the overlap matrix with elements representing the dot product between the pattern in the training set, namely,

$$Q_{\mu\nu} = \sum_i \xi_i^\mu \xi_j^\mu. \tag{8.27}$$

For linearly independent training vectors we can build the inverse of this matrix and use it in the learning rule in what is known as the *pseudo-inverse* method,

$$w_{ij} = \frac{1}{N} \sum_{\mu\nu} \xi_i^\mu (\mathbf{Q}^{-1})_{\mu\nu} \xi_j^\nu. \tag{8.28}$$

The inverse orthogonalizes the input pattern and the storage capacity is therefore $\alpha_c = 1$. It is interesting to see that we can include orthogonalization in the learning rule, although this learning is not very plausible biologically. However, we can still use this learning rule in simulations as long as we can argue that the orthogonalization can be implemented biologically, such as via expansion coding as we have seen before. Researchers are also looking for approximations of this learning rule that might be implemented in real nervous systems. Several other learning methods that can increase the critical load capacity of auto-associative memories have been proposed in the literature. However, the minimal Hebbian learning rule is often sufficient to store enough information in the brain, and the rule stands out for its simplicity.

8.5.4 The best storage capacity with a sparse pattern

There are many possible learning algorithms, some of which might be more biologically realistic than others. Each learning rule can lead to a different weight matrix. It is therefore interesting to ask what would be the best possible solution for a weight matrix. To be more precise, if we have a certain training set we can ask what the load capacity of the network, with a weight matrix that was produced with the optimal learning rule, is. To answer this question we have to try out all the possible weight matrices, which is, of course, a daunting task. However, methods from statistical physics can help us to get the answer to this question. This was proposed and worked out in some classical works by *Elizabeth Gardner*. She found that the maximal storage capacity of auto-associative networks with binary patterns that can have sparse representations is

$$\alpha_c = \frac{1}{a \ln(1/a)}. \tag{8.29}$$

This should be compared to the results in Hebbian networks with sparse patterns (eqn 8.26). This comparison shows that the simplest Hebbian rule comes close to giving the maximum value, another indication that this minimal mechanism is sufficient for many applications in the brain.

8.5.5 Control of sparseness in attractor networks

The investigation of the properties of attractor networks with sparse patterns is important when studying brain functions. However, training recurrent networks on patterns with sparseness a with the basic Hebbian covariance rule 7.11,

$$\delta w_{ij} = k(r_i - a)(r_j - a), \tag{8.30}$$

is not enough to ensure that the state that is retrieved also has sparseness of $a^{\text{ret}} = a$. Indeed, we will show that the retrieval sparseness in networks of sigma nodes with a sigmoidal transfer function that have zero firing offset (that is, $x_0 = 0$ in eqn 4.20) and are trained with the covariance rule 8.30 is always $a^{\text{ret}} = 0.5$.

Let us demonstrate this with binary patterns that have components 0 and 1, where 0 indicates no firing and 1 a firing of the node. The sparseness of the pattern is then given by the number of nodes that are active. In this case there are four possible combinations of pre- and postsynaptic firing and corresponding weight contributions of different patterns listed in Table 8.2. Hence, the components of the weight matrix after imprinting one pattern have three possible values with different probabilities (two for $a = 0.5$) and are hence trinomially (binomially for $a = 0.5$) distributed. The mean (expectation value) of these contributions is zero, that is,

$$\langle \delta w \rangle = \sum \delta w P(w) = a^2(1-a)^2 - 2a^2(1-a)^2 + a^2(1-a)^2 = 0, \tag{8.31}$$

Table 8.2 The contributions of the four possible firing patterns of pre- and postsynaptic firing rates to the Hebbian covariance matrix, and the probability of the occurrence of these patterns, for training sets with a sparse pattern of sparseness a.

r_i	r_j	δw	$P(w)$
0	0	a^2	$(1-a)^2$
0	1	$-a(1-a)$	$a(1-a)$
1	0	$-a(1-a)$	$a(1-a)$
1	1	$(1-a)^2$	a^2

and the variance is

$$\langle (\delta w)^2 \rangle = \sum (\delta w)^2 P(w) = a^2(1-a)^2. \tag{8.32}$$

Imprinting more patterns will increase the possible values of the weight components, and the values of the weight matrix are then distributed as a multinomial distribution of higher order. Instead of calculating these distributions for a given number of patterns we simplify our discussion by considering the case after imprinting many patterns, enough so that the central limit theorem holds. In this case we expect Gaussian-distributed weight values (as already stressed in the last chapter) with mean

$$\mu = \langle w \rangle = 0 \tag{8.33}$$

and variance

$$\sigma = \sqrt{\langle w^2 \rangle} = \sqrt{N^{\text{pat}}} a(1-a). \tag{8.34}$$

A weight matrix with Gaussian-distributed components means that the amount of inhibition is equal to the amount of excitation. This in turn forces half of the nodes to respond. We can state this more formally with the relative amount of inhibition

$$W^{\text{in}} = \int_0^\infty P(\delta w) \mathrm{d}(\delta w) \tag{8.35}$$

and the relative amount of excitation

$$W^{\text{ex}} = \int_{-\infty}^0 P(\delta w) \mathrm{d}(\delta w). \tag{8.36}$$

The number of firing nodes times the inhibition must match the amount of excitation of the other nodes, that is,

$$W^{\text{in}} a^{ret} = W^{\text{ex}} (1 - a^{ret}), \tag{8.37}$$

which determines the retrieval sparseness,

$$a^{ret} = \left(1 + \frac{W^{\text{in}}}{W^{\text{ex}}} \right)^{-1}. \tag{8.38}$$

This demonstrates that $a^{ret} = 0.5$ in the case of $W^{\text{in}} = W^{\text{ex}}$, which results from the standard Hebbian covariance rule.

There are several possible solutions to the problem of achieving the right retrieval sparseness. One is to adjust the firing thresholds of the nodes appropriately so that only a nodes can fire in the retrieval process. However, adjusting the thresholds of neurons with respect to the firing of all the nodes in the networks is a nonlocal operation that has to be carefully implemented to be biologically plausible. Instead, it is possible that the weight values themselves could be adjusted appropriately, something we have already discussed in the last chapter under the heading of gain control. Another solution that seems biologically realistic is to include an appropriate amount of *global inhibition*, mediated by inhibitory interneurons. It is possible that a combination of such effects is employed in the brain. In the following we will only outline the details of the latter proposal.

8.5.6 Global inhibition

Besides the Gaussian weight matrix 8.30 it is appropriate to include the effect of inhibitory interneurons in the networks. Inhibitory interneurons become activated by excitatory neurons in the network and in turn inhibit excitatory neurons. We can represent this activity-dependent global inhibition in a simple model by subtracting a constant c from the Hebbian weight matrix,

$$\bar{w}_{ij} = w_{ij} - c. \tag{8.39}$$

The inclusion of the constant shifts the probability distribution of the weight components as illustrated in Fig. 8.8A. The retrieval sparseness of the network with such shifted weight matrices is then given by

$$a^{\text{ret}} = \left(1 + \frac{\frac{1}{2} + W^{\text{dif}}}{\frac{1}{2} - W^{\text{dif}}}\right)^{-1} = \frac{1}{2} - W^{\text{dif}}, \tag{8.40}$$

where W^{dif} is the shaded area illustrated in Fig. 8.8A. For a weight matrix with Gaussian-distributed components the integral W^{dif} is given by the Gaussian error function (see Appendix B), so that in this case we can expect a retrieval sparseness of

$$a^{\text{ret}} = \frac{1}{2} - \frac{1}{\sqrt{2\pi}\sigma} \int_0^c e^{-\frac{t^2}{2\sigma^2}} \, dt = \frac{1}{2}[1 - \text{erf}(\frac{c}{\sqrt{2}\sigma})]. \tag{8.41}$$

We can evaluate this equation for the binary pattern with 0 and 1 components for which we have derived the variance above (eqn 8.34). The retrieval sparseness a^{ret} for various values of the pattern sparseness a as a function of the inhibition constant c is plotted in Fig. 8.8B. The retrieval sparseness matches the pattern sparseness for values of c given by

$$c = \sqrt{2}\sigma \text{erf}^{-1}(1 - 2a) \tag{8.42}$$

indicated by the thick line in Fig. 8.8B.

The curves specify what amount of global inhibition is necessary to keep the retrieval sparseness of the attractor network in the desired range. This seems to indicate a necessary fine tuning of the strength of the inhibition generated by inhibitory interneurons. However, we have not taken the attractor dynamics of the stored pattern

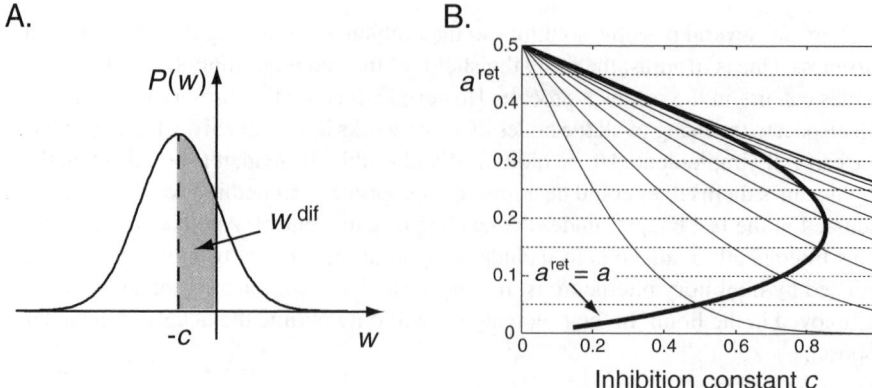

Fig. 8.8 (A) A Gaussian function centred at a value $-c$. Such a curve describes the distribution of Hebbian weight values trained on random patterns and includes some global inhibition with strength value c. The shaded area is given by the Gaussian error function described in Appendix B. (B) Theoretical retrieval sparseness a^{ret} as a function of global inhibition c is plotted as thin lines for different values of the sparseness of the pattern set a, from $a = 0.05$ (lower curve) to $a = 0.5$ (upper curve) in steps of $\Delta a = 0.05$. We assumed therein 40 imprinted patterns with Gaussian-distributed components of the weight matrix. The thick line shows where the retrieval sparseness matches the sparseness of the imprinted pattern ($a^{\mathrm{ret}} = a$).

into account. These will bias the networks to the desired sparseness once the inhibition from the inhibitory interneurons brings the network activity into the vicinity of the pattern sparseness. Simulation results, further detailed in Chapter 12, for a network trained on 40 patterns with sparseness $a = 0.1$ are shown in Fig. 8.9A. The plateau indicates the domain with sufficient attractor dynamics to stabilize the network to the sparseness of the stored pattern. For these values of the inhibition constant the network recalls the correct pattern as indicated by the Hamming distance also shown in this graph.

When inspecting the simulation results more carefully you can see that the values of the inhibition constants for which the plateau is achieved are less than the values indicated by the previous analysis shown in Fig. 8.8B for which eqn 8.42 would suggest a value of around $c = 0.73$. The discrepancy comes from the fact that an assumption of the above analysis, the applicability of the central limit theorem, is not met by the training of only 40 patterns. This can be seen from the normalized histogram of the weight values shown in Fig. 8.9B which is not well approximated by a Gaussian. The shift of this weight distribution by the inhibition constant therefore has a larger effect than the shift of a Gaussian used in the previous analysis. However, the distribution will eventually approach the Gaussian distribution for more patterns in the training set. As an example we show the distribution of weight components with 400 patterns in Fig. 8.9C. Fewer patterns are necessary to approach the Gaussian limit with larger values for the sparseness of the training patterns. This example should illustrate that it is important to verify that the conditions assumed in theoretical analysis are met under the experimental conditions.

We have seen that inhibition can control the retrieval sparseness of attractor networks. This might be an important functional role of inhibitory interneurons in the

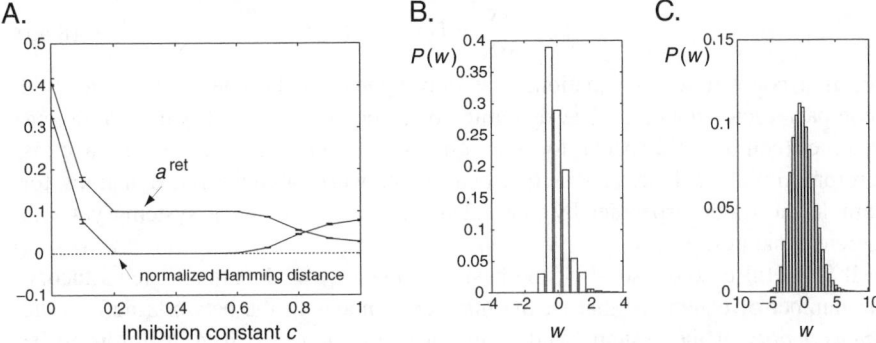

Fig. 8.9 (A) The simulated curve for pattern with sparseness $a = 0.1$. The plateau is due to the attractor dynamics not taken into account in the analysis that led to Fig. 8.8B. The lower curve indicates the average Hamming distance between the imprinted pattern and the network state updating the network. Correct recalls were indeed achieved for inhibition constants that coincide with the plateau in the retrieval sparseness. (B) Normalized histogram of weight components for 40 patterns trained with the Hebbian covariance rule. (C) Normalized histogram of weight components for 400 patterns trained with the Hebbian covariance rule.

brain, which is known to have many recurrent connections. There are, however, many more questions that have to be answered in the future. For example, the inhibition has to be very rapid if the inhibition is to be able to control the sparseness of a neural area in response to a sensory input, which seem to contradict the further delay that is introduced by the delays in the response of interneurons. Furthermore, we have adjusted the inhibition constant appropriately in the above examples. However, there is so far no evidence that the synaptic efficiency is modulated by Hebbian learning. The above scheme therefore has to be revised slightly so that the overall level of excitation matches the appropriate level of fixed inhibition.

8.6 Chaotic networks: a dynamic systems view

We end this chapter with a short excursion into the theory of dynamic systems. This will show us why we have called the memory states of auto-associative memories 'point attractors', and will give us further insight into the possible behaviour of recurrent networks. In particular, we will outline the conditions under which a recurrent network has point attractors. We will also show that recurrent networks with biologically more plausible non-symmetric weight matrices, in comparison to the symmetric weight matrices resulting from simplified Hebbian learning, frequently have properties similar to those of the Hebbian counterpart.

We have already stressed the dynamic nature of the models when recurrences are involved. In general, models with continuous dynamics are described in dynamical system theory by a set of coupled ordinary first-order differential equations, the so-called *equations of motion*,[22]

---

[22] Note that higher-order ordinary differential can be cast into a set of coupled first-order differential equations.

$$\frac{dx}{dt} = f(x). \tag{8.43}$$

This is a coupled set of equations that only appears to be one as we have used a compact vector notation. The dynamics of a recurrent network with continuous dynamics (eqn 8.1) is a special form of eqn 8.43, and an auto-associative network is therefore formally a dynamic system. Recurrent networks with discrete dynamics, for example, the systems specified by eqns 8.2 and 8.3, are still dynamic systems, yet with discrete dynamics.

It is useful to know some of the basic terminology of dynamic systems theory. The number of equations, that is, the number of nodes in the network, defines the *dimensionality* of the systems, and recurrent neural networks have therefore to be considered as high-dimensional dynamic systems. The vector **x** in eqn 8.43 is called a *state vector*, and a set of values for all components is called a *state*. A state describes a point within the high-dimensional *state space*, the space of all possible state values. The evolution of the state, defined by the equations of motion, describes a *trajectory*, a path within the state space.

8.6.1 Attractors

Dynamic systems can display a variety of different dynamic behaviour. We have already seen some examples of recurrent networks in which the networks converged to a fixpoint of the dynamic equations. This is a point in the state space, and it is for this reason that we called this point a *point attractor*. Other forms of attractors are possible in dynamic systems. For example, the attractor can be a loop within the state space, a so-called *limit cycle*, in which the system cycles through a continuous set of points. Such movements in the state space would appear in the components as oscillations, and oscillations in the brain may well be described by such attractors of neural networks. We can also define the dimensionality of an attractor. For example, a line attractor has a dimensionality of one, and a point in the state space has to be considered as having a dimensionality of zero. Higher-dimensional attractors are also possible in dynamic systems, although it is often difficult to find the corresponding regularities in the movement of the system.

We have so far only considered very regular movements of the dynamic systems. However, we know of examples of dynamic systems that display movements that are not completely regular but yet are also not completely stochastic (like noise). For example, a system can be attracted by two points in the phase space with irregular domination of these two attractor points. A popular example of a chaotic system is the *Lorenz system* defined by the equations of motion

$$\frac{dx_1}{dt} = a(x_2 - x_1) \tag{8.44}$$

$$\frac{dx_2}{dt} = x_1(b - x_3) - x_2 \tag{8.45}$$

$$\frac{dx_3}{dt} = x_1 x_2 - c x_3. \tag{8.46}$$

Note that we can write this system as a recurrent network of three nodes with sigma and sigma–pi couplings like

$$\frac{\mathrm{d}x_i}{\mathrm{d}t} = \sum_j w^1_{ij} x_i + \sum_{jk} w^2_{ijk} x_j x_k \qquad (8.47)$$

with

$$\mathbf{w}^1 = \begin{pmatrix} -1 & a & 0 \\ b & -1 & 0 \\ 0 & 0 & -c \end{pmatrix} \quad \text{and} \quad \mathbf{w}^2 = \begin{cases} w^2_{213} = -1 \\ w^2_{312} = -1 \\ 0 \quad \text{otherwise} \end{cases} . \qquad (8.48)$$

An example of a trajectory of the Lorenz system, showing the famous Lorenz attractor, is plotted in Fig. 8.10. With the above-mentioned definition of the dimensionality of an attractor we have to view two points having a dimensionality larger than zero but still less than one. It then becomes a question of how to define such rational dimensionalities, although we will not go into these discussions here. For our purposes it is enough to realize that there can be fractional dimensional attractors, from which the term *fractals*, often mentioned in the dynamic systems literature, is derived.

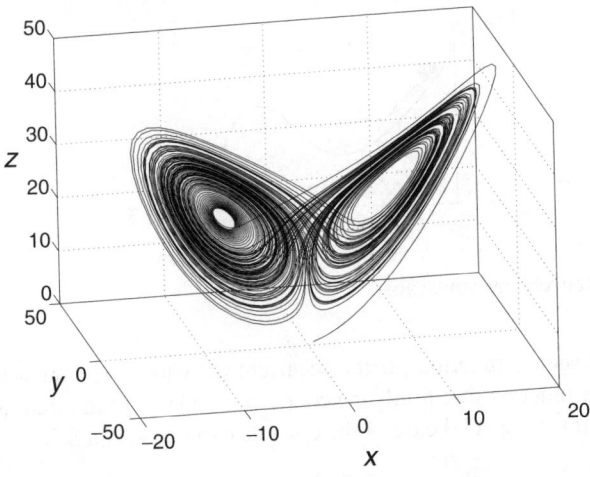

Fig. 8.10 Example of a trajectory of the Lorenz system from a numerical integration within the time interval $0 \le t \le 100$. The parameters used were $a = 10$, $b = 28$, and $c = 8/3$.

The movement around fractal attractors is very irregular, yet there is still some order. Such movements are therefore very different to random movements, such as movements dominated by noise. We call such behaviour *deterministic chaos*. It is deterministic because there is no noise in the equations and the future states of the systems are uniquely defined by the initial conditions of the system. In contrast, *stochastic* systems include noise so that systems that start with identical initial conditions can evolve in different ways.

8.6.2 Lyapunov functions

We have outlined in this chapter that point attractors of recurrent networks are useful as memories, and chaotic fluctuations in such systems are not desirable.[23] It is therefore of considerable interest to state under what conditions recurrent networks have point attractors, and under what conditions dynamic chaotic behaviour is expected.

From dynamic systems theory we know that a system has a point attractor if a *Lyapunov function* exists. The idea behind this statement can be illustrated with the help of Fig. 8.11. We illustrate there a 'landscape' in which a ball, driven by gravity and influenced by friction, can roll down a hill into a valley. The ball will ultimately come to a halt at the minimum (the valley) of the function describing the landscape. More formally, if there is a function $V(\mathbf{x})$ that never increases under the dynamics of the system,

$$\frac{dV(\mathbf{x})}{dt} \leq 0 \tag{8.49}$$

where the $\mathbf{x}$ is governed by the dynamic equations of the system (eqn 8.43), then there has to be a point attractor in the system (as long as the state space is bounded) corresponding to the minimum of the function V. If such a function exists with the required properties, then this function is called a *Lyapunov function* in dynamic systems theory. This function is also called an *energy function* in physics.

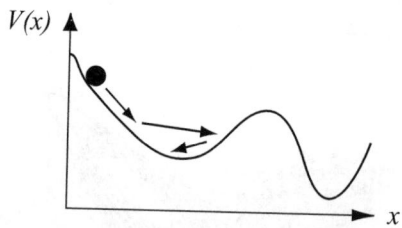

Fig. 8.11 A ball in an 'energy' landscape.

Can we find such a function for the recurrent networks? The answer is that we indeed know of a function that fulfils the conditions under certain circumstances. Let us illustrate this for the special case of the discrete dynamics, eqn 8.3,

$$h_i(t + 1) = \sum_j w_{ij} r_j(t) + I_i^{\text{ext}}(t). \tag{8.50}$$

For this system we propose to study the following function,

$$V(r_1, ..., r_N) = -\frac{1}{2} \sum_i \sum_j w_{ij} r_i r_j - \sum_i I^{\text{ext}} r_i. \tag{8.51}$$

---

[23] Chaotic fluctuations are not desirable within the basic auto-associative memory. However, chaotic networks may have useful properties that are employed by the brain. Indeed, EEG measurements of brain activity indicate a chaotic brain dynamic on a system level, while synchronized modes seem more devastating as they occur, for example, in epileptic seizures.

The change of this function in one time step is given by

$$\Delta V = V(t+1) - V(t)$$
$$= -\frac{1}{2}\sum_k\sum_j w_{kj}r_k(t+1)r_j(t+1) + \frac{1}{2}\sum_k\sum_j w_{kj}r_k(t)r_j(t)$$
$$- \sum_k I^{\text{ext}}[r_k(t+1) - r_k(t)]. \tag{8.52}$$

It is easiest to consider this model with sequential update. In this case, when the ith node is updated the other nodes stay constant, that is, $r_k(t+1) = r_k(t)$ for $k \neq i$. Only terms from the node i contribute to the change of the function V, and the change of this function in this time step is given by

$$\Delta V = -\frac{1}{2}r_i(t+1)\sum_{j\neq i} w_{ij}r_j(t) - \frac{1}{2}r_i(t+1)\sum_{k\neq i} w_{ki}r_k(t)$$
$$+ \frac{1}{2}r_i(t)\sum_{j\neq i} w_{ij}r_j(t) + \frac{1}{2}r_i(t)\sum_{k\neq i} w_{ki}r_k(t) - b_i[r_i(t+1) - r_i(t)]$$
$$= -[r_i(t+1) - r_i(t)][\sum_{j\neq i}\{\frac{1}{2}(w_{ij} + w_{ji})r_j(t)\} + I^{\text{ext}}]. \tag{8.53}$$

This difference is zero if $r_i(t+1) = r_i(t)$, but this means that we have already reached a stationary state. The interesting case occurs when $r_i(t+1) \neq r_i(t)$. Then we have to inspect the second term more carefully, and we notice that the result depends on the weight matrix. Let us first examine the case of Hebbian learning that results in a symmetric weight matrix, $w_{ij} = w_{ji}$. In this case we have $(w_{ij} + w_{ji})/2 = w_{ij}$ and the second factor in the last equation equals $h_i(t)$ according to eqn 8.50. We can therefore write eqn 8.53 as

$$\Delta V = -[r_i(t+1) - r_i(t)]h_i(t). \tag{8.54}$$

Let's consider again the system with binary states. If $r_i(t) = 1$ then it must be true that $h_i(t) < 0$ in order to have $r_i(t+1) = -1$. If $r_i(t) = -1$ then it must be true that $h_i(t) > 0$ in order to have $r_i(t+1) = 1$. In both cases we have $\Delta V < 0$ and hence a Lyapunov function. The system therefore always converges to a stable state, which corresponds to the trained patterns as we saw in the last section.

8.6.3 The Cohen–Grossberg theorem

Michael Cohen and *Steven Grossberg* studied more general systems with continuous dynamics of the form

$$\frac{dx_i}{dt} = a_i(x_i)\left(b_i(x_i) - \sum_{k=1}^{N}(w_{ik}f_k(x_k))\right). \tag{8.55}$$

This dynamic equation corresponds to the leaky integrator dynamics (eqn 8.1) with generalizations that include that the time constants and the activation functions can be different for each individual node in the network. Cohen and Grossberg found a Lyapunov function under the conditions that

1. **Positivity** $a_i \geq 0$: The dynamics must be a leaky integrator rather than an amplifying integrator.
2. **Symmetry** $w_{ij} = w_{ji}$: The influence of one node on another has to be the same as the reverse influence.
3. **Monotonicity** $\mathrm{sign}(\mathrm{d}g(x)/\mathrm{d}x) = \mathrm{const}$: The activation function has to be a monotonic function.

The statement that under these conditions a Lyapunov function exists has come to be known as the *Cohen–Grossberg theorem*. This theorem proves the existence of point attractors of the noiseless recurrent networks under the conditions just outlined.

8.7 Biologically more realistic variations of attractor networks

The synaptic weights between neurons in the nervous system cannot be expected to fulfil the conditions of symmetry of the weight matrix required to guarantee stable attractors in these networks. Also, we studied only weight matrices that have negative and positive values independently of each other. The positive weight values represent excitatory synapses, while the negative values represent negative synapses. Neurons receive a mixture of input from excitatory and inhibitory presynaptic neurons, but each class of neurons is bound to make only one type of synaptic contact, which defined the neuron type in the first place (Dale's principle). This corresponds to a weight matrix in which the signs within each column have to be consistent (all the elements in a column, corresponding to the synaptic efficiencies of one presynaptic neuron, have to be either positive or negative). This violates the symmetry condition of the weight matrix in the Cohen–Grossberg theorem as an inhibitory node could then only receive inhibitory connections and vice versa. However, the Cohen–Grossberg theorem only states the special case in which we can prove that networks have point attractors, while it is still possible that networks have point attractors even if the conditions for the Cohen–Grossberg theorem are not fulfilled. We will therefore test how networks behave that violate some of the conditions of the Cohen–Grossberg theorem.

8.7.1 Asymmetric networks

Let us start by studying a particularly simple case of non-symmetric weight matrices. Note that each matrix can be decomposed into a symmetric and an antisymmetric part,

$$\mathbf{w} = g^{\mathrm{s}}\mathbf{w}^{\mathrm{s}} + g^{\mathrm{a}}\mathbf{w}^{\mathrm{a}}, \tag{8.56}$$

with

$$w_{ij}^{\mathrm{s}} = w_{ji}^{\mathrm{s}} \tag{8.57}$$

$$w_{ij}^{\mathrm{a}} = -w_{ji}^{\mathrm{a}}. \tag{8.58}$$

We have thus introduced some factors in the decomposition to allow us to study the conditions with varying strengths of the symmetric and antisymmetric parts of the weight matrix. A simple example of symmetric and antisymmetric matrices is to set the magnitude of all components of $\mathbf{w}^s$ and $\mathbf{w}^a$ equal to one, e.g.

$$w_{ij} = \begin{cases} g^s + g^a & \text{for} \quad i > j \\ 0 & \text{for} \quad i = j \\ g^s - g^a & \text{for} \quad i < j \end{cases} \tag{8.59}$$

Let us study by simulations how networks with such matrices behave. We will use a network with dynamics specified by eqn 8.2 and a sigmoidal activation function. To get an indication in these simulations as to whether the networks have reached a steady state (point attractor) we measure the difference of the states of two consecutive time steps after the network has been allowed to evolve for a certain number of time steps. The difference between two consecutive time steps is then quantified by

$$d(t) = (|\mathbf{r}(t)| - |\mathbf{r}(t-1)|)^2 \tag{8.60}$$

This distance should be zero if the network converged to a point attractor, whereas this value is positive otherwise. The values of this convergence indicator are plotted on a grey scale for difference values of strength values g^s and g^a in Fig. 8.12A. The value is indeed zero for symmetric weight matrices as predicted by the Cohen–Grossberg theorem (horizontal line at $g^a = 0$). What is, however, interesting is that a stationary state is reached as long as the strength of the asymmetric part of the weight matrix is smaller than the strength of the symmetric part of the weight matrix. This indicates that only strong asymmetries destroy point attractors in the network.

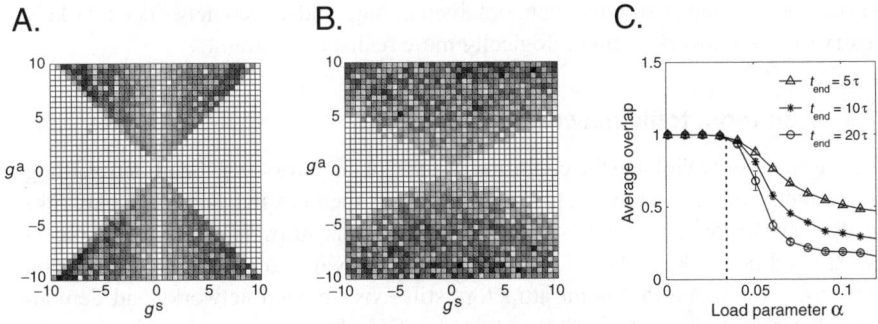

Fig. 8.12 (A) Convergence indicator for networks with an asymmetric weight matrix where the individual components of the matrix are chosen from the unit strength. (B) Similar to (A) except that the individual components of the weight matrix are chosen from a Gaussian distribution. (C) Overlap of the network state with a trained pattern in a Hebbian auto-associative network that satisfies Dale's principle.

8.7.2 Random and Hebbian matrices with asymmetries

In the previous example we used constant values with the same strength for all the components of the weight matrix, which is, of course, an oversimplification of a more realistic weight matrix. We will therefore give two more examples that incorporate further more biologically realistic details. In the first example, illustrated in Fig. 8.12B, we replaced, in experiments similar to the one just outlined, the weight components of strength $|w_{ij}| = 1$ with random variables. The random values were drawn from a Gaussian distribution, which we have found is typical for weight matrices trained with Hebbian rules as discussed in Chapter 7. The results are similar to the previous experiment in that the asymmetry has to be quite strong before the network displays signs of the onset of chaos. Strong asymmetries are only achieved if an excitatory connection from one neuron to another is countered by strong inhibition of the other neuron, or vice versa.

The next experiment was performed to test networks that comply with Dale's principle, for example, networks with separate excitatory and inhibitory nodes, while still utilizing the basic Hebbian imprinting. We therefore simulated networks with weight matrices chosen by the following procedure. We first generated a symmetric weight matrix through Hebbian learning. We also generated a matrix of the same size in which we randomly assigned excitatory and inhibitory nodes. We then deleted (set to zero) all the entries in the weight matrix that were either inconsistent with the nature of the node (to be inhibitory or excitatory) or which would result in a direct feedback between the nodes. This procedure generated a highly diluted weight matrix that is, however, consistent with biological constraints (while still being Hebbian). The overlap of the network state with a stored pattern after different durations of network cycles is shown in Fig. 8.12C. The results indicate that the storage capacity of this network coincides with the storage capacity of the Grossberg–Hopfield network with equivalent dilution, indicated by the vertical dashed line. This demonstrates that the theoretical results about the behaviour of recurrent networks that we discussed in the first part of this chapter and that were obtained in highly idealized networks are likely to carry over to networks with biologically more realistic constraints.

8.7.3 Non-monotonic networks

We have so far only violated the condition of the Cohen–Grossberg theorem regarding the symmetry of the weight matrix. However, the theorem indicates that networks can also behave chaotic when violating other constraints. *Masahiko Morita* and others have studied networks of Hebbian trained networks with non-monotonic activation functions. They found that point attractors still exist in such networks and demonstrated the profoundly enhanced storage capacities of those networks. Their results also indicate that the point attractors in these networks have basins of attraction that seem to be surrounded by chaotic regimes. The basins of attraction are like islands in a chaotic ocean. These chaotic regimes can have important functions. For example, they can indicate when a pattern was not recognized because it was too far from any trained pattern in the network. Non-monotonic activation functions seem, on a first inspection, biologically unrealistic. However, we have to keep in mind that nodes in these networks can represent collections of nodes, and a combination of neurons can produce a non-monotonic response. An effective non-monotonicity can also be the result of appropriate activation of excitatory and inhibitory connections, although such possibilities are largely unexplored. It is also possible that some neurons may have non-monotonic gain functions. It is certainly desirable to find more experimental support for such mechanisms in the brain.

Conclusion

It is well known that many neurons form collateral connections with neurons in the same brain area, and that bidirectional connections between brain areas are common. The brain is therefore not simply a feed-forward processor, but does utilize recurrences in the networks. We have seen that recurrent networks can have interesting properties that can be utilized in the brain in various ways. In particular, we outlined how auto-associative memories can be used to rapidly form memories from which the stored items can be recalled from partial information.

Recurrent networks add a temporal domain to the networks (in addition to the temporal nature of the signal processing in a single neuron) through the re-entry of signals into the network, and the temporal behaviour of recurrent networks can display complicated dynamics. However, Hebbian networks of the type found in the brain can display point attractors that can be utilized as associative memories, even if we relax the basic assumptions on which most theoretical work is based. The storage capacity of such networks can be quite large, and a sharp transition to an 'amnesic phase' can be expected if the information stored in the network exceeds a critical value that depends on the number of connections per neuron and the training algorithm used to imprint the pattern.

Further reading

Daniel J. Amit
>   *Modelling brain function: the world of attractor neural networks*, Cambridge University Press, 1989.
>   This book is dedicated to attractor networks, and many more details of such networks can be found in this book. Some parts are very mathematical.

John Hertz, Anders Krogh, and Richard G. Palmer
>   *Introduction to the theory of neural computation*, Addison-Wesley, 1991.
>   See chapters 2 and 3. A good overview of attractor networks and the analogy with spin models. This book is still one of the best introductions to the theory of artificial neural networks.

Edmund T. Rolls and Alessandro Treves
>   *Neural networks and brain function*, Oxford University Press, 1998.
>   See chapters 3, 6, and 7. This book concentrates on sparse attractor networks and the relations to brain functions. It gives many examples of brain areas that seem to implement auto-associative networks.

John J. Hopfield
>   Neural networks and physical systems with emergent collective emergent computational abilities, in *Proc. Nat. Acad. Sci., USA* 79: 2554–8, 1982, and Neurons with graded response have collective computational properties like those of two state neurons, in *Proc. Nat. Acad. Sci., USA* 81: 3088–92, 1994.
>   These are two of the original papers on attractor networks that lured many physicists into the study of neural networks.

M. A. Cohen and S. Grossberg

Absolute stability of global pattern formation and parallel memory storage by competitive neural networks, in *IEEE Trans. on Systems, Man and Cybernetics*, SMC-13: 815–26, 1983.

The original paper discussing the issue of stability as mentioned in this chapter.

M. Morita

Associative memory with nonmonotone dynamics, in *Neural Networks* 6: 115–26, 1993.

Masahiko Morita explains his ideas.

Michael E. Hasselmo and Christiane Linster

Neuromodulation and memory function, in *Beyond neurotransmission: neuromodulation and its importance for information processing*, Paul S. Katz (ed.), Oxford University Press, 1999.

Chapter 9, written by Hasselmo and Linster, reviews the roles of acetylcholine and noradrenaline in modulating learning. Other chapters discuss several additional interesting issues around neuromodulation.

9 Continuous attractor and competitive networks

The structure of continuous attractor neural networks discussed in this chapter is equivalent to that of the recurrent networks discussed in the last chapter. What is different is the representation of the pattern that we want to imprint in such networks. In particular we will be interested in storing patterns with continuous features. We will see that recurrent networks have interesting properties when trained on such continuous patterns as opposed to random patterns as in the last chapter. We will also discuss related models of self-organizing maps that suggest some possible mechanisms behind topographic maps in the brain.

9.1 Spatial representations and the sense of direction

In the last chapter we discussed auto-associative attractor models that we trained mainly on random patterns. The assumption of independent, random-like representations of states was motivated by the content of the representations. We had very general memory states in mind, such as the shape of objects, their smell, texture, or colour. The variety of objects with such different attributes might well be represented by independent vectors. In contrast to this type of memory represented by *point attractor neural networks* (PANNs) we will discuss in this chapter network models that rely on some forms of correlations between training patterns. We will call these networks *continuous attractor neural networks* (CANNs). Training patterns with a strong correlation make sense when the training patterns represent continuous or nearly continuous features. An example of a continuous feature is the spatial location of an object. Also, as mentioned in Chapter 4, representations in the brain are often topographic, characterized by the fact that similar features are represented in close physical proximity in the brain. We will discuss such issues in this chapter. We start with the representation of space within the brain while studying the formation of topographic maps in a later part of this chapter.

9.1.1 Head directions

Humans usually have to a certain degree a sense of direction. This suggests that we must have some form of spatial representation in our brain. The sense of direction can be tested with a simple experiment. If a subject is placed in a rotating chair with closed eyes while someone rotates the subject a certain amount, it can be shown that the subject's guess of the new direction is quite accurate. This demonstrates two important issues: first, we have to have a representation of body or head direction and, secondly, we have to have a mechanism to update this information without visual cues. Head directions are indeed represented on a neuronal level in the brains of mammals. This can be seen from cell recordings of neurons that respond maximally when a subject's head points in a particular direction. An example is shown in Fig. 9.1 for a neuron in the limbic system of a rodent. In this figure the firing rate of this neuron is shown when the subject was rotated in different directions. The neuron fires maximally for one particular direction and fires with lower firing rates to directions around the preferred direction of this neuron. Similar neurons with maximal firing rates for other directions can also be found.

There are several other experimental findings that are important to consider in the following models. For example, it was shown that the head-direction neurons continue to fire in the dark. This shows that the maintained firing rate cannot be explained by a continuous visual stimulus. Furthermore, different cells with corresponding preferred directions begin to fire after the rat is rotated in the dark. This shows that the update of the head direction representation can be achieved without visual cues as mentioned above. Also, the preferred direction of a cell is not fixed forever but can change in new environments. In the following we start by discussing models of head direction representations that can be maintained in the dark, while returning to the second issue, that of updating such information in the dark, in Section 9.4.

A. Head-direction cell in subiculus

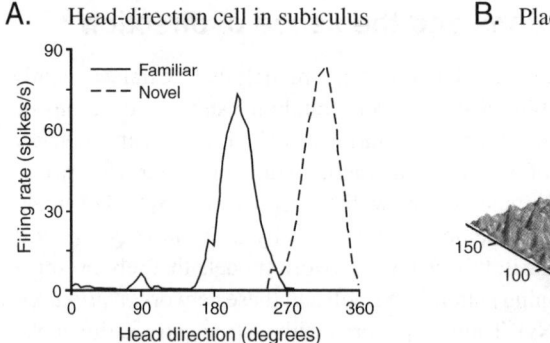

B. Place-field in hippocampus

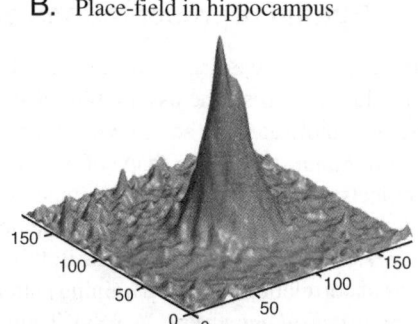

Fig. 9.1 (A) Experimental response of a neuron in the subiculum of a rodent when the rodent is heading in different directions in a familiar maze. The dashed line represents the new head properties of the same neuron when the rodent is placed in the new unfamiliar maze. The new response properties will normally be similar to the previous one, that is, head direction cells try to maintain approximately their response properties to specific head directions. However, the results shown were produced in experiments with a rodent that had cortical lesions that weakened the ability to maintain the response properties after the rodent was transferred into a new environment. [redrawn from Edward J. Golob and Jeffrey S. Taube, *J. Neurosci.* 19: 7198–211 (1999)]. (B) Neuronal response from many hippocampal neurons in a rodent that responded to the subject's location (places) in a maze. The figure shows the firing rates of the neurons in response to a particular place, whereby the neurons were placed in the figure so that neurons with similar response properties were placed adjacent to each other [A. Samsonowich and B. L. McNaughton, *J. Neurosci.* 17: 5900–20 (1997)].

9.1.2 Place fields

Head direction representations are a good example of the spatial representation of a one-dimensional feature space in the brain, and this example will guide most of the following discussion. However, the models outlined below apply equally to higher-dimensional representations. For example, neurons in the hippocampus of rats, the brain area that we discussed in the last section in connection with episodic memory, fire in relation to specific locations within a maze in which the rat is freely moving. In some experiments the activity of many neurons has been recorded while a rat could freely run in a maze. When the firing rates of the different neurons are plotted on a map reflecting their physical location in the hippocampus, the resulting firing pattern looks like randomly distributed code. A specific topography of neurons within the hippocampal tissue with respect to their maximal response to a particular place has not been found. However, if the plot is rearranged so that neurons that fire maximally in response to adjacent locations are plotted adjacent to each other, then a firing profile like the one shown in Fig. 9.1B can be seen. We will discuss in the next section why this rearrangement of the recorded data has to be made.

9.1.3 Spatial representations in network models

How could we represent spatial information in a neural network? For simplicity we will discuss as example head direction representations. The discussions of higher-dimensional spatial representations are similar. A possible solution to representing head directions in a network is illustrated in Fig. 9.2. There we have drawn a number of nodes that we have arranged on a circle to allude to the significance of node activity with respect to the head direction of a subject. This is done only for our convenience and the physical locations of the nodes are not relevant in the following. A possible form of head direction representation is given when we restrict the network activity to one active node. The active node then represents the head direction like a needle of a compass within the resolution of the network. With continuously varying values for neuron activities, as were found experimentally, it is possible to achieve a higher resolution if we represent the head direction as the centre of mass of an activity packet, [24] as long as we allow only one activity packet to be active. The angular resolution of the head direction representation can, of course, also be increased by increasing the number of nodes.

9.1.4 Graded winner-take-all models

In the representation of head direction proposed above we used the fact that only *one* node or *one* activity packet of nodes with a more graded response is active at any moment in time, so that the centre of this activity can be interpreted unequivocally as the direction of the subject. This is an example of a mechanism that is often simply described as *winner-takes-all*. This nomenclature is particularly appropriate in the model where only one node is active. A single activity packet can be viewed as a smooth generalization of the winner-take-all mechanism. To implement such a

---

[24] The centre of mass of the activity packet can be defined as the position where the areas under the packet left and right from this point are the same.

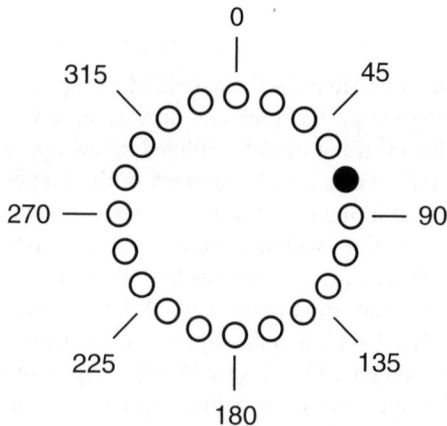

Fig. 9.2 A proposal as to how the activity of nodes, for clarity arranged into a circle, can represent head directions. With the 20 nodes of this model we can represent head directions with a resolution of 18 degrees when using a single binary node as a representation of a head direction. The single active node in the figure, represented as a solid circle, indicates a head direction of 72 degrees in this example.

mechanism it is necessary to include some information exchange between the nodes, and it is therefore obvious to again consider networks with collateral interactions. The general form of the dynamic equations of such recurrent networks have been introduced in the last chapter on auto-associative memory models. We will outline the models in the context of this chapter using the example of continuous dynamics (eqn 8.1); the treatments of the other discrete dynamics are analogous. We write the equations in the same form as in the last chapter, the only change being that we include a normalization of the weight values to the number of nodes, N, sometimes also included in auto-associative memory models. The dynamic equation for the networks considered in this chapter is thus

$$\frac{1}{\tau}\frac{dh_i(t)}{dt} = -h_i(t) + \frac{1}{N}\sum_j w_{ij}r_j(t) + I_i^{\text{ext}}(t), \qquad (9.1)$$

for each node i in the network. The interpretations of the symbols are the same as in Chapter 7. In the case of the model for the representation of head directions, as outlined in Fig. 9.2A, the increases of the number of nodes only increase the angular resolution of the head direction representation. In the limit of an infinite number of nodes, corresponding to an infinitesimal angular resolution, we can write the activity of all the nodes as a continuous quantity over the one-dimensional space of the ring, that is, $h_i(t) \rightarrow h(x,t)$. Such quantities are formally known as *fields*, values that depend on a spatial coordinate x and the time. The spatial coordinate can represent higher-dimensional spaces and should therefore be written more generally as a vector $\mathbf{x}$. With this field notation we can write the dynamic eqn 9.1 in the form of a *neural field equation*

$$\tau\frac{\partial h(\mathbf{x},t)}{\partial t} = -h(\mathbf{x},t) + \int_{y_1}\cdots\int_{y_d} w(\mathbf{x},\mathbf{y})r(\mathbf{y},t)dy_1\ldots dy_d + I^{\text{ext}}(\mathbf{x},t). \qquad (9.2)$$

This formulation is very convenient for analytical treatments and is a good approximation of a large discrete system where the properties of the model scale properly with the number of nodes. In the following we will outline the properties of the model in a one-dimensional space, although the properties we discuss can be directly generalized to higher-dimensional models. In one dimension the dynamic field equation simplifies to

$$\tau \frac{\partial h(x,t)}{\partial t} = -h(x,t) + \int_y w(x,y)r(y,t)\mathrm{d}y + I^{\mathrm{ext}}(x,t). \qquad (9.3)$$

To return from this continuous space notation to a model with discrete nodes we have simply to follow the discretization rules,

$$x \;\; \to i\Delta x \qquad (9.4)$$

$$\int \mathrm{d}x \to \Delta x \sum. \qquad (9.5)$$

Notice the scale factor in the last mapping equation, which can easily be forgotten when replacing the integral with a sum. We recover the equations for the individual nodes when we identify the correspondences $h(i\Delta x, t) = h_i(t)$ and $\Delta x = 1/N$. We can switch between the continuous and discrete notations easily and will do so frequently for convenience. There is nothing new in the equations above compared to the recurrent models of auto-association. The only difference will be the nature of the patterns we want to imprint in these networks.

9.2 Learning with continuous pattern representations

To enable recurrent networks to represent a continuous set of patterns we again use Hebbian learning to adjust the weights between the nodes. In contrast to the point attractor networks studied in the last chapter we do not use the covariance rule 8.30 but use a simpler correlation rule as we will only train the excitatory connections and adjust the sparseness of the network activity with global inhibition as discussed before. The rules also become equivalent in the $a \to 0$ limit. In the neural field representation we can write this Hebbian rule for the excitatory weights w^{E} as

$$w^{\mathrm{E}}(x,y) = \sum_\mu r^\mu(x,t)r^\mu(y,t). \qquad (9.6)$$

The firing rate r^μ is the firing rate of the neural field while dominated by the training example of a pattern μ presented to the network. This is again analogous to the learning phase in auto-associative networks discussed in the last chapter. The inhibition from inhibitory interneurons will again be represented by subtracting a constant from the excitatory weights,

$$w = w^{\mathrm{E}} - c. \qquad (9.7)$$

The constant c is adjusted in the following so that the pool of inhibitory neurons (modelled with this constant) is able to keep the sparseness of the network at a desired level.

9.2.1 Learning Gaussian head direction patterns

The major difference to the point-attractor networks described in the last chapter is the training set we use. On the basis of the firing profiles of the head direction cells we will choose a Gaussian profile around a preferred direction for each cell. The external input to a node i is then given as

$$r_i = e^{-\delta_i^2/2\sigma^2}.\tag{9.8}$$

The displacement δ_i for each cell between the head direction α^{HD} provided by the external input and the optimal firing direction of the cell α_i is defined as

$$\delta_i = \min(|\alpha_i - \alpha^{\text{HD}}|, 360 - |\alpha_i - \alpha^{\text{HD}}|),\tag{9.9}$$

where we have measured the directions in degrees. We can think of these firing patterns in the way that a given head direction cue excites a particular node in the recurrent network and also neighbouring nodes. The strength of these connections to the neighbouring nodes falls off with a Gaussian-shaped function.

What will be the weights if we train the network on all possible directions of the subject? As the firing rates have Gaussian profiles during the learning phase, we get a contribution to each weight component for a given training example of magnitude

$$\delta w_{ij}^{\text{E}} = r_i r_j = e^{-(\delta_i^2 + \delta_j^2)/2\sigma^2}.\tag{9.10}$$

Thus, for a node i with a preferred direction equal to that of the training example, that is, $\alpha^{\text{HD}} = \alpha_i$, we get a contribution of

$$\delta w_{ij}^{\text{E}}(\alpha^{\text{HD}} = \alpha_i) = e^{-(\alpha_i - \alpha_j)^2/2\sigma^2}.\tag{9.11}$$

Similarly, node i with a preferred direction different from the direction of the training example, $\alpha^{\text{HD}} = \alpha_{i+n} = \alpha_i + n\Delta\alpha$, where $\Delta\alpha$ is the angular resolution of the model, will be updated with the value of

$$\delta w_{ij}^{\text{E}}(\alpha^{\text{HD}} = \alpha_i + n\Delta\alpha) = e^{-[(\alpha_i + \alpha_j)^2 - (\alpha_i - \alpha_j)n\Delta\alpha + (n\Delta\alpha)^2]/2\sigma^2}.\tag{9.12}$$

For infinite resolution of the model, $\Delta\alpha \to 0$, and a training covering all possible locations of head directions,[25] we get a weight matrix

$$w_{ij}^{\text{E}} = N e^{-\frac{(\alpha_i - \alpha_j)^2}{2\sigma^2}}.\tag{9.13}$$

The weight matrix therefore has a Gaussian shape and the same width as the receptive fields of the nodes.

The weight matrix (eqn 9.13) depends only on the distance between nodes. We can write this in the continuous notation as

$$w^{\text{E}}(x, y) = w^{\text{E}}(|x, y|).\tag{9.14}$$

This regularity of the weight matrix has important consequences as we will see. Furthermore, the inhibition constant shifts the Gaussian by a certain amount vertically.

---

[25] We will see later that we can modify the model slightly to achieve similar results with a very small training set instead.

The resulting effective weight matrix describes a model with short-distance excitation and long-distance inhibition. We have derived the weight matrix for Gaussian receptive fields. However, for most of the characteristics of CANN models it is sufficient to have weight matrices with the features just mentioned, shift invariance, short-distance excitation, and long-distance inhibition.

9.2.2 Gaussian interaction profiles in the brain

There is a variety of examples from cell recordings that suggests an effective interaction structure with short-distance excitation and long-distance inhibition in the brain. Such evidence has, for example, influenced the *Cowan* and *Wilson*'s important model of the neocortex that includes such interaction structures on the level of columnar organizations in the cortex. Such models are therefore frequently studied when studying the formation of orientation columns in the primary visual cortex. We will return to this issue later in this chapter. Another example was found in the intermediate layer of the superior colliculus, a midbrain area that is an important integration stage for many cortical and subcortical pathways that guide the direction of gaze. An example of the interaction structures within the superior colliculus from cell recordings in monkeys is shown in Fig. 9.3. In this figure we plotted the results for two sample neurons. The figure shows the influence of activity in other parts of the colliculus on the activity of each neuron. This influence indeed has the characteristics of short-distance excitation and long-distance inhibition. A CANN model with these characteristics was able to reproduce many behavioural findings for the variations in the time required to initiate a fast eye movement as a function of various experimental conditions.

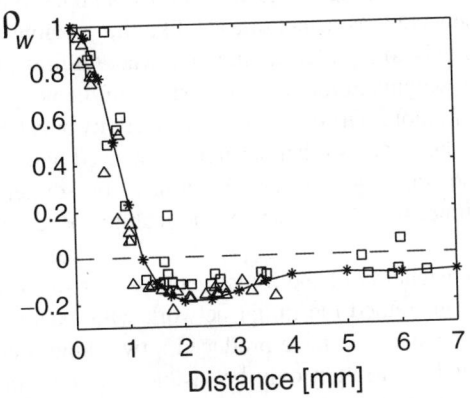

Fig. 9.3 Data from cell recordings in the superior colliculus in a monkey that indicate the interaction strength ρ_w between cells in this midbrain structure. The solid line displays the corresponding measurement from simulations of a CANN model of this brain structure [from Trappenberg *et al.*, *J. Cogn. Neurosci.* 13: 256–71 (2001)]. .

Similar models have also been used in a variety of other models of brain functions, which is an indication that such mechanisms might indeed be employed for various information-processing needs in the brain. Such an interaction structure, for example, reflects receptive fields in the *lateral geniculate nucleus* (LGN), which have been termed *on centre–off surround*. Many studies that employ such interaction structures thus incorporate this form explicitly without training the networks. This is a valid strategy when studying the consequences of such interactions. The regularity of the weight profiles also makes it possible for the brain to implement this type of interaction structure by other means than synaptic plasticity, for example, by genetic coding. However, such 'hardwired' connectivity patterns demand some regularity and topographic organizations of the response properties of the neurons. We will see later that the strategy can be inverted in the sense that a given lateral interaction structure of the proposed form can guide the self-organization of topographic maps in the brain.

9.2.3 Self-organized interaction structures in CANNs

We have already mentioned that the hippocampus does not display a regularity in the interaction structure in the physical space of the brain tissue, so that a hardwired (for example, genetically coded) interaction structure is not likely to be employed in this structure. However, note that the only reason we saw the regularities in the above model is that we organized the nodes according to the feature values they represented, labelling each node with numbers representing a value proportional to the topography of the feature space. Discovering this organization in the interaction structures would have been difficult otherwise. This is demonstrated in Fig. 9.4, where we have labelled the nodes and indicated the relative strengths of weights between the different nodes with lines of different widths corresponding to the relative strengths of the connections. Before learning, illustrated in Fig. 9.4A, we can assume that all nodes have equal weights relative to each other.[26] The dimensionality of this model can be regarded as high when we take the number of neighbours, the number of nodes to which each node is relatively strongly connected, as a measure. Unlike the previous model we have assigned each node randomly to a preferred direction where it fires maximally. After training we therefore get weights in the 'physical' space of the nodes as indicated in Fig. 9.4B that look rather random. The order in the connectivity only becomes apparent when we finally reorder the nodes so that strongly connected nodes are adjacent to each other. After doing so (Fig. 9.4C) we see that it has a one-dimensional structure which reflects the one-dimensional structure of the feature space that was used for training this network.

We reduced the dimensionality of the initial network to a one-dimensional connectivity pattern. If we had trained the initial network with examples from a two-dimensional feature space we would have produced a two-dimensional structure in the weight matrix, although this might only be visible after rearranging the nodes accordingly. The rearrangement of nodes we used is equivalent to the rearrangement of hippocampal neurons discussed in the last section (Fig. 9.1B) and explains why this had to be done to see the 'regularity' in the firing pattern. From this we can see that the network self-organizes to reflect the dimensionality of the feature space that is conveyed to it by the characteristics of the receptive field. The network therefore 'discovers' the dimensionality of the underlying problem. It is good to keep this point

---

[26] We started the training algorithm in this chapter with zero initial weights between the nodes, so that all weights have an equal relative strength. With some forms of weight decay it is also possible to get similar results with random initial weights.

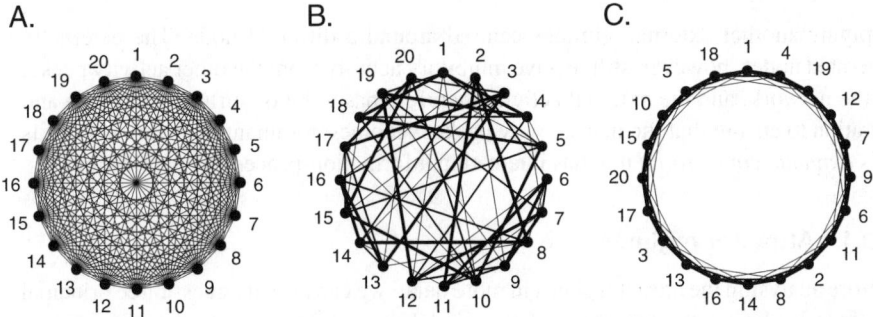

Fig. 9.4 A recurrent associative attractor network model, similar to the model shown in Fig. 9.2, where the nodes have been arbitrarily placed in the physical space on a circle. The relative connection strength between the nodes is indicated by the thickness of the lines between the nodes. Each node responds during learning with a Gaussian firing profile around the stimulus that excites the node maximally. Each node is assigned a centre of the receptive field randomly from a pool of centres covering the periodic training domain. (A) Before training all nodes have the same relative weights between them. (B) After training the relative weight structure has changed with a few strong connections and some weaker connections. (C) The regularities of the interactions can be revealed when reordering the nodes so that nodes with the strongest connections are adjacent to each other.

in mind as we will discuss a complementary model of so-called 'self-organizing maps' later in this chapter. There we organize the topography of receptive fields within the constraint of a network with a given dimensionality in its interaction structure.

9.3   Asymptotic states and the dynamics of neural fields

What are the asymptotic states when we train the auto-associative memory networks on continuous patterns instead of the discrete random patterns used in the last chapter? The behaviour of the network depends, of course, on the weight matrix. We derived in the last section that the weight matrix is *shift invariant* after training the network on continuous Gaussian patterns, that is, it just depends on the distance between two nodes regardless of their absolute position. Furthermore, the weights are positive (excitatory) for short distances between the nodes and negative (inhibitory) for large distances between nodes, when the distance is measured in the feature space. We therefore have a model with local *cooperation* and global *competition*. The local cooperation encourages the ongoing firing of a collection of nodes that are close to each other, while the long-range inhibition tends to reduce the firing of remote nodes. This can enable a collection of nodes to be active. We will call this the *activity packet* in the following.

   Due to the shift invariance of the weight matrix, the activity packet can be stabilized at any location in the network depending on an initial external stimulus. The attractor is therefore a continuous manifold in the neural field limit. For this reason we call these models continuous attractor neural network (CANN) models, a special form of attractor neural network (ANN) models such as the point attractor networks discussed in the last section. A new activity packet can be established at a new location by applying another external stimulus centred around a different node. The externally activated nodes, however, still receive inhibitory activity from the other activity packet in the network, and the external activity therefore has to be of sufficient strength and duration to ensure that the new activity packet becomes dominant in the network. It is this *dynamic competition* that has interesting information-processing applications.

9.3.1 Attractor regimes

Before analysing the attractor states in more detail we can already guess three principal regimes in the parameter space of the model with different attractor types in the network. The different regimes in the CANN model depend on the level of inhibition c (once we have specified the activation function and other parameters in the model such as the excitatory weights by training). The regimes can be categorized as follows.

1. **Growing activity:** When the inhibition is weak compared to the excitation between nearby nodes, so that the model is dominated by excitation, the dynamics of the model are governed by positive feedback. The activity of all neurons will then increase to the maximum possible firing rate, and information processing by the system is not possible.

2. **Decaying activity:** When the inhibition is strong compared to the excitation, then the dynamics of the model are dominated by negative feedback. The neurons will respond to sufficiently strong input, but the activity of all the nodes will decay rapidly after the removal of the external input. In this mode we can expect useful transient information processing such as that produced by competition between external inputs.

3. **Stable activity packet:** In an intermediate range of the strength of inhibitions relative to that of the excitations, which we will specify further below, we can reach a stable packet of active nodes. This mode can be utilized in different kinds of information processing, in particular to represent spatial information for a long time.

The regimes can easily be explored numerically as detailed in Chapter 12. An example of a CANN simulation is shown in Fig. 9.5A where we plotted the firing rates of the nodes in a network of 100 nodes on a grey-scale during the evolution of the system in time. All nodes are initialized to have medium firing rates, and a strong external stimulus to nodes number 30–70 was applied at time $t = 0\ \tau$. The stimulated nodes respond with an increase in their firing rate due to the excitatory weights for nearby nodes while the nodes far from the stimulated site even decrease their firing rates slightly due to inhibition from the nodes that have increased activity. The external stimulus was removed at $t = 10\tau$. The overall firing rates in the network decreased slightly following this removal of the external stimulus, and the activity packet became smaller. However, the most interesting effect is that a group of neighbouring nodes stayed active asymptotically. The firing rate profile at time $t = 20\ \tau$ is shown as a solid line in Fig. 9.5B.

In these simulations we used an inhibition constant that was equal to three times the square of the average firing rate of a node when driven by a Gaussian external input used for training the network, that is, $c = 3\langle r \rangle^2$. The firing rate profile of the nodes in a simulation with reduced inhibition ($c = \langle r \rangle^2$) at time $t = 20\tau$ is shown as a dashed

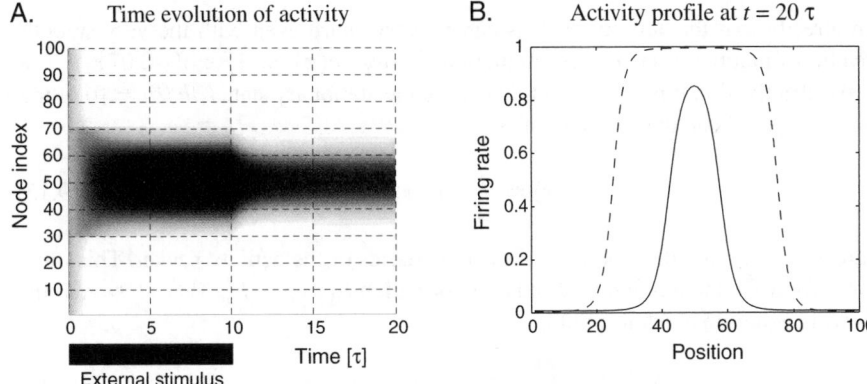

Fig. 9.5 (A) Time evolution of the firing rates in a CANN model with 100 nodes. Equal external inputs to nodes 30–70 were applied at $t = 0$. This external input was removed at $t = 10\tau$. The inhibition was set to three times the average firing rate of a node when driven by a Gaussian external input like that used for training the network. (B) The solid line represents the firing rate profile of the simulation shown in (A) at $t = 20\tau$. The dashed line corresponds to the firing rate profile in a similar simulation with reduced (by a factor of three) inhibition.

line in Fig. 9.5B. With such an inhibition, which would correspond to the value when the network was trained with the Hebbian covariance rule (eqn 7.11) instead of the Hebbian variance rule (eqn 9.6), it turns out that the activity packet takes up roughly half of the node in the network. This is indeed what we expect from the analysis in the previous chapter where we saw that recurrent networks with a global inhibition of $c = \langle r \rangle^2$ result in a retrieval sparseness of $a^{\text{ret}} = 0.5$.

The size of the activity packet shrinks with increasing levels of inhibition, which correspond to lower values for the retrieval sparseness of the network. At the same time we find that the firing rate profile becomes more Gaussian in shape. In this case, the firing rates of the neurons do not come close to the maximum possible firing rates given by the sigmoidal activation function. Similar profiles can thus also be expected with threshold linear activation functions that resemble much more closely the effective activation functions discussed in Chapter 4. With low inhibition we find that the excitation within the activity packet is so strong that the firing rates of the neurons deep inside the activity packet nearly reach their firing limit. The activity packet therefore has relatively sharp boundaries. We will utilize this finding in the following analysis.

9.3.2 Formal analysis of attractor states

The numerical simulations can guide us through some more detailed analysis on the nature of the asymptotic states (attractors) in CANN models following the classical analysis of *Shun-ichi Amari*. For simplicity, we assume a threshold activation function, $g(x) = \theta(x)$, that makes the firing rates within the activity packet equal to one while setting the firing rates outside the activity packet equal to zero. This choice is convenient as we will see. However, the results with this example are expected to be a good approximation for networks with smoother gain functions because we have seen already that the activity packets can be very sharp even with the very smooth activation functions used in the simulations above ($g(x) = 1/\exp(-0.07x)$). The threshold activation function is useful because the stationary state ($\partial h/\partial t = 0$) of the dynamic eqn 9.3 can then be written as

$$h(x) = \int_{x_1}^{x_2} w(x,y)dy, \tag{9.15}$$

where x_1 and x_2 are the positions of the boundaries of the activity packet. This must be true for all x including the boundaries for which $h(x_1) = h(x_2) = 0$, so that the following equation must also hold,

$$\int_{x_1}^{x_2} w(x_1,y)dy = 0. \tag{9.16}$$

For the weighting function $w = w^{\mathrm{E}} - c$ with excitatory Gaussian components (eqn 9.13) the above equation yields

$$\sqrt{\frac{\pi}{2}}N\sigma\mathrm{erf}(\frac{x_2 - x_1}{\sqrt{2}\sigma}) = c(x_2 - x_1). \tag{9.17}$$

This equation can be solved graphically as illustrated in Fig. 9.6 and has solutions with a finite extent of the activity packet as long as $0 < c < N$. The regime $c \geq N$ corresponds to the above-mentioned regime with decaying activity (asymptotic size of activity packet $x_2 - x_1 = 0$), and $N \leq 0$ corresponds to the regime with growing activity (asymptotic size of activity packet $x_2 - x_1 = \infty$).

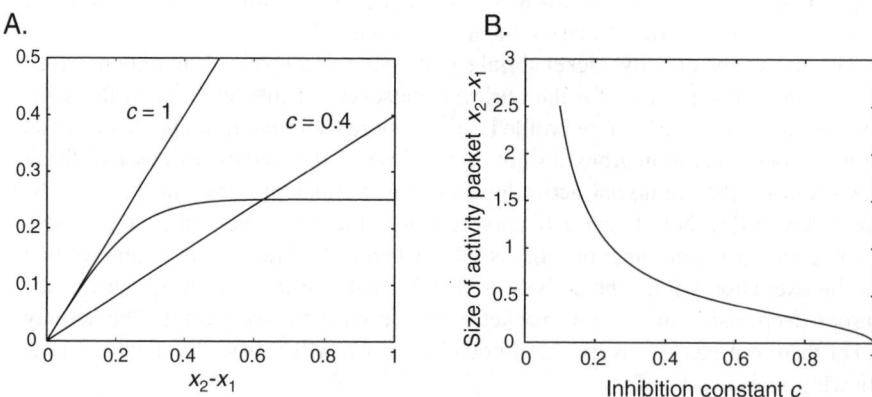

Fig. 9.6 (A) Plot of the functions $\sqrt{\frac{\pi}{2}}\sigma\mathrm{erf}(\frac{x_2 - x_1}{\sqrt{2}\sigma})$ and two linear functions with slope $c = 1$ and $c = 0.4$. The intersection of the functions (other than at $x_2 - x_1 = 0$) gives the solutions we are seeking of eqn 9.17. (B) The solutions of eqn 9.17 as a function of the inhibition constant.

9.3.3 Stability of the activity packet

What about the stability of the activity packet with respect to its movements? To answer this question we calculate the velocity of the boundaries. A movement of the activity packet without external input can only result from some forces generated by the shape of the activity packet itself. The force is then proportional to the gradient of the activity packet, and the velocity is

$$\frac{\mathrm{d}x}{\mathrm{d}t} = -\Delta x \frac{\partial h}{\partial t}, \tag{9.18}$$

where we also took a possible explicit time dependence of the activity packet into account. To get the velocity of the boundaries we have to substitute $x = x_1$ or $x = x_2$ into this equation.[27] We can then substitute eqn 9.3 into this equation to get a formula for the velocity of the boundary. The velocity of the centre of the activity packet is actually more interesting, where we define the centre as

$$x_c(t) = \frac{1}{2}(x_1(t) + x_2(t)). \tag{9.19}$$

It is then straightforward to get an expression for the velocity of the centre of the activity packet, namely

$$\frac{\mathrm{d}x_c}{\mathrm{d}t} = -\frac{1}{2\tau\Delta x_1} \int_{x_1}^{x_2} w(x_1, y) \mathrm{d}y - \frac{1}{2\tau\Delta x_2} \int_{x_1}^{x_2} w(x_2, y) \mathrm{d}y. \tag{9.20}$$

The expression on the right side is zero when the weighting function is symmetric and shift-invariant, and the gradients of the activity packet at the boundaries are the same except for sign. These conditions are illustrated in Fig. 9.7. For the Gaussian weighting function, which is symmetric and shift-invariant, we see that the integral from x_1 to x_2 of the Gaussian centred around x_1 is the same as that for the Gaussian centred around x_2. Also, once we have a symmetric activity packet, it is clear that the gradients at the boundaries are equal except for sign as illustrated in Fig. 9.7B. The velocity of the centre of the activity packet for a symmetric weighting function is therefore zero ($\mathrm{d}x_c/\mathrm{d}t = 0$), and the activity packet stays centred around the location where it was initialized.

9.3.4 Drifting activity packets

We have formally derived that the activity packet in models with shift-invariant and symmetric weight matrices is stable. Equation 9.20 also tells us when the activity packet is not stable and can hence drift. For example, the velocity of the centre of the activity packet is not equal to zero when the shift invariance of the weight matrix is broken. The weight matrix generated by Hebbian learning on random patterns, as discussed in the last section, is not shift-invariant. Such associative memory networks therefore drift away from initial conditions toward a point attractor as we discussed in the last chapter.

Another important factor is noise in the system, which we always have to take into account when discussing brain mechanisms. Noise breaks the symmetry as well as the shift invariance of the weighting functions when the noise is independent for

---

[27] The gradients are formally not defined at the boundaries with the threshold activation function. The activity packet is smooth with sigmoid activation functions, while the onset can still be sharp enough to define a boundary, particularly if we are considering the limit of increasing slope in the sigmoidal activation function.

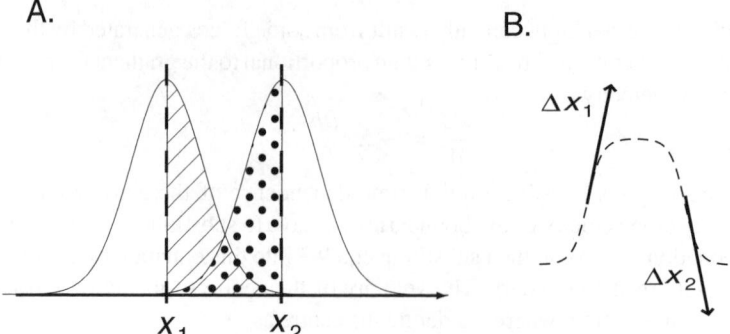

Fig. 9.7 (A) Two Gaussian bell curves centred around two different values x_1 and x_2. The striped and dotted areas are the same due to the symmetry of the bell curve. The integrals from x_1 to x_2 over the two different curves are therefore the same. This is not true if the two curves are not symmetric and have different shapes. (B) The dashed line outlines the shape of an activity packet from a simulation. The symmetry of this activity packet makes the gradients of boundaries equal except for a sign.

each component of the weight. The activity packet in such networks therefore drifts to some points where the shifting forces compensate each other. This leads to a *clustering* of end states. An example is shown in Fig. 9.8A. Each curve in the plot corresponds to the time evolution of the centre of gravity of the activity packet after the network was initialized with an activity packet centred around a different node in the network. This drift caused by noise has some interesting consequences for the operation of the brain. For example, the drift makes an accurate representation of locations over a long time impossible. Indeed, experiments show that the sense of direction in the dark diminishes after some period of time. Note that the drift of the activity packet due to noise decreases with the size of the system because the noise components can average out when the weighting function is the result of the interaction of many noisy nodes.

Another source of asymmetries in the weighting function is an irregular or partial training of the network. In the previous examples we have always trained the network for the same amount of time with activity packets centred at each node in the network. This not only requires long training sessions, but is also unrealistic in biological terms as the subject would have to explore a new environment in a very regular manner. The effects of partial training are illustrated with the simulations shown in Fig. 9.8C-D. There we have trained the network with activity packets centred around only 10 different nodes in the network, but not the remaining nodes. A partial view of the resulting weight matrix is shown in Fig. 9.8B. The values of the weight matrix are largest at the locations that were used for training. The centre of the activity packet with the resulting weight matrix, not including any noise, also shows a drift for initial states around the trained locations (Fig. 9.8C).[28] We thus basically recover a point attractor network.

---

[28] The locations precisely in between the trained locations are also stable, but only marginally so. A small deviation would result in a drift to the trained locations.

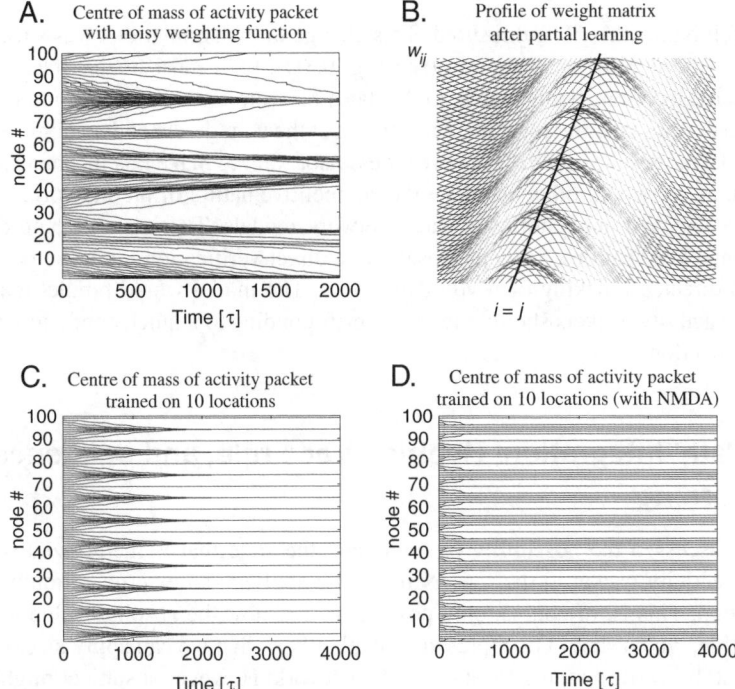

A. Centre of mass of activity packet with noisy weighting function

B. Profile of weight matrix after partial learning

C. Centre of mass of activity packet trained on 10 locations

D. Centre of mass of activity packet trained on 10 locations (with NMDA)

Fig. 9.8 (A) Noisy weight matrix. Time evolution of the centre of gravity of activity packets in CANN model with 100 nodes. The model was trained with activity packets on all possible locations. Each component of the resulting weight matrix was then convoluted with some noise. (B) Irregular or partial learning. Partial view of the weight matrix resulting from training the network with activity packets on only a few locations. (C) Time evolution of the centre of gravity of activity packets in CANN model with 100 nodes after training the network on only 10 different locations. (D) 'NMDA'-stabilization. The network trained on the 10 locations was augmented with a stabilization mechanism that reduces the firing threshold of active neurons.

9.3.5 Stabilization of the activity packet

The drift in the activity packet can be stabilized by a small increase in the excitability of neurons once they have been recently activated. This corresponds to a voltage-dependent nonlinearity in the postsynaptic neuron that could help to maintain the firing in a recurrent network. Such voltage-dependent nonlinearities could, for example, be implemented by NMDA receptors. We have already mentioned in Chapter 2 that NMDA receptors are blocked by magnesium ions when the neuron is at rest. This blockade is removed after an increase in the membrane potential. Thus, we can excite the neuron the next time much more easily if the time necessary to block the channel again is long relative to the time of the next incoming spike.

We can simulate such a voltage-dependent nonlinearity with a relatively long time constant by altering the threshold in the activation function of the model neurons. In the simulations shown in Fig. 9.8D we have changed the threshold value α from the value $\alpha = 0$ used in all previous simulations to a lower value of $\alpha = -10$ for neurons that exceeded 50% of their maximal firing rates. The value was reset to $\alpha = 0$ when the neurons fell below this 50% threshold. This change in the simulations was sufficient to increase the number of attractors (see Fig. 9.8D), with most of the states close to the trained locations being stable. Only states far from the trained location drifted initially to the closest attractor state. An increase of the voltage-dependent nonlinearity would make more states stable. However, we do not want to make this mechanism too dominant, as we would otherwise lose the competitive nature of the network, which is relevant for most applications of these network models. There is so far no direct experimental verification of this proposal, but a direct verification might be possible with well-directed blocking of NMDA receptors. The model would predict that the drift in the activity packets should increase, corresponding to a quick confusion in the sense of direction.

9.4 'Path' integration, Hebbian trace rule, and sequence learning

We have discussed the possibility of 'updating' the state that is memorized with a particular activity packet in the CANN model by applying an external stimulus to a new location. This is, of course, only possible if we know the absolute value for the new location, which should be represented explicitly so that we can apply an external stimulus at the corresponding location in the network. However, a subject might not have such an absolute value available, for example, we rotate a subject with closed eyes. This is like driving a blindfolded person around in a city and asking him where we are after some time. To solve this problem we have to 'calculate' the new position from the old position and the changes we made (velocity information including rotation and forward speed) over this time period. This 'calculation' is called *path integration*, and we will adopt this terminology for the generic situation of calculating a new state representation from an initial state representation and signals that indicate the change of the state.

9.4.1 Path integration with asymmetric weighting functions

We saw in the last section that asymmetries in the weighting functions of CANN models lead to a movement of the activity packet. A proposal for solving the path integration problem involves using such asymmetries in a systematic way. To do this we have to find a way to relate the strength of the asymmetry to the velocity of the movement. A velocity signal can be generated by the subject itself, and we call therefore such information *idiothetic cues*, where 'idiothetic' means self-generated. Examples are inputs from the *vestibular system* in mammals that can generate signals indicating the rotation of the head and *proprioceptive* feedback from muscles that signal the change in their position. Such signals will be the input to our following models, where we will again concentrate on head direction as an example.

9.4.2 Idiothetic update of head direction representations

A proposal as to how idiothetic velocity signals can be used in CANN models of head direction representations to update the system is shown in Fig. 9.9. For simplicity we have only shown three nodes of the recurrent network representing the head direction of a subject, and have only included collateral connections to the neighbouring nodes. In addition to these head direction nodes we included two other nodes (which can also represent a collection of neurons), which we will call *rotation nodes*. We assume for simplicity that the firing rate of these nodes is directly proportional to the velocity of the head movement. The principal idea behind the model is that these rotation nodes can *modulate* the strength of the collateral connections within the continuous attractor model. This modulatory influence makes the effective weighting function within the attractor network in one direction stronger than in the other direction, thus enabling the activity packet to move in a particular direction with a speed that is determined by the firing rate of the rotation nodes.

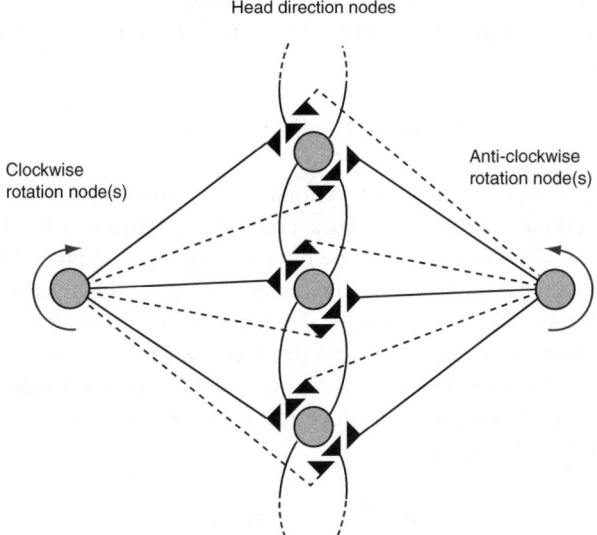

Fig. 9.9 Model for path integration in CANNs. The central nodes are part of the network with collateral connections as used to represent head directions (Fig. 9.2). The rotation nodes represent collections of neurons that signal rotation velocities proportional to their activity. The afferents of these rotation cells can modulate the collateral connections within the head direction network. We symbolized this with synapses close to the synapses of the collateral connections. Each rotation cell can synapse on to each synapse in the head direction network. The separation of the connections, as indicated by the solid and dashed lines in the figure, is self-organized during learning.

The effect of rotation node activity on the attractor network has to be modulatory (that is, multiplicative) as opposed to additive because the latter case would produce only an equal external input to all nodes that could not shift the activity packet. The modulatory effect can, for example, be implemented with the sigma–pi node discussed in Chapter 4. This in turn has several possible implementations on the neuronal level. For the following discussion it is sufficient to think about two physically close synaptic terminals that can interact to produce such nonlinear modulation affects.

9.4.3 Self-organization of a rotation network

The major problem we have to solve to make this model biologically realistic is to find a way to self-organize the network. If we simply assume that rotation nodes modulate all synapses equally we cannot move the activity packet. Instead, the network has to learn that the firing in a rotation node that indicates, for example, clockwise rotations should modulate only the appropriate 'clockwise synapses' in the network. Thus, the network has to learn that synapses have strong weights only in response to the appropriate weights in the recurrent network as indicated by solid lines in Fig. 9.9. The influence of the opposite synapses, indicated by dashed lines, has to at least be weaker. To achieve this we need a learning rule that can associate the recent movement of the activity packet with the firing of the appropriate rotation node. We therefore need to have a trace, a 'short-term memory', in the nodes that is related to the recent movement of the activity packet. An example of such a *trace term* (indicated by a bar over the firing rate) is,

$$\bar{r}_i(t+1) = (1 - \eta)\bar{r}_i(t) + \eta r_i(t). \tag{9.21}$$

This is a discrete version of a leaky integrator and can also be written in a differential form. This trace term represents the sliding average of recent firing, with an exponential sliding window of width characterized by the parameter η. The precise form of this trace term is not essential for the following mechanisms, and other trace terms can also be utilized. With this trace in the firing of the nodes in the recurrent network we can associate the co-firing of rotation cells with the movement of the activity packet in the recurrent network. The weights between rotation nodes, which we indicate with the superscript rot, and the synapses in the recurrent network, which have no superscript, can be formed with a Hebbian rule,

$$\delta w_{ijk}^{\text{rot}} = k r_i \bar{r}_j r_k^{\text{rot}}. \tag{9.22}$$

The rule strengthens the weights between the rotation node and the appropriate synapses in the recurrent network. As before we can form these weights during the learning phase where the firing of the nodes is determined by the firing of external input. This learning phase corresponds to the exploration of an environment by a subject using visual cues.

9.4.4 Updating the network after learning

After the weights have been learned we can update head directions of a subject without external input. The dynamics of the model are given by

$$\tau \frac{\partial h(x, t)}{\partial t} = -h(x, t) + \int_y w^{\text{eff}}(x, y, r^{\text{rot}}) r(y, t) r_i^{\text{rot}}(t) \mathrm{d}y, \tag{9.23}$$

where the index i labels the group of nodes of either clockwise or anti-clockwise rotation cells. The weight matrix $\mathbf{w}^{\text{eff}}$ depends on the activity of the rotation nodes, and describes the effective weighting function within the recurrent network from the collateral connections and the modulatory influence of the idiothetic cues. The modulatory nature of these influences can, for example, be expressed by

$$w_{ij}^{\text{eff}} = (w_{ij} - c)(1 + w_{ijk}^{\text{rot}} r_k^{\text{rot}}), \qquad (9.24)$$

though other forms of modulatory functions are possible and generally lead to results similar to the ones outlined below.

The behaviour of the model when trained on examples of one clockwise and one anti-clockwise rotation with only one rotation speed is demonstrated in Fig. 9.10. Details of the simulations are included in Chapter 12. An external position stimulus was applied initially for 10τ to initiate an activity packet, and this activity packet is stable after the removal of the external stimulus when the rotation nodes are inactive. Between $20\tau \leq t \leq 40\tau$ we applied a clockwise rotation activity corresponding to the activity used during learning. The activity packet then moved in the clockwise direction linearly within this time. The movement stops immediately after the rotation cell firing is abolished at $t = 40\tau$. During $50\tau \leq t \leq 70\tau$ we applied an anti-clockwise firing rate of the anti-clockwise node that was twice the value used during learning. The activity packet moved at nearly twice the speed in the anti-clockwise direction, demonstrating that the network can generalize to other rotation speeds.

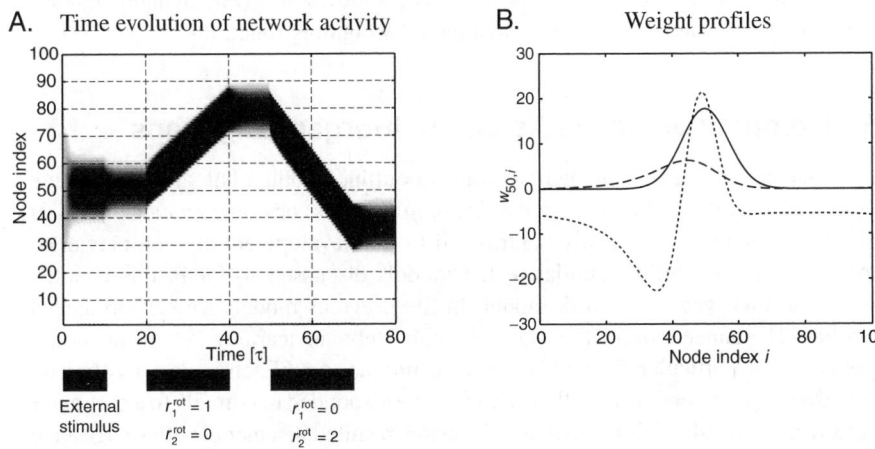

Fig. 9.10 (A) Simulation of a CANN model with idiothetic updating mechanisms. The activity packet can be moved with idiothetic inputs in either clockwise or anti-clockwise directions, depending on the firing rates of the corresponding rotation cells. (B) The different weighting functions from node 50 to the other nodes in the network after learning. w, solid line; w^{rot}, dashed line; w^{eff}, dotted line.

Examples of the weighting functions after learning are shown in Fig. 9.10. The solid line represents symmetric collateral weighting values $w_{50,i}$ between node 50 and the other nodes in the network. The clockwise rotation weights $w_{50,i,1}^{\text{rot}}$ are shown as a dashed line in the figure. Their functional form is not symmetric as expected. The corresponding anti-clockwise rotation weights $w_{50,i,2}^{\text{rot}}$, not shown in the figure, are a mirror-image of the dashed line. The resulting effective weighting function (eqn 9.24) is shown as a dotted line in the figure. All the examples are within a feature space with an intrinsic one-dimensional topography. However, path integration can be achieved in analogous ways also with feature spaces in higher dimensions as the network self-organizes.

9.4.5 Sequence learning

We can also apply the generic mechanisms of asymmetric weighting functions to other applications such as *sequence learning*. The Hebbian rule 8.5 produced symmetric weights with random patterns. The weights are not shift-invariant and therefore led to the drift of activity towards point attractors as we have mentioned. However, the symmetry of the weight matrix made the activity in the network stable for the point attractors. *John Hopfield* suggested including a trace term (in pattern space) in the canonical learning rule. This results in a new form of the learning rule that includes an asymmetric component,

$$w_{ij} = \frac{1}{N} \sum_{\mu} (\xi_i^{\mu} \xi_j^{\mu} + \lambda \xi_i^{\mu+1} \xi_j^{\mu}). \tag{9.25}$$

For a sufficient strength of the asymmetric component, which can be controlled using the strength parameter λ, the network is able to jump between the patterns in the sequences that was used to train the network. The abilities of several dynamic models in the literature are due to effectively asymmetric weighting functions.

9.5 Competitive networks and self-organizing maps

In this final section of this chapter we want to outline another, but strongly related, class of models, often called *self-organizing maps* (SOMs) or *Kohonen networks* after the Finnish scientist who greatly contributed to the development of such networks. Some form of competition, similar to the models discussed so far in this chapter, is central to this type of network model. In the previous models we self-organized the collateral connections in the network with Hebbian learning while the nodes responded to a particular feature vector in a unique way. Each node was thereby assigned a unique receptive field, that is, each node responded maximally to a particular feature vector. Together with global inhibition this resulted in a model with short-range cooperation and long-range competition between the nodes in the recurrent network. In the following model we will do the reverse in the sense that we fix the interaction structure between the nodes representing the feature values with an activity packet while self-organizing the receptive fields of the nodes.

9.5.1 Two-dimensional SOM

We will outline the ideas using a two-dimensional feature space and discuss generalizations below. With two-dimensional feature vectors,

$$\mathbf{r}^{\text{in}} = \begin{pmatrix} r_1^{\text{in}} \\ r_2^{\text{in}} \end{pmatrix}, \tag{9.26}$$

it is natural to consider a two-dimensional sheet of nodes when we want to represent features with those two-dimensional feature values with an activity packet of node activity similar to that used before. Such a sheet of nodes is illustrated in Fig. 9.11. The collateral interactions within the network are fixed (in contrast to the previous models), but have again the general structure of short-distance excitation and long-distance inhibition. These connections are not shown in Fig. 9.11. The major difference from the previous model is that we initially allow any external feature value to reach every node in the network. We therefore have included an additional set of weights w_{ij}^{in} in the model as shown in the figure. The input connections in the previous model were fixed. By self-organizing these input weights we can explore how they are organized in response to some training examples while the network is still governed by competitive dynamics.

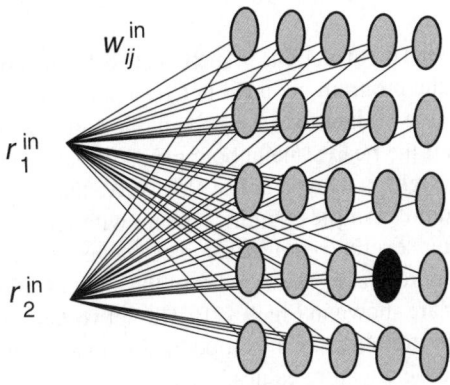

Fig. 9.11 Architecture of a two-dimensional self-organizing map. Each of the two input values r_1^{in} and r_2^{in}, each representing one of two feature components, is mapped on the map network with individual weight values $\mathbf{w}^{\text{in}}$. The nodes in the map network are arranged in a two-dimensional sheet with collateral connections (not shown) corresponding to the distances between nodes in this two-dimensional sheet.

9.5.2 Simplifying winner-take-all description

We will make some simplifications in the dynamics of the recurrent model that will enable us to describe the principal ideas more easily. We have already seen that an external stimulus will evoke an activity packet in the recurrent network, and the strongest packet will inhibit other activities in the network. We will therefore describe the response of the network as a Gaussian firing rate around the node that receives the strongest input. Let us call this node the *winning node* and label it with a '*'. The firing rates of the other nodes in the network are then given by,

$$r_{ij} = e^{-((i-i^*)^2 + (j-j^*)^2)/2\sigma^2}, \tag{9.27}$$

where i and j label the node in the corresponding row and column components of the two-dimensional feature layer. With this response property of the nodes in the feature layer we can train the input weights of the network with a Hebbian learning rule

$$\delta w_{\mathbf{x}i}^{\text{in}} = k r_{\mathbf{x}} (r_i^{\text{in}} - w_{\mathbf{x}i}^{\text{in}}), \tag{9.28}$$

where the index vector $\mathbf{x}$ has two components indexing the position of the node in the two-dimensional network. The index $i = 1, 2$ is the index of the input node. We includ a weight decay term to keep the weight values bounded. The change of a weight component is zero when the value of the weight component equals a specific component of an input vector. If, in addition, the weights for each node are normalized, then the weight vector of the winning node $\mathbf{w}_{\mathbf{x}^*}^{\text{in}}$ is closest to the corresponding input vector, that is,

$$|\mathbf{w}_{\mathbf{x}^*}^{\text{in}} - \xi| \leq |\mathbf{w}_{\mathbf{x}}^{\text{in}} - \xi| \quad \text{for all } \mathbf{x}. \tag{9.29}$$

A node is therefore maximally excited for a feature value represented by the weights to this node.

The development of the input weights in such a network is demonstrated in Fig. 9.12 from simulations with the program included in Chapter 12. In all of these figures the values of input weights are plotted for each node in the network at the intersection of the grid that connects all the nearest-neighbour pairs. We started the simulation with random values for the weights as shown in Fig. 9.12A. We then exposed the network to equally distributed random examples within a square with feature components in the range $0 \leq r_i^{\text{in}} < 1$. After several training examples we get a relatively homogeneous representation of the feature space. The weights of the network after updating the network on 1000 examples are shown in Fig. 9.12B. The network has learned to represent the feature values of the training set in a topographic map in, which neighbouring nodes represent neighbouring feature values.

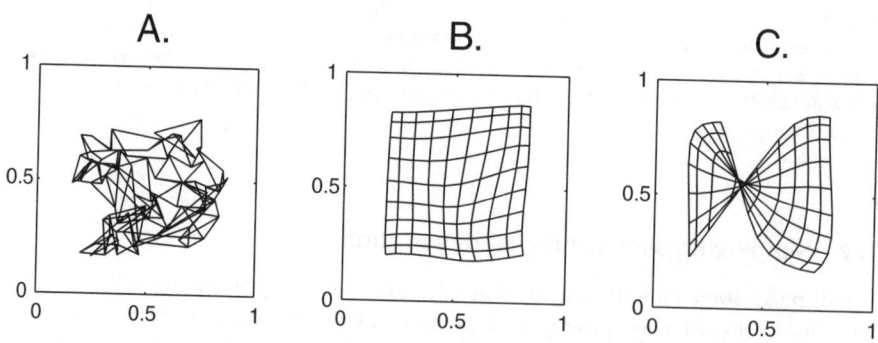

Fig. 9.12 Experiment with two-dimensional self-organizing feature maps. (A) Initial map with random weight values. (B) and (C) Two examples of the resulting feature map after 1000 random training examples with different random initial conditions. These simulations are discussed further in Chapter 12.

The form of the feature space representation can dramatically depend on the initial conditions and the precise sequence of examples. In Fig. 9.12C we show another example of the same simulations with a different set of random variables for the initial weights and the training examples. A twist or fold in the representation, not uncommon in such simulations, occurs in this example. It can take a long time to unwind such *topological defects*. A method to avoid such topological defects is to use a broad neighbourhood function (large σ in eqn 9.27) at the beginning of the simulation and to reduce the width with time. Topological defects are also present in cortical maps and might have similar origins.

Another example of a self-organized feature map is shown in Fig. 9.13. We exposed there a two-dimensional Kohonen network to random training vectors in the unit square as before. We started the simulations with a perfect representation to avoid the topological defects just mentioned (see graph for $t = 0$). The representation did not change much after 1000 updates with the random training examples in the unit square; only some small distortions were introduced caused by the updates of the weight matrix with the training examples. After this set of training examples we changed the training vectors presented to the network. These new training examples consisted of random values similar to the previous ones, and, in addition, training examples with components $1 \leq \xi_i < 2$, with random examples from each of the two quadrants. The simulation results after 100 and 1000 training steps with these new training examples are shown in the last two graphs in Fig. 9.13, respectively. The topographic map branched out to reach representations of the additional feature values, although the representation even after some considerable time is not as good as the representation of the initial training set.

The latter simulations demonstrate that SOM networks can learn to represent new domains of feature values after some training time, although the representation seems less fine-grained compared to the initial feature domain. The simulation results lead to some interesting conclusions if we speculate that the formation of cortical representation is driven by similar mechanisms. For example, it seems best to be exposed early in life to training examples from a broad feature space, including examples from all the features for which a fine-grained representation is desirable. This suggests, for example, that it is important to be exposed to different languages early in life to be able to achieve some high-level phonetic sound discrimination in different languages. Another example is face discrimination from people with different racial features. This is often difficult until one is exposed much more frequently to those people, for example, when moving to a different part of the world.

9.5.3 Other competitive networks

The competitive networks discussed above are only examples of networks that rely on some form of competitive mechanism, and many more examples can be found in the literature. Indeed, most of the models in computational neuroscience incorporate some form of competition that is commonly achieved through some form of global inhibition, or at least some inhibition between subpopulations of nodes. The specific ability of SOM to form self-organized topographic maps was mainly built on the fact that we trained not only the winning node, but also, to a lesser extent, the neighbouring nodes. Specifying what we mean by 'neighbouring nodes' introduces a set of topographic relations in the category space spanned by the output nodes. Competitive networks without such topographic relations (for example, when only the winning node is trained and the rest of the nodes are inhibited equally) can still learn to categorize training examples. The categorization in winner-take-all networks is achieved by the fact that a specific input vector evokes the firing of a specific output node that represents a specific category. More graded winner-take-all networks, in which several output nodes are active, can represent categories within a more distributed representation,

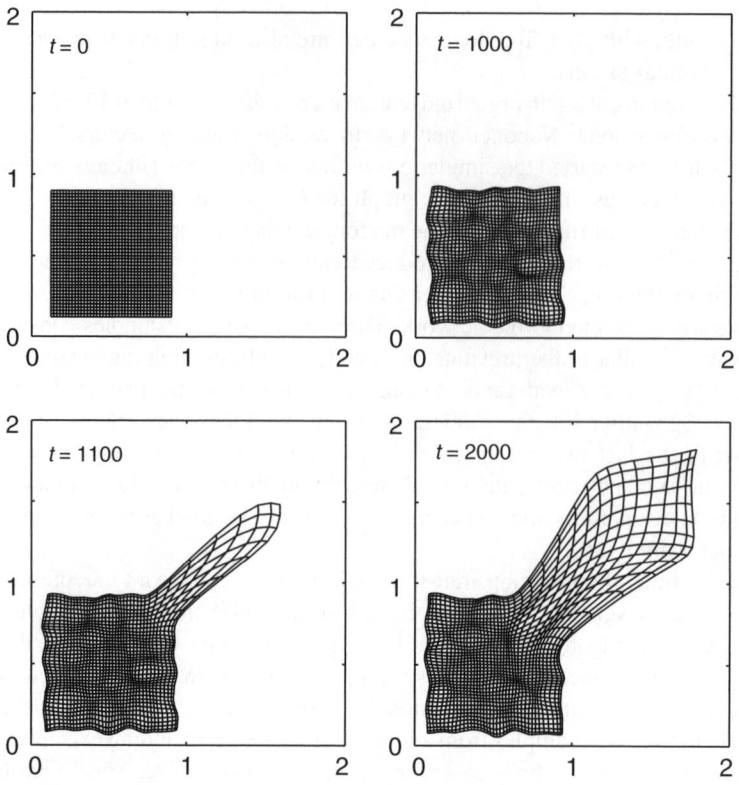

Fig. 9.13 Another example of a two-dimensional self-organizing feature map. In this example we trained the network on 1000 random training examples from the lower left quadrants. The next training examples were then chosen randomly from the lower-left and upper-right quadrant. The parameter t specifies how many training examples have been presented to the network.

but the basic mechanism of categorization is the same. In categorization we replace all the input vectors that evoke a specific output node with the same output vector ('prototype'). Such a functionality is sometimes called *vector quantization* and is illustrated in Fig. 9.14.

Note that without weight normalization in a competitive network we would enlarge the projections to some of the nodes (in the extreme case only one), which in turn would increase their chance even further to be a winning node and hence would continue to increase their weight values with all subsequent training examples. Only one node (the always-winning node) would continue to change its preferred direction by following the latest example (continuous plasticity), and other nodes would quickly lose the ability to get updated. To prevent this we included a weight decay term in the Kohonen network.

A related general problem of learning systems was pointed out by *Steven Grossberg* who coined the phrase *plasticity–stability dilemma*. Plasticity, which means the continuous ability of the networks to learn, would counter the desire for stability, in which we can build on the learned abilities of the networks. A possible solution to this

A.   Before training   B.   After training

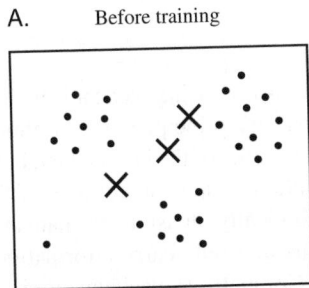 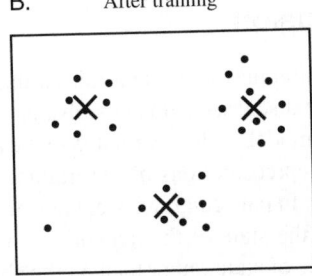

Fig. 9.14 Example of categorization (vector quantization) of two-dimensional input data (two-dimensional training vectors). The training data are represented as dots, and the input vector that would best evoke a response of one of the three output nodes is represented by a cross. (A) Before training there is no correspondence between the group of input data and the output node representing a category. (B) After training we have a 'preferred vector' for each node that corresponds to each of the clusters in the training data set.

problem within competitive networks was developed by *Gail Carpenter* and *Steven Grossberg* with a framework they called *adaptive resonance theory*. The principal idea is to employ a threshold in the degree of similarity a training vector must have with the preferred direction of a node in order to update the weights to this node. A close match is like a resonance, which enables some form of stability in the system. A training example that does not elicit a resonance with a class already represented by the nodes in the network can be used as a starting value for a new class represented by a previously unused (or a newly created) node. This enables a continuous plasticity in response to novel training examples.

Competitive mechanisms are important for many behavioural abilities in biological systems as many models in the literature demonstrate. A neural implementation of global competition requires long-range inhibition which is not clearly documented in the anatomy of the brain. Biologically faithful network models should therefore only include local forms of inhibition when the models are based on a fine-grained neuronal level. However, there are several ways in which local forms of inhibition can lead to effective global competitive mechanisms in networks. For example, multilayer models with short-range inhibition within each layer can spread this inhibition to larger scales if the feed-forward connections are divergent as is often observed in the brain. An interesting possible source of long-range inhibition was pointed out by *John Taylor* and *Farrukh N. Alavi* from King's College in London. They speculated that the thalamic-cortical loops that reach through the reticular nucleus of the thalamus (NRT) (which is a structure with dominating inhibitory neurons) could effectively lead to long-range inhibitory effects in the cortex. They demonstrated that such a structure could be modelled effectively with an interaction structure in the form of CANN models.

Conclusion

Competition plays a crucial role in many information-processing mechanisms in the brain. We discussed some models that rely heavily on such principles. The continuous attractor model is closely related to the point attractor model of the last chapter, but the specific representations of the training patterns resulted in specific abilities of these networks. In particular, we were able to develop biologically plausible mechanisms of updating the state of the network with neural signals that only carry information on the change of state rather than on the absolute state. Competitive mechanisms are also though to be crucial in many other mechanisms, such as the formation of topographic maps, which are common in the brain.

Further reading

Shun-ichi Amari

Dynamic pattern formation in lateral-inhibition type neural fields, in *Biological Cybernetics* 27: 77–87, 1977.

A wonderful paper on the dynamical states in CANN models. This includes additional possible states in such networks not mentioned in this chapter.

Simon M. Stringer, Thomas P. Trappenberg, Edmund T. Rolls, and Ivan E.T. de Araujo

Self-organizing continuous attractor networks and path integration I: One-dimensional models of head direction cells, to appear in *Networks*, 2002.

This paper includes the proposal of self-organization of CANN models and path integration with idiothetic cues utilizing self-organized asymmetric weighting functions.

Thomas P. Trappenberg, Michael Dorris, Raymond M. Klein, and Douglas P. Munoz

A model of saccade initiation based on the competitive integration of exogenous and endogenous signals in the superior colliculus, in *J. Cognit. Neurosci.* 13: 256–71, 2001.

Example of CANN model of the superior colliculus that combines modelling with physiological and behavioural data.

John Hertz, Anders Krogh, and Richard G. Palmer

Introduction to the theory of neural computation, Addison-Wesley, 1991.

See chapter 9 for a good introduction to competitive networks and self-organizing maps.

T. Kohonen

Self-organization and associative memory, Springer Verlag, 3rd edition, 1989.

A classic book that includes most of the ideas behind associative memory and self-organizing maps.

10 Supervised learning and rewards systems

One of the basic processes the brain is involved in is the accurate guidance of purposeful limb movements. In this chapter we will discuss the control of motor actions within control networks that can be implemented in the brain. Such models utilize supervised learning of which we will discuss several variants, including the basic delta rule, the biologically unrealistic but widely employed error-back-propagation algorithm, as well as reinforcement training, which employs only supervision with simple reward contingencies.

10.1 Motor learning and control

In the previous chapters we discussed fast associative learning, which we think is central in forming memories and associations. This can be contrasted with other types of learning in which we try to discover regularities in training data in a statistical sense. This type of learning therefore acts on a large number of training data without the intention of storing all the specific examples. An example is the learning of motor skills, such as catching a ball or using the hand to bring a glass of water to the mouth. Such motor learning can take some time; just think about how long it takes to play the piano. It seems important for the survival of a species to achieve a certain precision in limb movements, and motor control is therefore an important task for our central nervous system.

When reaching for a cup it helps to look at it. Visual guidance can help to achieve the task; hence the brain must be able to direct the control system for arm movements with visual signals. It is a fascinating experience when we perturb this system, for example, by using prism glasses that shift the visual appearance of a target systematically away from its actual location, or by changing the dynamics of an arm with a computer controlled guidance system. Reaching for a target with altered parameters of the controlled system we will typically fail within the first few trials. However, we are commonly able to adapt to the changed environment within only a few additional trials.

10.1.1 Feedback controller

The control of mechanical or electrical devices is, of course, a major challenge faced regularly by engineers, and it is therefore not surprising that engineers have contributed a lot of ideas on how limb movements could be controlled by the nervous system. A very simple, yet often very efficient, form of control is *feedback control*, which is illustrated in Fig. 10.1. In the example we want to follow in this section we designate a signal specifying the desired state as input to the system. This input signal is then converted by a specific module into a motor command that drives the device we want to control. We will therefore call this module *motor command generator*. The generated motor command could, for example, be the signal to activate specific motor neurons in the brainstem that in turn stimulate muscle fibres. The motor command generator can be viewed as an *inverse model* of the dynamics of the controlled object as it takes a state signal and should produce the right motor command so that the controlled object ends up in the desired state. The motor command thus causes a new state of the object that we have labelled 'actual state' in Fig. 10.1.

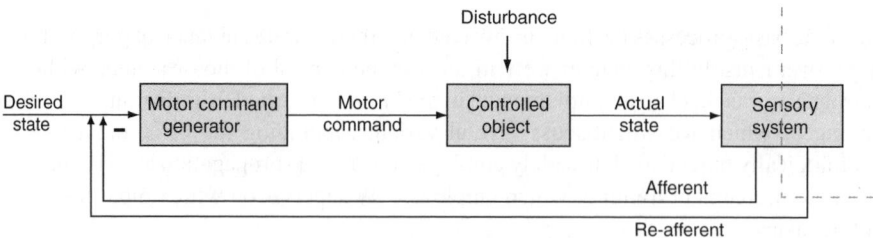

Fig. 10.1 Negative feedback control and the elements of a standard control system.

It is, of course, a major question how to find and implement an appropriate and accurate motor command generator. In principle we can utilize a feed-forward mapping network that is trained on examples of the motor actions. Whatever method we use, if the motor command generator is a perfect inverse model of the controlled object and can thus produce the right motor commands, and the controlled object responds perfectly, then we have reached the goal. However, in reality this almost never happens as many circumstances can prevent such idealized motor movements. We have indicated this in the figure by a disturbance signal to the controlled object, which can include internal noise, external influences such as different loads on the system, ageing of the system, different temperatures altering physical properties, etc. In such cases we might not reach the desired state and we have to initiate a new movement to compensate for the discrepancy between the desired and actual state of the controlled object. It is then necessary to estimate this discrepancy, which demands a sensory system. For arm movements we could, for example, use visual feedback as mentioned above. The sensory feedback is then generated outside of our body, and we include a dotted line running through the sensory system box to indicate sensory cues originating inside and outside of our body. The visual input about the position can be converted into a feedback signal that we call *re-afferent*. Sensory feedback can also originate from internal sensors in the body. For example, neural signals are generated by the contraction of muscles that are fed back to the central nervous system, so-called *proprioceptive feedback*. We labelled this type of sensory feedback *afferent* in Fig. 10.1.

The sensory feedback can be used in various ways to regulate the system. A simple example, that of a negative feedback controller, is illustrated in Fig. 10.1. The sensory signal that indicates the actual state of the system is therein subtracted from the signal that specifies the desired state. This produces a new desired state (*motor-error map*). The feedback thus yields a new motor-error signal for the correction movement that generates a new movement that will probably be closer to the desired state. It is clear that the sensory feedback has to be converted into the right reference frame to be used by the motor control system.[29] The basic feedback control systems often works very well and are commonly used in mechanical and electrical devices. There are, however, several factors that make this simple scheme unrealistic for biological subjects. For example, the sensory feedback is usually too slow for many control tasks, and the necessary accuracy is also not without problems. Proprioceptive feedback can be faster but is often inaccurate. The simple feedback controller can therefore not be the whole story in the brain.

10.1.2 Forward controller

Two refined schemes for motor control with slow sensory feedback are illustrated in Figs. 10.2 and 10.3. The first one employs some subsystems that mimic the dynamic of the controlled object and the behaviour of the sensory system. These subsystems are called *forward models*. If these models are good approximations of the real systems they are modelling, then we can use the output of these systems, instead of the slow sensory response, as feedback signal. The models have, of course, to be measured against the real systems. So they have to get some form of sensory feedback that can be used to change their behaviour. This scheme works as long as the changes in the systems that they mimic are much slower than the time scale of the movement that is controlled, which is a condition that is often fulfilled in biological motor control.

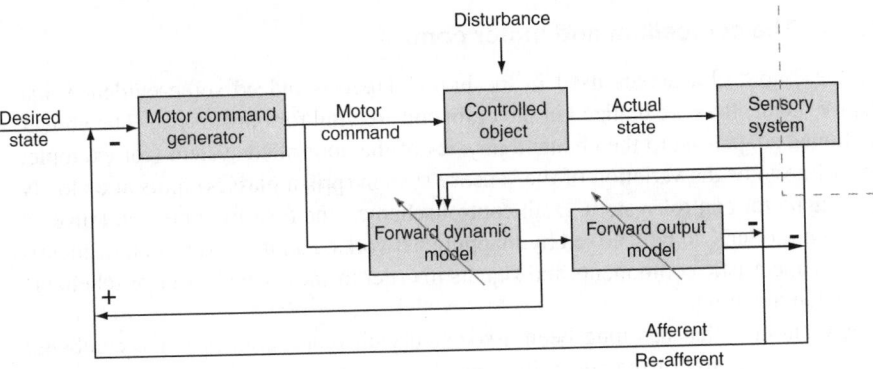

Fig. 10.2 Forward model controller.

10.1.3 Inverse model controller

The second scheme, shown in Fig. 10.3, employs an inverse model instead of the forward model in the previous scheme and is therefore called an *inverse model controller*. The inverse model, which is incorporated as side-loop to the standard feedback controller, learns to correct the computation of the motor command generator. The reason that this scheme works is similar to that for the previous example. The sensory feedback can be used to make the inverse model controller more accurate while it

---

[29] Such a mapping is not included in the figure.

provides the necessary correction signals in time to be incorporated into the motor command. Both control systems, the forward model controller and the inverse model controller as shown in Figs. 10.2 and 10.3, are robust controllers and could well be implemented in the brain as discussed below. However, these are not the only control systems that might be relevant for brain processes, and we will discuss another one, the adaptive critic, later in this chapter.

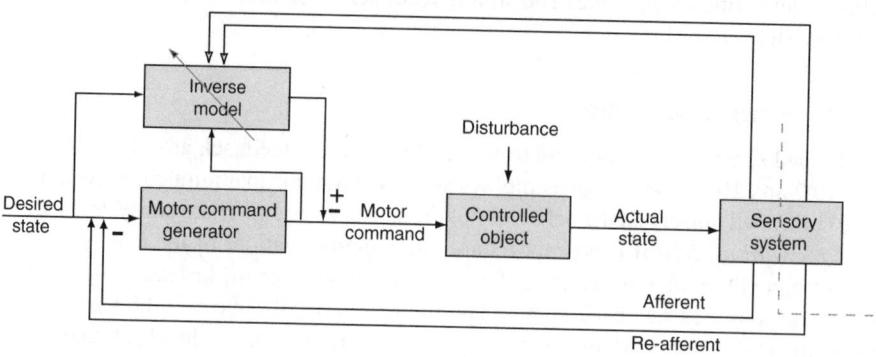

Fig. 10.3 Inverse model controller.

10.1.4 The cerebellum and motor control

Are these control schemes used in the brain? There is indeed some evidence that adaptive controllers are realized in the brain and are vital for our survival. The above-mentioned adaptation to the changed physics of the controlled system (for example, arm stiffness) or the variation of the sensory system (prism glasses) hints at a slowly adapting motor control system. Both control schemes, the forward model and inverse model controller, can be realized by mapping networks, and it is a question of identifying characteristic components and signals in order to show which control scheme is realized in the brain.

A brain area that has long been associated with motor control is the cerebellar cortex. The anatomy of the cerebellum displays great regularity and is summarized schematically in Fig. 10.4. Inputs from different sources enter the cerebellum through *mossy fibres* that contact *granule cells* and *Golgi cells*. Granule cells are probably the most numerous cells in the brain, estimated to be in the order of 10^{10}–10^{11}. The granule cells exceeds that of mossy fibres by a few hundred times, which makes this architecture a candidate for expansion recoding as discussed in Chapter 8. The granule cells, and some other intermediate neurons, provide a major input to *Purkinje cells* through *parallel fibres*, each Purkinje cell receiving as many as 80,000 inputs from different granule cells. Purkinje cells in turn provide the (inhibitory) output of the cerebellum.

The Purkinje cells also receive input from so-called *climbing fibres* from the *inferior olive*, each cell with its own climbing fibre. This one-to-one architecture is unique in the brain, and the climbing fibres have long been speculated to provide a teaching signal to the Purkinje cells that enables the mapping network of the cerebellum to learn to

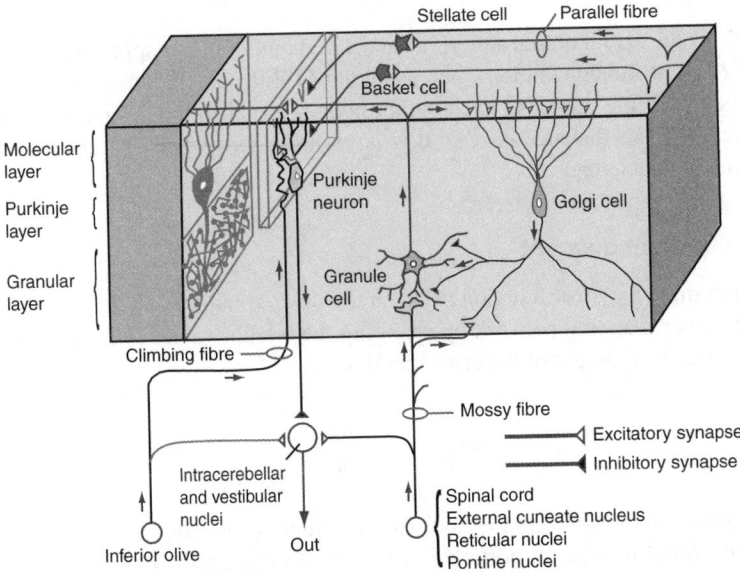

Fig. 10.4 Schematic illustration of some connectivity patterns in the cerebellum. Note that the output of the cerebellum is provided by the Purkinje neurons that make inhibitory synapses. Climbing fibres are specific for each Purkinje neuron and are tightly interwoven with their dendritic tree [adapted from Albus, *Math. Biosci.* 10: 25–61, 1971].

control motor functions as a *side-loop controller*. In this chapter we will therefore discuss in more detail supervised learning algorithms for mapping networks, some of which might be utilized in the cerebellum, and some of which have implications for other adaptation schemes in the brain.

10.2 The delta rule

Forward and inverse models can be implemented by feed-forward mapping networks as mentioned above. In this section we will specify how such mapping networks can be trained. Our objective in the following is to minimize the mean difference between the output of a feed-forward mapping network and a desired state provided by a teacher. We can quantify our objective with an *objective function* or *cost function*, labelled E in the following, that measures the distance between the actual output and the desired output. There are many possible definitions of a distance between two states. Formally, each definition defines the metric of the state space as long as the distance measure has certain obvious characteristics such as that the distance is zero only if the output of the network is equal to the desired state, and the distance is positive otherwise. Some examples of distance measures are listed in Appendix A. A simple and often used cost function is the *mean square error* (MSE)

$$E = \frac{1}{2} \sum_i (r_i^{\text{out}} - y_i)^2, \tag{10.1}$$

where r_i^{out} is the actual output and y_i is the desired output of a mapping network. The factor $1/2$ is a useful convention when using the MSE as an objective function as we will see below. If the output of each output node is equal to the desired value y_i for each node, then this distance is zero. It is larger then zero when one or more of the components do not agree.

10.2.1 Gradient descent

We can minimize the error function between the desired output and the actual output of a single-layer mapping network by changing the weight values to the output layer along the negative gradient of the error function,

$$\delta w_{ij} = -k\frac{\mathrm{d}E}{\mathrm{d}w_{ij}}. \tag{10.2}$$

This is called a *gradient descent* method. The error function is monotonically decreased by this procedure in steps proportional to the gradient along the error function. The constant k is a *learning rate*. An example is shown in Fig. 10.5 for a hypothetical network with only one weight value w. By starting with some random weight value for this 'network' it is likely that this network shows poor performance with a large error. The weight value is then changed in the direction of the negative gradient, which is a vector pointing in the direction of the maximal slope of the objective function from this point, and has a length proportional to the slope in this direction. Changing the weight values of the network to new values in this direction therefore guarantees that the network with the new weights has a smaller error as before. When approaching a minimum the gradient of the error function is decreasing and the change in the weight values slows down. This method often results in a rapid initial decrease of the network error, and the convergence can also be proven for small learning rates.

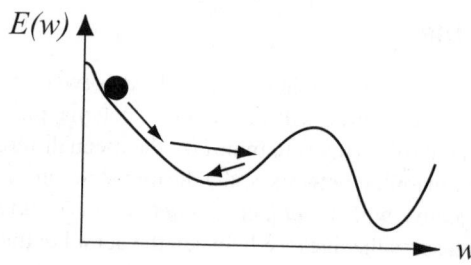

Fig. 10.5 Illustration of error minimization with a gradient descent method on a one-dimensional error surface $E(w)$.

We can easily derive the particular algorithm to implement the gradient descent method for a single-layer feed-forward mapping network (simple perceptron) when using the MSE as a measure of the distance between desired output y_i and actual output r_i^{out} of the perceptron. The gradient of the error function is given by

$$\frac{\partial E}{\partial w_{ij}} = \frac{1}{2}\frac{\partial}{\partial w_{ij}}\sum_i (g(\sum_j w_{ij}r_j^{\text{in}}) - y_i)^2$$

$$= f'(h_i)(\sum_j w_{ij}r_j^{\text{in}}) - y_i)r_j^{\text{in}}. \tag{10.3}$$

$f'(x) = \frac{\partial f(x)}{\partial x}$ is the derivative of the activation function that enters the formula by using the *chain rule* for calculating differentials

$$\frac{\partial f(g(x))}{\partial x} = \frac{\partial f(g)}{\partial g}\frac{\partial g(x)}{\partial x}. \tag{10.4}$$

For a linear perceptron with activation function $f(x) = x$, this is simply given by $f' = 1$. The weight change for a linear perceptron is therefore

$$\delta w_{ij} = k(y_i - r_i^{\text{out}})r_j^{\text{in}}. \tag{10.5}$$

This learning rule is called the *delta rule* as the change of the weight values is proportional to the difference between the desired and actual output often expressed as δ. Following this procedure, each weight value w_{ij} is increased ('learning') by an amount proportional to the product of the presynaptic node (input value) r_j^{in} and the desired postsynaptic value y_i, and at the same time decreased ('unlearning') by the product of the input value r_j^{in} and the actual postsynaptic value r_i^{out}. When the desired output is equal to the actual output, then the change of weight values is equal to zero as required. The difference between the actual and desired output of the network can be measured with the aid of sensory information in the control problems discussed in the last section, and the delta rule can provide the necessary adaptation of the mapping network of the motor system.

We have just outlined the delta rule for a linear perceptron. In the case of a nonlinear activation function of the output node we have to include the derivative of the activation function in the calculation of the delta term that is used to update the weights. The algorithm of the delta rule is summarized in Table 10.1.

10.2.2 Batch versus online algorithm

In technical applications of mapping networks, in which we have a set of examples of a function we want to represent with the mapping network, we can do the repetitions mentioned in the summary (Table 10.1, item 6) in various ways. A pattern can be learned very accurately with the gradient descent method if we train the network only on one pattern for a long time before switching to the next pattern. However, when a second pattern is trained consecutively in a similar way, then the network unlearns the previous pattern because the weights are adjusted entirely to represent the new pattern. This is not of course what we intended. In practical applications of the delta rule it is often (though not always) the best strategy to represent all pattern in a training in a way that is known as a *batch algorithm*. In this strategy we first apply all pattern in a training set to the network and calculate the desired changes for each pattern. Only after all the patterns have been applied do we change the weights according to the mean value over all patterns before we repeat this procedure. Batch algorithms often

Table 10.1 Summary of delta-rule algorithm

1. Initialize weights to random values
2. Apply a sample pattern to the input nodes $$r_i^0 = r_i^{\text{in}} = \xi_i^{\text{in}}$$
3. Calculate rate of the output nodes $$r_i^{\text{out}} = g(\textstyle\sum_j w_{ij} r_j^{\text{in}})$$
4. Compute the delta term for the output layer $$\delta_i = g'(h_i^{\text{out}})(\xi_i^{\text{out}} - r_i^{\text{out}})$$
5. Update the weight matrix by adding the term $$\Delta w_{ij} = k\delta_i r_j^{\text{in}}$$
6. Repeat steps 2–5 until error is sufficiently small

work well in practice if there are not many redundancies in the training data, though batch algorithms are a bit more prone to get stuck in local minima.

However, batch algorithms are not plausible in biological systems, and it is more realistic to assume that the sensory information is processed directly. *Online learning algorithms* are therefore very important in computational neuroscience. These algorithms seem at first more restrictive and not as effective as batch algorithms as we can not repeatedly utilize past experiences. However, batch algorithms also have severe problems when applying them to many realistic applications. This becomes apparent if we think about training a network on realistic data, for example, on visual images. The data stream generated by the eyes is enormous, and even in artificial systems, such as systems that employ a digital camera with a much lower resolution than that of human eyes, the amount of data generated is still enormous. The amount of storage necessary to keep the images present for the batch algorithm will rapidly overload any reasonable system. Online algorithms are therefore not only more interesting in computational neuroscience, but also for technical applications.

10.2.3 Supervised learning

The delta learning rule depends on knowledge of the desired output. We therefore need a *teacher* who supplies the network with the desired response, the training signal. This class of learning algorithms is called *supervised learning*. This is different from Hebbian learning rules, which change the synaptic strength between nodes purely on the basis of the actual firing rates of the nodes involved, irrespective of the reason that determined the firing of the nodes. Hebbian learning rules are therefore categorized as *unsupervised learning*. However, note that the δ-rule (eqn 10.5) for the weight changes still takes the form of a correlation rule between an error factor $\delta_i = y_i - r_i^{\text{out}}$ at the output node and the firing rate of the input node r_j^{in}. The biological mechanisms underlying synaptic plasticity can therefore be used to implement such a learning rule, provided that an error signal reaches the node. The climbing fibre in the cerebellum could very well supply such an error signal to the Purkinje cells.

10.2.4 Supervised learning in multilayer networks

We have so far discussed supervised learning in single-layer mapping networks (perceptrons). In Chapter 6 we indicated that perceptrons without hidden layers have limitations in the kind of functions that can be represented by them. However, it is straightforward to generalize the delta rule to multilayer mapping networks. The resulting algorithms is known as the *error-back-propagation algorithm* or *generalized delta rule*. This algorithm dominates most of the applications of multilayer feed-forward mapping networks (multilayer perceptrons), such as connectionist modelling. Due to its wide use we summarize this algorithm in the next section. However, we will also discuss some difficulties in connecting the computational steps with brain processes. Some alternative schemes closely resembling the basic scheme of error-back-propagation while claiming to be more biologically realistic have been proposed in the literature, although direct verifications have not yet been established. These difficulties in the biological realism of the implementations of this algorithm are in addition to the problematic applications of mapping networks with strongly restricted number of hidden nodes to achieve good generalization as discussed in Chapter 6.

Before outlining the back-propagation algorithm we note that there might not be the need in the brain to train multilayer mapping networks with supervised learning algorithms with the generalized delta rule. Single-layer networks can represent complicated functions, even nonlinear separable functions, if we allow some more advanced processing in the single nodes compared to that accomplished by the very much simplified sigma nodes. Another example that enables single-layer networks to cope with linear non-separable functions is *expansion recoding*, an increase of the representation dimensionality of the input vectors as discussed in Chapter 8. This is indeed what might be used in the cerebellum. The mossy fibres contact a large number of granule cells, which exceeds by far the number of mossy fibres. With this expanded representation of the input vectors it is possible to solve a mapping problem with a single-layer perceptron in which the error signal is only supplied to the output nodes of this network. Finally, modular networks, which we will discuss in the next chapter, can break the problems down in various other ways, so that simplified learning rules can be used in each subsystem.

10.3 Generalized delta rules

In this section we outline some of the learning rules that dominate many of the technical applications of artificial neural networks and connectionist models in cognitive science. The learning ability of multilayer mapping networks has been studied predominantly in the context of this algorithm. Although we will stress some problems in finding signs that the error-back-propagation algorithm is used directly in biological networks, it is nevertheless instructive to study this learning algorithm as it can help us to get a deeper understanding of learning procedures in general, and the frequent use of the back-propagation algorithm in the literature makes it necessary to get at least a basic knowledge of it. Furthermore, a discussion of this algorithm will allow us to outline some of the problems regarding its biological plausibility.

The delta rule (eqn 10.5) cannot be applied directly to each layer of a multilayer mapping network as this would require a teaching signal for each node in the network.

The desired value of the output node can be supplied by a teacher, but proper values of the hidden nodes are not known *a priori*. It is, however, straightforward to generalize the delta rule in the gradient descent formalism. It is therefore surprising that this generalization was widely recognized only from the early 1980s onward, although the algorithm was known and used for training neural networks long before that. Let us illustrate the algorithm for a multilayer feed-forward network with one hidden layer denoted by the superscript 'h'. The gradient of the MSE error function with respect to the output weights is given by

$$
\frac{\partial E}{\partial w_{ij}^{\text{out}}} = \frac{1}{2} \frac{\partial}{\partial w_{ij}^{\text{out}}} \sum_i (r_i^{\text{out}} - y_i)^2
$$

$$
= \frac{1}{2} \frac{\partial}{\partial w_{ij}^{\text{out}}} \sum_i (f^{\text{out}}(\sum_j w_{ij}^{\text{out}} r_j^{\text{h}}) - y_i)^2
$$

$$
= f^{\text{out}\prime}(h_i^{\text{h}})(\sum_j w_{ij}^{\text{out}} r_j^{\text{h}} - y_i) r_j^{\text{h}}
$$

$$
= \delta_i^{\text{out}} r_j^{\text{h}}, \tag{10.6}
$$

where we have defined the delta factor

$$
\delta_i^{\text{out}} = f^{\text{out}\prime}(h_i^{\text{h}})(\sum_j w_{ij}^{\text{out}} r_j^{\text{h}} - y_i)
$$

$$
= f^{\text{out}\prime}(h_i^{\text{h}})(r_i^{\text{out}} - y_i). \tag{10.7}
$$

Equation 10.6 is, of course, just the delta rule as before because we have only considered the output layer. The calculation of the gradients with respect to the weights to the hidden layer again requires the chain rule as they are more embedded in the error function. Thus we have to calculate the derivative

$$
\frac{\partial E}{\partial w_{ij}^{\text{h}}} = \frac{1}{2} \frac{\partial}{\partial w_{ij}^{\text{h}}} \sum_i (r_i^{\text{out}} - y_i)^2
$$

$$
= \frac{1}{2} \frac{\partial}{\partial w_{ij}^{\text{h}}} \sum_i (f^{\text{out}}(\sum_j w_{ij}^{\text{out}} f^{\text{h}}(\sum_k w_{jk}^{\text{h}} r_k^{\text{in}})) - y_i)^2. \tag{10.8}
$$

After some battle with indices (which can easily be avoided with analytical calculation programs such as MAPLE or MATHEMATICA), we can write the derivative in a form similar to that of the derivative of the output layer, namely

$$
\frac{\partial E}{\partial w_{ij}^{\text{h}}} = \delta_i^{\text{h}} r_j^{\text{in}}, \tag{10.9}
$$

when we define the delta term of the hidden term as

$$
\delta_i^{\text{h}} = f^{\text{IN}\prime}(h_i^{\text{in}}) \sum_k w_{ik}^{\text{out}} \delta_k^{\text{out}}. \tag{10.10}
$$

The error term δ_i^{h} is calculated from the error term of the output layer with a formula that looks similar to the general update formula of the network, except that a signal

Table 10.2 Summary of error-back-propagation algorithm

1. Initialize weights to small random values

2. Apply a sample pattern to the input nodes

$$r_i^0 := r_i^{\text{in}} = \xi_i^{\text{in}}$$

3. Propagate input through the network by calculating the rates of nodes in successive layers l

$$r_i^l = g(h_i^{l-1}) = g(\sum_j w_{ij}^l r_j^{l-1})$$

4. Compute the delta term for the output layer

$$\delta_i^{\text{out}} = g'(h_i^{\text{out}})(\xi_i^{\text{out}} - r_i^{\text{out}})$$

5. Back-propagate delta terms through the network

$$\delta_i^{l-1} = g'(h_i^{l-1})\sum_j w_{ij}^l \delta_j^l$$

6. Update weight matrix by adding the term

$$\Delta w_{ij}^l = k\delta_i^l r_j^{l-1}$$

7. Repeat steps 2–7 until error is sufficiently small

is propagating from the output layer to the previous layer. This is the reason that the algorithm is called *error-back-propagation algorithm*. The algorithm is summarized in Table 10.2. An example of using this basic algorithm to solve the XOR problem as mentioned in Chapter 6 is included in Chapter 12.

10.3.1 Biological plausibility

The back-propagation of error signals is probably the most problematic feature in biological terms. Some form of information exchange between postsynaptic and presynaptic neurons might be possible as discussed before. However, a wide use of such mechanisms for a back-propagation of errors through the whole network introduces several other problems. A major problem is the non-locality of the algorithm in which a neuron has to gather the back-propagated errors from all the other nodes to which it projects. This not only raises synchronization issues, but also has disadvantages for true parallel processing in the system. The inclusion of derivative terms in the delta signals is also problematic. The back-propagation of inaccurate derivative terms can quickly lead to inaccurate updates of the weights in the network. Finally, it has never been resolved how a forward propagating phase of signals can be separated effectively from the back-propagation phase of the error signals.

10.3.2 Advanced algorithms

The basic error-back-propagation algorithm has many problems, in particular with its convergence performance. A lot of effort was devoted to the improvement of the basic version of this gradient descent method. This effort established a deep understanding of the learning in the form of statistical learning theories and led to considerable improvements over the basic algorithm. These techniques include the smart choice of initial conditions, different error functions, various acceleration techniques, and hybrid methods. We will only provide a short overview of some of those techniques without many details because they are often of minor biological significance and the subject of more technologically oriented publications. In the discussion we will stress some of the limitations of the basic error-back-propagation algorithm and demonstrate that there are many alternative learning strategies.

10.3.3   Momentum method and adaptive learning rate

In applications of the basic gradient descent method we can typically find an initial phase where the average error over the training examples is rapidly decreasing. However, this is unfortunately often followed by a phase of very slow convergence, often caused by a shallow part of the error function. Many solutions have been proposed to overcome this problem. One of the oldest is to use a *momentum* term that 'remembers' the change of the weights in the previous time step,

$$\Delta w_{ij}(t+1) = -k\frac{\partial E}{w_{ij}} + \alpha\Delta w_{ij}(t). \tag{10.11}$$

The momentum term has the effect of biasing the direction of the new update vector towards the previous direction (hence the name 'momentum'), which is often a good guess for improved weight values. Another method is to increase the learning rate when the gradient becomes small. Several methods with adaptive learning rates were proposed and are used in technical applications. Acceleration of the gradient descent method is very important as the convergence to an acceptable error level can take thousands of learning steps.

10.3.4   Different error functions

Shallow areas in the error function depend on the particular choice of the error function on which the gradient descent method is based. An acceleration of the learning process can often be achieved with error functions other than the MSE. A particularly interesting choice is the *entropic error function*,

$$E = \frac{1}{2}\sum_{\mu,i}[(1+y_i^\mu)\log\frac{1+y_i^\mu}{1+r_i^{\text{out}}} + (1-y_i^\mu)\log\frac{1-y_i^\mu}{1-r_i^{\text{out}}}], \tag{10.12}$$

which is a proper measure for the information content (or entropy) of the actual output of the multilayer perceptron given the knowledge of the correct output. Although this error function looks computationally more demanding than the MSE, the application in gradient descent methods can be computationally less demanding. For example, the delta term of the output layer with an activation function $g(x) = \tanh(x)$ reduces to $\delta_i = y_i - r^{\text{out}}$. It is not always obvious which error functions should be used, and a general strategy for choosing the error function can unfortunately not be given.

10.3.5   Higher-order gradient methods

The basic line search algorithm of gradient (or *steepest*) decent is known for its poor performance with shallow error functions. However, the training algorithms discussed here are based on the minimization of an error function, and we can employ many other advanced minimization techniques to achieve this goal. Several of such techniques, common in technical minimization procedures, take higher-order gradient terms into account. These can also be viewed as including curvature terms describing the curvature of the error surface in the weight change calculations. Such methods typically involve the calculation of the inverse of the Hessian matrix, which can be numerically time-consuming. In the context of statistical learning theory we have to evaluate a proper distance measure such as the *Fisher information*. A superior gradient method, called *natural gradient algorithm*, has been proposed by *Shun-ichi Amari* based on such considerations.

The standard *Levenberg–Marquardt method* is also based on higher-order gradient information, and has been used to train feed-forward mapping networks. This method is also included in the MATLAB Neural Networks toolbox from The MathWorks, Inc. Simulations of training mapping networks with the natural gradient method and with the Levenberg–Marquardt method have demonstrated that these methods can overcome the problems of shallow minima (and even turning points in the error function), and that the convergence times for training the networks to a specified training error are many orders of magnitude less then those of other learning methods. The drastically improved convergence times can outnumber the increased simulation time due to the increased algorithmic complexity. Such algorithms should therefore always be considered when applying mapping networks to technical problems. The relations of such algorithms to biological learning are, however, so far unclear.

10.3.6 Local minima and simulated annealing

A general limitation of pure gradient descent methods is the possibility that the network gets trapped in a local minimum of the error surface. The system is then not able to approach a global minimum of the error function as anticipated. This is illustrated in Fig. 10.5 where the global minimum on the right was actually not reached. A solution to this problem is to include some stochastic processes that enable random search. Several methods with random components are very successful in training multilayer perceptrons, most notably a method called *simulated annealing*. This method adds some noise to the weight values during the update of the weights on top of a deterministic algorithm such as gradient descent. This noise helps to escape shallow local minima. The noise level is then gradually reduced to ensure convergence.

10.3.7 Hybrid methods

A variety of methods utilize the rapid initial convergence of the gradient descent method and combine it with global search strategies. For example, we can employ a method in which, after the gradient descent method slows down below an acceptable level, a new starting point is chosen randomly. From this new starting point in the configuration space a few gradient descent steps are performed on the network. Only when the error value after these few gradient descent steps is lower than that at the previous level do we accept these new weight values for further gradient descent steps. Such *hybrid methods* combine the efficient local optimization capabilities of gradient descent methods with the global search abilities of stochastic processes. Genetic algorithms,

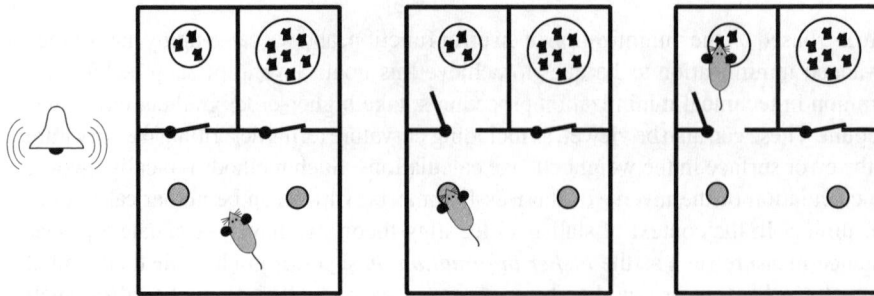

Fig. 10.6 Classical conditioning and temporal credit assignment problem. A subject is required to associate the ringing of a bell with the pressing of a button that will open the door to a chamber with some food reward. In the example the subject has learned to press the left button after the ringing of the bell. This is an example of a temporal credit assignment problem. It is difficult to devise a system that is still open to possible other solutions such as a bigger reward hidden in the right chamber.

mentioned in Chapter 6, use similar combinations of deterministic minimization and stochastic components.

Most of the algorithms mentioned in the last few paragraphs are incorporated in neural network simulation libraries and software environments such as the MATLAB Neural Networks Toolbox. The biological implausibility of such algorithms together with the caution needed when using multilayer mapping networks as stressed in Chapter 6 should, however, be taken very seriously when applying such networks as models for the brain.

10.4 Reward learning

In the supervised learning that we outlined above we always assumed that a teacher supplies the exact desired state of each output node in the network. The teacher therefore has to supply very detailed vector information, which seems unrealistic in common learning situations. More realistic are teaching procedures where a teacher provides only limited feedback such as only an indication of the performance of a 'student'. In the extreme case the indication of the student's performance might only be binary as in 'correct' or 'incorrect', or 'good' and 'bad'. In animal experiments such a training signal is often supplied with a food reward. An absence of a food reward can indicate a 'false response' to the animal.

10.4.1 Classical conditioning and temporal credit assignment problem

Learning with reward signals has been known to psychologists for many years under the term *conditioning*. An example of classical conditioning is shown in Fig. 10.6. In the illustrated experiment we place a rodent in a box that has three chambers, two doors, and two buttons. The rodent is placed in the chamber with the two buttons, and the doors to the other two chambers, which contain food, are initially closed. A press of a button can open a door to the corresponding food chamber. However, the system is prepared in such a way that a press of the button only opens a door within a small time interval after the time indicated by an auditory signal such as the ringing of a bell. The rodent can run around freely in the button chamber, but initially does not know the conditions under which the door will open. However, it may happen that the rodent will press by chance the left button in the required time after the bell rings. The door then opens and the rodent gets some small food reward. The rodent will then quickly associate the ringing of the bell with the motor action it has to take, pushing the button to get a reward. Note that this implies that the ringing of the bell has to be associated with the occurrence of the reward later in time. This condition has been termed the *temporal credit assignment problem*. Neural mechanisms have to be able to solve this problem.

10.4.2 Stochastic escape

We included in the experiment another chamber with a larger food reward that the rodent would certainly prefer. The rodent in our example was, however, conditioned by chance to open the left door after the ringing of the bell. If the rodent always stuck to the initial conditioned situation it would never learn about the existence of the larger food reward. However, if the rodent is running around randomly in the button chamber before the bell rings it could still happen that it presses the right button before running to the left button. The opening of the right door and the subsequent larger food reward then changes the association of the auditory signal to a new motor action, that of pressing the right button. This is an example of *stochastic escape* that can balance *habit* versus *novelty*. Such a balance seems to be an important ingredient in the survival of a species.

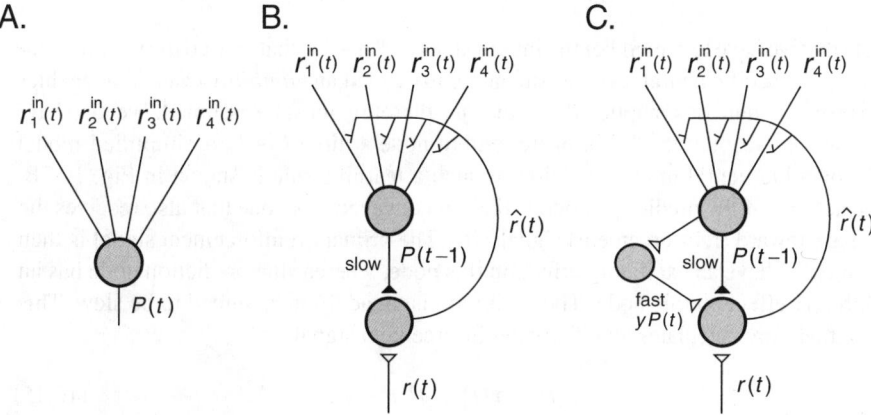

Fig. 10.7 (A) Linear predictor node. (B) Neural implementation of temporal delta rule. (C) Neural implementation of temporal difference learning.

10.4.3   Reinforcement models

To describe the implementation of a system that learns from reward signals within neural architectures we can consider a single node as illustrated in Fig. 10.7A (the generalization of the schemes to networks is straightforward). The inputs to this node represent a certain input stimulus such as the ringing of the bell. We suppose that the node gets activated under the right conditions and is therefore able to predict the future reward. To stress this predictive nature of the firing of the node we will label the activity of the node with $P(t)$. We assume here a simple linear sigma node for which the activity of the node at time t is calculated from the inputs to the node at time t by

$$P(t) = \sum_i w_i(t) r_i^{\text{in}}(t), \tag{10.13}$$

where we have labelled the activity of the presynaptic nodes with r_i^{in} and the weight values as usual with w_i. We included a time dependence in the weight values because we want to change them over time as a function of reinforcement signals.

10.4.4   Temporal delta rule

Let us assume that a reward is given at time $t+1$, which is indicated by the system with a scalar value $r(t+1)$. To solve the temporal credit assignment problem we have to be able to associate this future reward value with the specific input stimulus that occurred reliably before the reward was given. This can be done with a temporal version of the delta rule in the form

$$w_i(t) = w_i(t-1) + (r(t) - P(t-1)) r_i^{\text{in}}(t-1). \tag{10.14}$$

Note that we have to remember the input values ($r_i^{\text{in}}(t-1)$) that caused the reward value at $r(t)$. A specific neural mechanism, generally called an *eligibility trace*, that enables the proper update of synaptic efficiencies w_i therefore must be assumed. We also have to take into account the firing of the output node at time $t-1$. A simplified model of a possible neural implementation of such a learning rule is shown in Fig. 10.7B. The activity of the prediction node is therein conveyed to a node that also receives the primary reward (reinforcement) signal $r(t)$. The primary reinforcement signal is then assumed to have an excitatory effect on this node, whereas the prediction node has an inhibitory effect on this node. The inhibition is, in addition, assumed to be slow. This node therefore calculates an effective reinforcement signal

$$\hat{r}(t) = r(t) - P(t-1). \tag{10.15}$$

This effective reinforcement signal has to be conveyed to all the synapses of the incoming signals where it has to be correlated with an eligibility trace to produce the appropriate weight change proposed in eqn 10.14. This model of reward learning is known as the *Rescorla–Wagner theory* in classical conditioning. The model can produce one-step ahead predictions of a reward signal. The effective reinforcement signal is zero when the activity of the prediction node matches the primary reinforcement signal.

10.4.5 Reward chains

Learning in the previous model is restricted to the prediction of reward in the next time step. Much more important for the survival of an individual is the ability to predict future reward at different time steps or even whole series of rewards. To extend the above scheme to this situation we define a reinforcement value $V(t)$ that takes all the future rewards into account. For example, we can define the reinforcement value

$$V(t) = \alpha_1 r(t+1) + \alpha_2 r(t+2) + \alpha_3 r(t+3) + ..., \qquad (10.16)$$

where we have included some parameters α_i that allow us to specify the weights we give to the rewards at different times. For example, we can give A reward in the close future more weight than a reward far in the future. A simple realization of such a model is to include a discount factor $0 \le \gamma < 1$. By setting $\alpha_i = \gamma^{i-1}$ we define the reinforcement value to be

$$V(t) = r(t+1) + \gamma r(t+2) + \gamma^2 r(t+3) + ... \qquad (10.17)$$

Our aim is to predict this reinforcement value. Note that the Rescorla–Wagner model corresponds to the case $\gamma = 0$.

10.4.6 Temporal difference learning

The prediction of this reinforcement value seems to be a hopeless task as we do not know the future rewards. In the case of the temporal delta rule the system had to remember the events in the previous time steps to solve this problem, which was done with the help of an eligibility trace. It seems at first that using additional eligibility traces for different time frames would demand an infinite memory of all previous events. There is however a wonderful trick to solve this problem, which was named *temporal difference learning* by *Richard Sutton* and *Andrew Barto* who developed most of the theory of this advanced reinforcement learning. To derive the learning rule let us assume that we already have a system that can predict the reinforcement value at time t correctly. Then we have

$$P(t) = r(t+1) + \gamma r(t+2) + \gamma^2 r(t+3) + ... \qquad (10.18)$$

This also implies that we were able to predict the correct reinforcement value at the previous time step, that is,

$$P(t-1) = r(t) + \gamma r(t+1) + \gamma^2 r(t+2) + ...$$
$$= r(t) + \gamma[r(t+1) + \gamma r(t+2) + ...]. \qquad (10.19)$$

The expression in square brackets of eqn 10.19 is equal to the expression on the right-hand side of eqn 10.18 so that we can replace this with $P(t)$. We therefore have the condition

$$P(t-1) = r(t) + \gamma P(t), \qquad (10.20)$$

which must be true if we have a perfect prediction. If we don't have a perfect prediction then the equation does not hold. *Richard Sutton* proposed to minimize the *temporal difference* error

$$\hat{r}(t) = r(t) + \gamma P(t) - P(t-1). \tag{10.21}$$

Such an algorithm allows the network to learn to predict future rewards. Note that this scheme includes the temporal delta rule (eqn 10.15) as special case by setting $\gamma = 0$. Figure 10.7C shows a possible neuronal implementation of the learning method that is similar to the previous case (Fig. 10.7B) but includes a fast side-loop with a decay factor that conveys the value $\gamma P(t)$ to the node that calculates the primary reinforcement signal.

10.4.7 Adaptive critic controller

Temporal difference learning is a method of learning to predict future reward contingencies. *Richard Sutton* and *Andrew Barto* have incorporated this scheme into a powerful control method, called *adaptive critic*, which is illustrated in Fig. 10.8. Note the similarities of this control scheme to the inverse model controller outlined in Fig. 10.3. The critic is, however, designed to predict the correct motor command for accurate future actions and can thus supervise the motor command generator. The motor command generator is often called *actor* within this framework, and the scheme is also known as the *actor–critic scheme*. This scheme is very useful in engineering applications such as controlling elevators or adjusting the parameters of a petroleum refinery's operation. Neural implementations have been suggested and it has been noted that there are some similarities with specific architectural features in the brain as we will outline next.

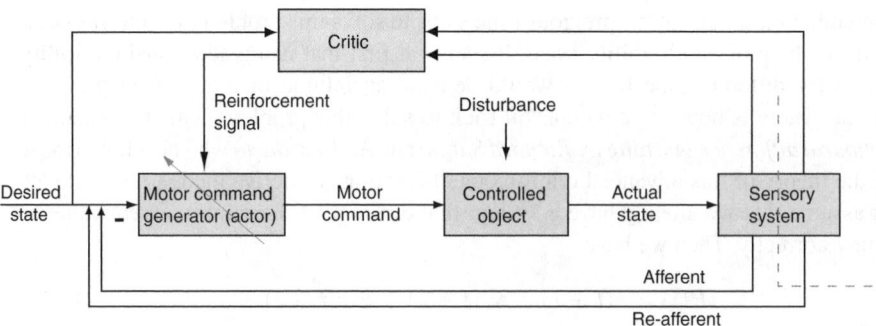

Fig. 10.8 Adaptive critic controller.

10.4.8 The basal ganglia in the actor–critic scheme

There are several speculations as to how the adaptive critic control scheme might be realized in the brain. For example, it was hypothesized that the *basal ganglia*, which are known to be instrumental in the initiation of motor commands, have many of the structural components required to implement an adaptive critic and to supervise actors that can control the initiation of motor movements. The basal ganglia are a collection of five subcortical nuclei as illustrated in Fig. 10.9A. They receive cortical and thalamic input mainly through the *putamen* and the *caudate nucleus*, which are hence thought to comprise the input layer of the basal ganglia. The information stream then runs through the *globus pallidus* (with an internal and external segment) to the major output layer, the *substantia nigra pars reticulata*. The internal side-loop from the globus pallidus via the *subthalamic nucleus* is also important for our next discussion. In addition, note that the *substantia nigra pars compacta* projects back to the striatum, the input layer of the basal ganglia that is made up of the caudate nucleus and the putamen.

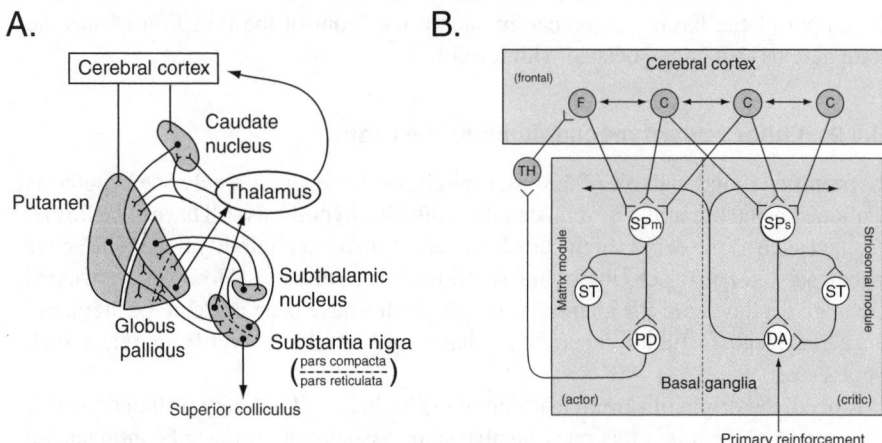

Fig. 10.9 (A) Anatomical overview of the connections within the basal ganglia and the major projections comprising the input and output of the basal ganglia [adapted from Eric R. Kandel, James H. Schwarz, and Thomas M. Jessell, *Principles of neuroscience* (3rd edition), Elsevier (1991)]. (B) Organizations within the basal ganglia are composed of processing pathways within the striosomal and matrix modules reflecting an architecture that could implement an actor–critic control scheme. C, cerebral cortex; F, frontal lobe; TH, thalamus; ST, subthalamic nucleus; PD, pallidus; SPm, spiny neurons in the matrix module; SPs, spiny neurons in the striosomal module; DA, dopaminergic neurons [adapted from Houk, Adams, and Barto, in *Models of information processing in the basal ganglia*, Houk, Davies, Beiser (eds.), MIT Press (1995)].

The information streams within the basal ganglia are thought to be segregated (to a certain extent) within interleaved modules that are called the matrix modules and the striosomal modules, respectively. A very simplified proposal for the specific architecture within the basal ganglia, which reflects a neural architecture that could implement an actor–critic scheme, is shown in Fig. 10.9B. The input layer of the basal ganglia is rich in *spiny neurons* (SP), which receive massive cortical (C) connections. The spiny neurons in the striosomal module (SPs) also receive projections from dopaminergic neurons (DA) in the substantia nigra pars compacta (SNpc) that synapse on to the spines of the spiny neurons in the caudate and putamen. It is possible that the dopaminergic input is thereby able to alter the efficiencies of specific cortical inputs that are marked with an eligibility trace. The neurons shown in the basal ganglia are also inhibitory so that the dopaminergic neurons in the SNpc are inhibited by SPs activity. The subthalamic side-loop in contrast disinhibits the DA, which can result in some excitation of DA neurons proportional to the inputs from this side-loop and a primary reinforcement signal. If, in addition, the side-loop is faster than the direct

SPs–DA influence, then it is possible that the striosomal module implements the critic that minimizes the temporal difference as discussed above (compare Fig. 10.7C).

The matrix module in turn could implement the actor. The dopaminergic neurons project to these spiny neurons (SPm) and could therefore alter the C–SPm weights in a fashion similar to that in the C–SPs connections. The only difference is that these neurons project to the internal segment of the pallidus and the substantia nigra pars reticulata (SNpr), which are thought to be the major output layers of the basal ganglia. The output of the basal ganglia can in such a way control the initiation of specific motor actions that are associated with reward.

10.4.9 Other reward mechanisms in the brain

The proposed functional role of the basal ganglia discussed here is only one hypothesis mentioned in the literature. Several details within this hypothesis still have to be further specified such as the details of the biochemical nature of an eligibility trace. The above hypothesis so far also lacks important experimental verifications. However, we wanted here to give a flavour of the interesting proposals that have been made in the literature, which may lead to future experiments that might be able to verify or reject such hypotheses.

Indeed, the origin of reward learning in the brain is still not very well understood. What seems clear is that this must involve some association of reward contingencies with specific motor actions in the brain. Some other subcortical and cortical areas are thought to be instrumental in making such reward associations. For example, the sub-cortical area called the *amygdala* receives projections from many information streams of many senses as well as the prefrontal cortex, and is therefore placed strategically to form associations between different modalities and reward contingencies. Indeed, bilateral damage of the amygdala is known to considerably impair such associations. Furthermore, it has been shown that neurons in the *orbitofrontal cortex* adapt their response to stimuli after changing their reward associations. Dopaminergic neurons also project into the frontal cortex so that reward mechanisms could originate predominantly in the frontal cortex as opposed to the hypothesis of the basal ganglia discussed above. Further research is needed to learn more about these important issues.

Conclusion

In the previous chapters we discussed some basic networks that utilized Hebbian learning on the neuronal (or node) level. In this chapter we explored adaptive mechanisms that are thought to be relevant on a larger scale (system level) of the brain. Our discussions focused in particular on control systems that are crucial for basic motor operations in primates. The learning algorithms discussed in this chapter have in common that they are supervised with varying degrees of specification, ranging from the mere utilization of a reward signal to the specification of the desired firing pattern of output nodes. More research is needed to relate such algorithms to neural mechanisms and to understand their biochemical implementation.

Further reading

Daniel M. Wolpert, R. Chris Miall, and Mitsuo Kawato
  Internal models in the cerebellum, in *Trends Cogn. Sci.* 2: 338–47, 1998.
  A good review of the control schemes and the cerebellum by leading researchers in this area.
Edmund T. Rolls and Alessandro Treves
  Neural networks and brain function, Oxford University Press, 1998.
  See chapter 9 for a discussion of the involvement of the cerebellum and the basal ganglia in motor control, and chapter 7 for some discussions of reward associations in the amygdala and the orbitofrontal cortex.
Richard S. Sutton and Andrew G. Barto
  Reinforcement learning: an introduction, MIT Press, 1998.
  The book is by the specialists who invented temporal difference learning.
John Hertz, Anders Krogh, and Richard G. Palmer
  Introduction to the theory of neural computation, Addison-Wesley, 1991.
  See chapters 5 and 6.
James C. Houk, Joel L. Davis, and David G. Beiser (eds.)
  Models of information processing in the basal ganglia, MIT Press, 1995.
  A good collection of ideas about information processing in the brain. This includes a highly readable introduction to the actor–critic scheme and temporal difference learning, as well as speculations about the relations of these schemes to the basal ganglia.

11 System level organization and coupled networks

In order to understand how complex mental tasks can be solved by the brain we have to understand the specifics of modular networks reflecting large-scale organizations of the brain. In this chapter we briefly review some examples of system level organizations in the brain. We then explore the advantages and specific features of modular networks, where we simplify the discussions by considering basic modular networks that consist of combinations of some of the basic networks discussed in the previous chapters. These include the mixture of experts, the product of experts, and coupled attractor networks. We will then outline as examples of system level models three specific models of brain functions, namely one of working memory, one of visual object recognition that takes attention into account, and one of the Stroop task. The latter is a specific example of a more general framework that proposes a neural workspace for a flexible recruitment of processors in a modular network.

11.1 System level anatomy of the brain

In the previous chapters we reviewed some important biochemical mechanisms for specific information processing in single neurons and discussed various principal architectures of neural networks that can be used in the brain. These are important building blocks of the nervous system. The nature of these information-processing principles can explain some characteristics of mental processes as we have seen in various examples. However, the brain is more than just a big neural network with completely interconnected neurons as we must recognize that it displays a lot of structure. This alludes to specific information-processing mechanisms for specific tasks in the brain, which we want to explore. It is clear that we have to combine the basic networks discussed in the previous chapters, such as associative and competitive networks, into more global architectures reflecting large-scale organizations of the brain if we want to understand how specific tasks are solved by the brain. The control systems in the last chapter are examples of such global organizations. In this chapter we concentrate on a more formal description of modular networks resulting from combining the basic networks discussed in the previous chapters, and we give some specific examples of simplified system level models of specific brain functions.

We consider modular networks generally to be networks that display some structure within their architecture as opposed to completely interconnected networks. Such networks can be viewed as special cases of completely interconnected networks with small or vanishing weight values between classes of nodes. Within this view we can consider modular networks as large-scale networks with constraints. Having constraints is biologically realistic and it is therefore important to understand how large-scale networks behave under specific constraints. An example of a purely physical constraint is the following. Consider a completely interconnected network with a number of elements (nodes) of the order of the neurons in the brain (say conservatively 10^{11}) and with individual interconnecting axons of 0.1 μm radius. Placing the nodes on the surface of a sphere and allowing the interior of the sphere to be densely packed with the axons would result in a sphere with a diameter of more than 20 km. The constraints of the physical extent of interconnecting 'wires' therefore demands some more economic solutions. We will see in this chapter that there are many important advantages of modular structures within large systems.

11.1.1 Large-scale anatomical and functional organization in the brain

The brain displays organization on many levels as mentioned in Chapter 1 (see Fig. 1.1). We discussed some of the connectivity patterns in the cortex in Chapter 4, and throughout the book have mentioned several examples of specialized functions in which specific brain areas display adequate structural components. For example, in the previous chapter we saw that the cerebellum is well equipped with neuronal organizations that can be used in the control of motor actions, and we have seen in Chapter 7 that the CA3 region of the hippocampus has many collateral connections that can be used to store auto-associative memories. It is known that there are anatomical and functional organizations on much larger scales in the brain, such as the visual system, which takes up most of the primate's brain. Some large-scale brain organizations can be revealed by brain imaging techniques such as *positron emission tomography* (PET) and *functional magnetic resonance imaging* fMRI, which can highlight what areas are involved in certain mental tasks, and such studies established clearly that different brain areas do not work in isolation. On the contrary, many specialized brain areas have to work together to solve complex mental tasks.

In order to understand how different brain areas work together we have to establish the anatomical and functional connectivity between brain areas in more detail. Anatomical connections are not easy to establish as it is extremely difficult to follow the path of a stained axon through the brain in brain-slices (including the branches that can often have different pathways). This seems indeed to be a daunting task, though it has been done in isolated cases. There are other methods of establishing connectivities in the brain. These include the use of chemical substances that are transported by the neurons to target areas or from target areas to the origin as mentioned in Chapter 4. Functional connectivity patterns, in which we are particularly interested when studying how brain areas work together, can also be established with simultaneous stimulations and recordings in different brain areas. Such experiments show correlations in the firing patterns of neurons in different brain areas if they are functionally connected.

The modularity of the brain is already indicated by electrophysiological recordings, which show specific response properties of neurons in different areas, and by functional brain imaging, which shows that different brain areas are active in different mental tasks. A large database of experimental results has been studied by *Claus-C Hilgetag*, *Mark A. O'Neill*, and *Malcolm P. Young* following the earlier examples of various groups. In contrast to earlier work they considered an algorithm that would evaluate many possible configurations, and they found a large set of possible connectivity patterns in the visual cortex satisfying most of the experimental constraints. An example is shown in Fig. 11.1. Interestingly, most solutions of this optimization problem displayed some consistent hierarchical structures. All solutions found violated some of the experimental constraints (dashed line in Fig. 11.1), which is probably based on the inaccuracy of some of the experimental results.

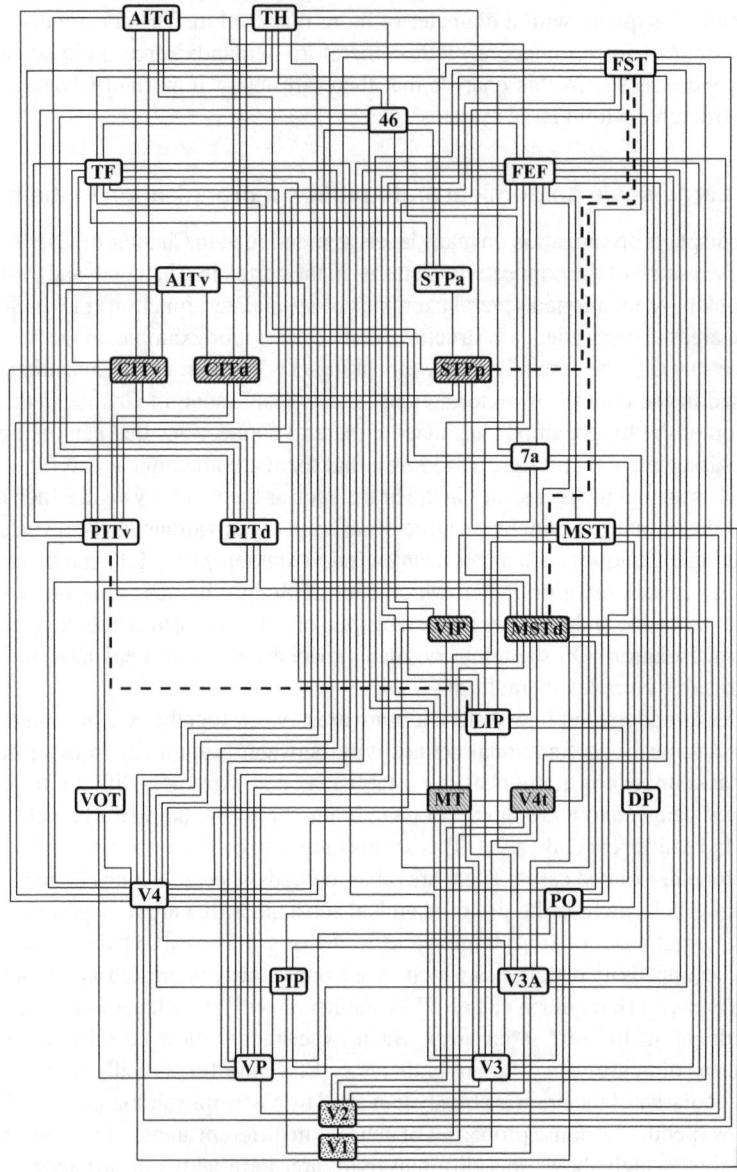

Fig. 11.1 Example of a map of connectivities between cortical areas involved in visual processing [reprinted with permissions from Claus-C. Hilgetag, Mark A. O'Neill, and Malcolm P. Young, *Phil. Trans. R. Soc. Lond.* B355: 71–89 (2000)].

Each box in Fig. 11.1 represents a cortical area that has been distinguished from other areas on different grounds, typically anatomical and functional ones. The solid pathways between these boxes represent known anatomical or functional connections. The order from bottom to the top indicates roughly the hierarchical order in which these brain areas are contacted in the information-processing stream, from primary visual areas establishing some basic representations in the brain to higher cortical areas that are involved in object recognition and the planning and execution of motor actions. The authors also took two basic visual processing pathways into account, the *ventral visual pathway* running from the occipital lobe to the temporal lobe (with brain areas shown on the left side) and the *dorsal visual pathway* running from the occipital lobe to the temporal lobe (with brain areas shown on the right side). We will comment on these pathways later in this chapter. But note that there are also interactions within these pathways.

Note that the connections indicated are not unidirectional. It is well established that a brain area that sends axon to another brain area also receives back-projections from the structures it sends to. Such back-projections are often in the same order of magnitude as the forward projections. Interesting examples, not included in Fig. 11.1, are so-called *corticothalamic loops*. The subcortical structure called the *thalamus* was initially viewed as the major relay station through which sensory information projects to the cortex. However, it is becoming increasingly clear that the notion of a pure relay station is too simple as there are generally many more *back-projections* from the cortex to the thalamus compared to the forward projections between the thalamus and the cortex. Some estimates even indicate a number of back-projections that exceed the forward projections tenfold. The specific functional consequences of back-projections between the thalamus and the cortex as well as within the cortex itself are still not well understood. However, such structural features are consistent with reports of the influence of higher cortical areas on cell activities in primary sensory areas, for example, attentional effects in V1.

Two final remarks on cortical maps such as the one illustrated in Fig. 11.1. The connections between brain modules are simply indicated with lines in such figures. However, note that the nature of the connections can be rather complicated. For example, the connections could be specific to certain neurons in the receiving structure, or there could be different types of functional connections between brain areas such as topographic or non-topographic mappings. The wiring diagram shown as an example in Fig. 11.1 is therefore only limited to a few connections to give an impression of the modularity in the brain. Also, keep in mind that the visual system illustrated in Fig. 11.1 interacts with other areas that are not only dedicated to visual processing.

11.1.2 Advantages of modular organizations

In the light of our earlier discussion of completely connected neural networks as universal function approximators (see Chapter 6) it is obvious to ask why modular specialization is used in the brain. Some suggestions of the functional significance of modular specialization in visual processing have been outlined by the British psychologist *Alan Covey*. He speculated that, if the cortex uses inhibition to sharpen various visual attributes, such as colour, edges, or orientations, then it is possible to use local inhibition, which can be implemented with inhibitory interneurons within retinotopic maps, as long as the different attributes are kept separate. Such a specialization would also allow the separate attentional amplifications of separate features. Many other advantages of task decompositions and modular architectures have been outlined in the literature. These include advantages in learning speed, generalization abilities, representation capabilities and task realizations within hardware limitations. We will explore some examples that illuminate the reasons behind such advantages in this chapter.

Several ideas about modular architectures and the way in which modules can interact and organize themselves have been discussed in the literature. For example, *Marvin Minsky* discussed some of his thoughts in his book, *The society of mind*. Another example is the *committee machine* first described by *N.J. Nilson* in 1965, and the *pandemonium* suggested by *O.G. Selfridge* in 1958. To get a deeper understanding of such information-processing architectures within the neural network framework we have to study how neural networks interact. We will thus review some properties of *modular networks* that are comprised of some of the basic networks we discussed in previous chapters. The whole system can still be called a neural network, but the fact that we can distinguish subsystems in those networks makes them modular for our discussion's purposes. The emphasis in the following discussions is therefore on the interaction between basic neural networks such as *modular mapping networks* and *modular attractor networks*. To understand specific mental abilities of the brain we must, of course, study much more specific modular architectures reflecting actual brain organization. Some examples will be given in the second half of this chapter, while the first half is dedicated to highlighting some general properties and advantages of modular networks.

11.2 Modular mapping networks

11.2.1 Mixture of experts

We can think of many possible architectures for modular networks, and several specific designs have been discussed in the literature. We will concentrate here on some fundamental examples that can give us a feeling for specific properties of modular networks. In this section we start by combining feed-forward mapping networks in various ways. An example of such an architecture, called the *mixture of experts*, is illustrated in Fig. 11.2. In this architecture we have a column of parallel working modules or *experts*. Each of these experts receives potentially the same input. Another module, called the *gating network*, also receives some input from the general input stream and has at its purpose weighting the outputs of the expert networks. The outputs of the expert networks are then combined by the *integration network*, which sums the weighted output of the experts to form the output vector of the system. Note that the integration network can be represented by sigma–pi nodes because the output of the gating network has modulatory effects on the inputs to the integration network.

11.2.2 Divide-and-conquer

What are the benefits of using such an approach rather than using a single large mapping network, especially as we know that a large mapping network is a universal

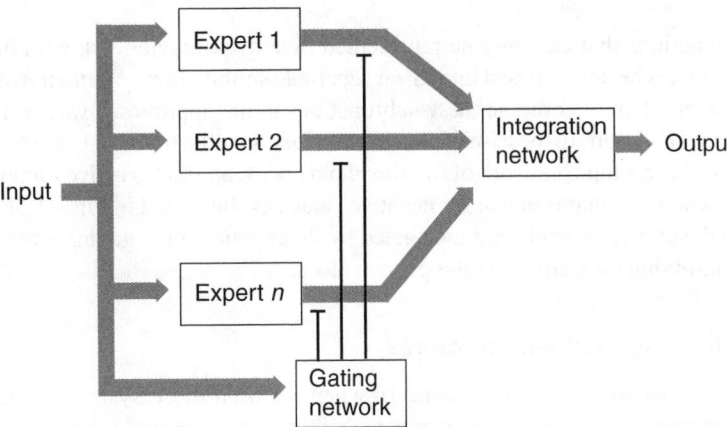

Fig. 11.2 An example of a type of modular mapping network called mixture of experts. Each expert, the gating network, and the integration network are usually mapping networks. The input layer of the integration network is composed of sigma–pi nodes, as the output of the gating network weights (modulates) the output of the expert networks to form the inputs of the integration network.

function approximator and can thus solve any mapping task (in principle). An example can illustrate some of the benefits of such a modular architecture. In Fig. 11.3A we plotted the absolute function $f(x) = |x|$. It is difficult to approximate this piecewise linear function with a single mapping network due to the sharp discontinuity at $x = 0$. However, a single linear node can represent each branch of the function. By dividing the problem into these subproblems we can employ a simple solution for the subproblems.

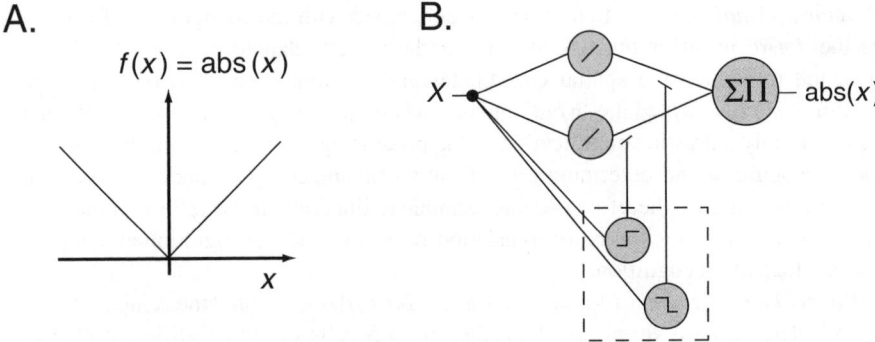

Fig. 11.3 (A) Illustration of the absolute function $f(x) = |x|$ that is difficult to approximate with a single mapping network. (B) A modular mapping network in the form of a mixture of experts that can represent the absolute function accurately.

This modular network is an example of the *divide-and-conquer* strategy often used in technical applications, which seems also to be a strategy that we use frequently in our daily lives in order to solve complex problems. By dividing a problem into subproblems it is possible that each subproblem can be solved by simple means. For example, functions that can only be represented by a mapping network with hidden nodes can always be decomposed into linear separable subfunctions. We then still have to solve the problem to combine these subfunctions in the appropriate way, which is often not trivial in itself. In the case of the absolute function illustrated in Fig. 11.3A we can solve this by a simple network of two threshold nodes, one that is active for positive input values and one that is active for negative values as illustrated in Fig. 11.3B. The outputs of the expert networks are then gated by the outputs of the gating network by simply multiplying the outputs by the gating values.

11.2.3 Training modular structures

Training such networks is a major concern when we want those systems to be able to solve specific tasks in a flexible manner. Training the experts alone now has two components: one is to assign the experts to particular tasks, and the other is to train each expert on the designated task. However, even with trained experts we still have to train the gating network, which is a form of credit-assignment problem. Note that the division of the training into a task assignment phase and an expert training phase may be useful in solving the training problem, but in biological systems we might not be able to divide the learning into these separate steps. We will see below an example where the different steps are indeed combined. Several training methods have been proposed in the neural network and machine learning literature. We will not discuss these here in detail as their relevance for brain processes is still not clear. Instead we will follow an instructive example.

11.2.4 The 'what-and-where' task

Several studies have shown that the brain has two partly distinguished visual pathways, the *ventral visual pathway* that is mainly concerned with the recognition of objects and the *dorsal visual pathway* that is particularly well adapted to selecting objects for action for which the spatial coordinates are important. The two pathways are sometimes simply termed the '*what*' and the '*where*' pathways, respectively, although this is certainly a drastic simplification of the processing within the brain. Performing object recognition and determining the location of objects in a single network are difficult tasks, and we therefore use this example to illustrate the benefits of a modular network with separate experts for translation-invariant object recognition and object-invariant location recognition.

Robert Jacob, *Michael Jordan*, and *Andrew Barto* demonstrated the principal ideas in 1991. They used a simple model retina of 5×5 cells on which different objects represented by different 3×3 patterns can be placed in nine different locations (see the example in Fig. 11.4A). A network that can solve this 'what-and-where' task can be designed to have 26 input channels, consisting of 25 for the retinal input and 1 for the task specification. We can represent the output with 18 nodes, the first 9 to specify 9 objects and the second 9 specify the locations. The representation is therefore local, that is, only one node is active at any time (grandmother cells). *Jacobs* and co-workers have demonstrated that a single network with a hidden layer of 36 nodes can solve the task after training the multilayer mapping network with back-propagation learning. However, the authors also demonstrated improved performance with modular networks

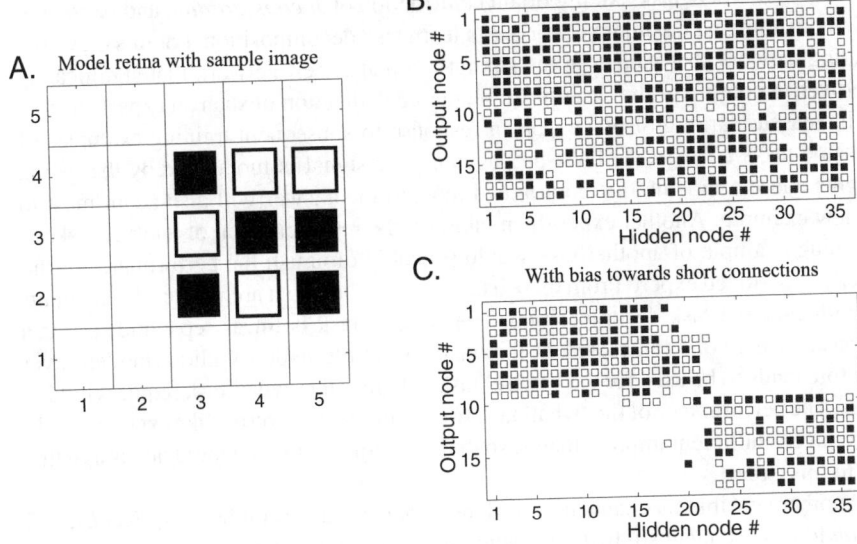

Fig. 11.4 Example of the 'what-and-where' tasks. (A) 5×5 model retina with 3×3 image of an object as an example. (B) Connection weights between hidden and output nodes in a single mapping network. Positive weights are shown by solid squares, while open squares symbolize negative values. (C) Connection weights between hidden and output nodes in a single mapping network when trained with a bias towards short connections [adapted from Jacobs and Jordan, *J. Cogn. Neurosci.* 4: 323–36, 1992].

for the reasons summarized next. Note that the use of back-propagation learning in this circumstance is acceptable as it is only used to find an example of a network that can solve the problem.

11.2.5 Temporal and spatial cross-talk

A general problem that diminishes learning in single networks is conflicting training information. Conflicting training information can cause different problems. For example, the network will quickly adapt to reasonable performances of the 'what' task if we train the network first entirely on this task. The representations of the hidden layers will however change in a subsequent learning period on the 'where' task, which is likely to conflict with the representation necessary for the 'what' task. This *temporal cross-talk* is generally a problem in training sets with conflicting training pattern. In addition, there can be conflicting situations within one training example due to the distributed nature of the representations. This is called *spatial cross-talk*. The division of the tasks into separate networks can abolish both problematic cross-talk conflicts.

11.2.6 Task decomposition

The above discussion suggests that modular networks have important merits, though we haven't yet solved the problem of establishing a modular network that can solve the what-and-where task. An important contribution of *Jacobs*, *Jordan*, and *Barto* was to demonstrate that modular networks can learn task decomposition. For this they used a gating network that increased the strength to that expert network that significantly improved the output of the system. An increased dedication of such an expert ensured the increased learning of this expert in response to subsequent training examples of the same tasks because the back-propagated error signal is modulated by the gating weights and the module that contributed most to the answer will also adapt most to the new example. Another expert then takes on the representation of another task, as a training example of another task would probably diminish the performance of the already specialized expert. From these examples it is clear that architectural constraints can influence the task decomposition. The 'where' task is linear separable so that a single-layer network can represent this task. A simple expert without hidden layer therefore tends to be used for the 'where' task, whereas the 'what' task requires hidden nodes. The performance of the 'what' task with a simple perceptron-like expert is likely to produce insufficient improvements, so that the gating network would not assign this task to this expert.

Another possible mechanism for task decomposition was outlined by *Jacobs*, and *Jordan* in a subsequent study. In this study they considered a physical location of the nodes in a single mapping network and used a distance-dependent term in the objective function, which leads to a weight decay favouring short connections. The multilayer network was trained on the objective (or error) function

$$E = \frac{1}{2} \sum_i (r_i^{\text{out}} - y_i)^2 + \lambda \sum_{ij} \frac{d_{ij} w_{ij}^2}{1 + w_{ij}^2}. \tag{11.1}$$

The first term is the common mean square error between the actual and desired output of the network (as in eqn 10.1), which enforces the correct functioning of the network. The second term is yet another weight decay term that makes solutions with large weight values between distant nodes unfavourable. Figure 11.4B and 11.4C indicates the connections between the hidden nodes and the output nodes trained on the what-and-where task as outlined above with a single 25–36–18 feed-forward mapping network. In Fig. 11.4B the network was trained entirely on the MSE objective function (that is, $\lambda = 0$), whereas Fig. 11.4C shows an example of the results when trained with the bias term that enforces short connections. As can be seen, the connections from the hidden layers to the output nodes, and therefore the hidden nodes themselves, were separated in relation to the two separate tasks. More hidden nodes were thereby allocated to the more difficult 'what' task. A learning procedure with such physical constraints therefore leads to a modularization of the network as demonstrated in the example.

11.2.7 Product of experts

We mentioned in Chapter 6 that the normalized output of a feed-forward mapping network with a competitive output layer can be interpreted as the probability that an input vector has a certain feature or belongs to a certain class symbolized by the output node. The previously discussed modular networks can easily be generalized to allow similar interpretations if the summed outputs of the expert networks are normalized.

We can thus also view the mixture of experts as a collection of experts whose weighted opinion is averaged (added and re-normalized) to determine the probability of the feature value. However, if we are considering probabilities it is much more natural to consider products of probabilities that would determine the joined probability of independent events. This was pointed out by *Geoffrey Hinton*, and he proposed an architecture that he named *product of experts*. He not only stressed some of the advantages of such networks but also alluded to their biological plausibility when compared to empirical data.

Hinton gave an illustrative example that illuminates the advantages of a product of experts relative to the weighted mean calculated by the mixture of experts. Consider the recognition of a face. Forensic scientists know too well that many features are necessary to determine a face uniquely. Face recognition is therefore a high-dimensional problem (which is typical of mental processes that have to be solved by the brain). A weighted sum of experts typically results in a wide distribution function that would not determine a particular face with much certainty. A product of experts can produce much sharper probability functions and more closely resembles face recognition abilities in humans. The reason for this is that even a large probability assigned by one expert can be largely suppressed by low probabilities assigned by other experts. A large probability is only indicated if there is some agreement between the experts. This scheme allows the individual experts to assign unreasonably large probabilities to some event as long as other experts represent such events more accurately with low probabilities. [30]

How to train products of experts is not obvious, and some research is currently being pursued in this area. Hinton and colleagues have already proposed generalized gradient descent methods for training such networks. It is likely that advanced applications will soon be developed using such networks, and further research in relating such methods to brain functions will hopefully be pursued.

11.3 Coupled attractor networks

In the examples of modular networks of the previous section we coupled mapping networks to an overall feed-forward scheme. In this section we will discuss the combination of recurrent networks. We can also view such a system as a subdivided recurrent network. An example is illustrated in Fig. 11.5 where we have divided the nodes in the whole network into two groups. We can thus distinguish between connections (or weights) between nodes in the same group and connections (or weights) between nodes of different groups. The whole system is, of course, still a recurrent network and thus basically still one large dynamic system. However, it is very enlightening to study the behaviour of such systems with strongly coupled subsystems and weak interaction between them as we will see in this section. Note that neither the nodes within the groups nor the nodes between the nodes have to be completely connected. However, we will simplify the discussion by considering fully connected attractor networks of the types discussed in Chapter 8.

---

[30] Training a network to faithfully represent low probabilities is difficult when learning from examples as low probabilities correspond to rare events. With the mixtures of experts we could allow some experts to concentrate on rare events and even allow them to assign to them relatively large (even overestimated) probabilities. Other experts can then adjust for the magnitudes of the rare events.

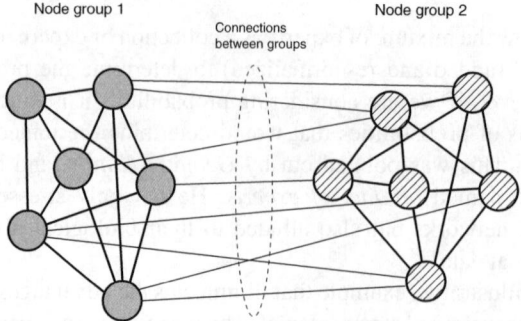

Fig. 11.5 Coupled (or subdivided) recurrent neural networks. The nodes in this example are divided into two groups (the nodes of each group are indicated with different shadings). There are connections within the nodes of each group and between nodes of different groups.

11.3.1 Imprinted and composite patterns

Let us first compare the extremes of coupled attractor networks, one single point attractor network as discussed in Chapter 8 versus two single point attractor networks. As an example we want to imprint into this system all objects that can be described by two independent feature vectors representing, for example, shape and colour. We consider therefore three different shapes {*circle, square, triangle*} and three different colours {*red, blue, green*}. The possible training set should consist of all nine possible combinations {*red circle, blue circle, ...*}. In general, if we have m possible feature values for each feature of an object we can build m^2 possible objects as long as there are no correlations between the features that would make certain combinations more likely than others. A network with 1000 nodes can store around 138 patterns (feature values) when random representations of the feature values are trained with the Hebbian rule for which the storage capacity was found to be $\alpha_c \approx 0.138$ (see Chapter 8). A network with two such independent subnetworks, each carrying the information of one feature, could store

$$P = (\alpha_c N/m)^m = 138^2 = 19,044 \qquad (11.2)$$

objects represented by all possible feature combinations, where m is the number of features or the number of modules with an equal number of nodes N/m. The number of patterns that can be stored in a single network is only

$$P = \alpha_c N. \qquad (11.3)$$

To store all the 19,044 possible objects in one attractor network we need a network with at least $N = 19,044/\alpha_c = 138000$ nodes. The saving of resources of the modular network in terms of nodes is very impressive (2000 versus 138,000), and the savings are even more impressive if we consider the number of weights we have to use ($1000^2 + 1000^2 = 10^6$ versus $138,000^2 = 210^{10}$).

Why then would we ever want to use large single networks? The answer is that we have to use them if we want to represent correlations of features and remember specific objects. For example, if we want to remember a *green square* and a *blue triangle*, then we have to imprint these two objects with the specific combination of the features, and not the other possible combinations. If we do this we can recall the objects from partial information; for example, *green* can trigger the memory of *green square*. In the case of separate networks the input of *green* cannot trigger any memory in the 'shape' network, and we would potentially recall all possible combinations of features even if they were not part of the set of objects we used for training.

11.3.2 Signal-to-noise analysis

The previous discussion suggests that a certain intermediate level of coupling between attractor neural networks is likely to be necessary for more complex (and realistic) processing in the brain. We can get some further insight into the behaviour of coupled attractor networks using a signal-to-noise analysis along the lines of that outlined in Chapter 8. We consider an overall network of N nodes that can be divided into

$$m = \frac{N}{N'} \tag{11.4}$$

modules, each having the same number of nodes N'. The weights are trained with the Hebbian rule

$$w_{ij} = \frac{1}{N} \sum_{\mu=1}^{P} \xi_i^\mu \xi_j^\mu. \tag{11.5}$$

However, we modulate the weight values between the modules with a factor g. Formally, we can define a new weight matrix with components

$$\tilde{w}_{ij} = g_{ij} w_{ij}. \tag{11.6}$$

The components of a modulation matrix **g** are given by

$$g_{ij} = \begin{cases} 1 & \text{if nodes } i, j \text{ are within the same module} \\ g & \text{otherwise} \end{cases}. \tag{11.7}$$

To shorten the statement 'if nodes i, j are within the same module' we can define a set $m(i)$ that lists all the node numbers that are in the same module as node number i.

11.3.3 Imprinted pattern

The next step in the signal-to-noise analysis is to evaluate the stability of the imprinted pattern. First we will evaluate an imprinted pattern that is made up of the subvectors of each module that were imprinted together. Without loss of generality we can evaluate again the first pattern $\mu = 1$, that is, $s_i(t) = \xi_i^1$. We can then separate the signal terms from the noise terms in the updating rule as outlined in Chapter 8. This gives

$$s_i(t+1) = \text{sign}(\frac{N'}{N}\xi_i^1 + g\frac{N-N'}{N}\xi_i^1 + \frac{1}{N}\sum_{j\in m(i)}\sum_{\mu\neq 1}\xi_i^\mu\xi_j^\mu\xi_j^1 + \frac{g}{N}\sum_{j\notin m(i)}\sum_{\mu\neq 1}\xi_i^\mu\xi_j^\mu\xi_j^1). \tag{11.8}$$

From this we can identify the signal and noise terms. The strength S of the signal term is the factor in front of the desired signal ξ_i^1, and the variance of the noise term is determined by the remaining part analogously to the procedure described in Chapter 8. In summary we get

$$\text{signal: } S = \frac{N'}{N} + g\frac{N-N'}{N} = \frac{1}{m} + g(1 - \frac{1}{m})$$

$$\text{noise: } \sigma = \sqrt{\frac{P-1}{N}}\sqrt{\frac{1}{m} + g^2(1 - \frac{1}{m})}$$

$$\approx \sqrt{\frac{P}{N}}\sqrt{\frac{1}{m} + g^2(1 - \frac{1}{m})} = \sqrt{\alpha z_1}$$

$$(11.9)$$

We introduced in the last line a shorthand notation z_1 that encapsulates the difference between the modular networks ($m > 1$) and the standard single network ($m = 1$) for which $z_1 = 1$. The parameter α is again the load parameter defined in eqn 8.12. Further discussions on the load capacity depend again on the error bound we impose on the network (see eqn 8.13). To simplify the formulas we introduce the shorthand notation,

$$z_2 = [\text{erf}^{-1}(1 - 2P_{\text{error}})]^2. \tag{11.10}$$

With these terms we can write the capacity bound (eqn 8.14), generalized to modular networks started in a trained pattern, as

$$\alpha < \frac{S^2}{2z_1 z_2}. \tag{11.11}$$

We called the limiting case the critical load parameter or the storage capacity, α_c, which is plotted for $P_{\text{error}} = 0.001$ and different values of g and m in Fig. 11.6A. With only one module we, of course, get the same results as in Chapter 8. However, with more than one module ($m > 1$) the results depend on the relative strength of the intramodular weights relative to the intermodular weights that we parametrized with g. With values of $g < 1$ we get less support from the signals of the other modules so that the load capacity of the networks becomes diminished.

11.3.4 Composite pattern

In contrast to the previous case we can start the network from states that correspond to different subpatterns in the different modules. We can consider all kinds of combinations of patterns in different modules, but for the following discussion there are two most interesting and limiting cases. The first one occurs when all the subpatterns in the modules are chosen to be the first training pattern except for the module to which the node under consideration belongs. For this module we choose a random initial state and ask how strong the intermodular coupling strength has to be in order for the other modules to trigger the corresponding state in the module with the random initial state. We thus study the network with the starting state

$$s_j(t) = \begin{cases} \eta_j & \text{for } j \in m(i) \\ \xi_j^1 & \text{otherwise} \end{cases}, \tag{11.12}$$

where the variable η_j is a random binary number. The state of the network after one update is then

$$s_i(t+1) = \text{sign}(g\frac{N-N'}{N}\xi_i^1 + \frac{1}{N}\xi_i^i \sum_{j \in m(i)} \xi_j^1 \eta_j + \text{other noise-terms for } \mu \neq 1). \tag{11.13}$$

The corresponding strength of the signal is determined by the first factor. The second term is an additional noise term that has to be added to the noise term of the previous case (eqn 11.9). However, this noise term contributes only with $1/(mN)$ and, since we consider large networks, we can ignore this small contribution. In summary we get

$$\text{signal: } S = \frac{N-N'}{N}g = (1 - \frac{1}{m})g$$
$$\text{noise: } \sigma \approx \sqrt{\alpha z_1} \qquad (11.14)$$

We can insert these expressions into eqn 11.11 from which we can derive a lower bound on the relative intracoupling strength g (for a given number of modules m and load parameter α) that we have to impose if we want to be able to trigger the first pattern in the module that we started in a random state. This is given by

$$g^2 > \frac{2\alpha z_2}{(m-1)(1 - \frac{1}{m} - 2\alpha z_2)}. \qquad (11.15)$$

This lower limit on g is plotted in Fig. 11.6B as the line restricting the white area from below (for $P_{\text{error}} = 0.001$ and $\alpha = 0.01$). The curve confirms the intuitive conjecture that the g-factor can be reasonably small if the number of modules increases because an increasing number of modules results in an increasing relative number of nodes that are pointing in the desired direction.

A.   Load capacity   B.   Bounds on intermodular strength

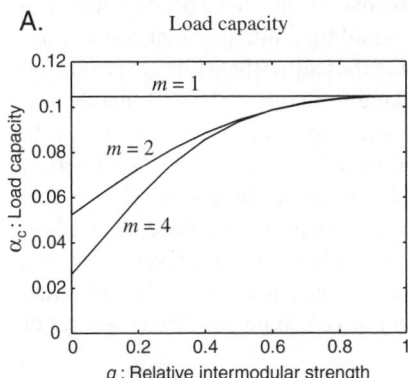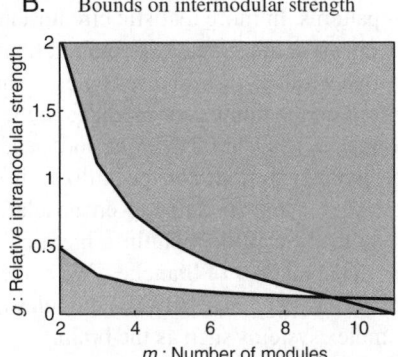

Fig. 11.6 Coupled attractor neural networks: results from signal-to-noise analysis. (A) Dependence of the load capacity of the imprinted pattern on relative intermodule coupling strength g for different numbers of modules m. (B) Bounds on relative intermodule coupling strength g. For g values greater than the upper curve the imprinted patterns are stable. For g less then the lower curve the composite patterns are stable. In the narrow band in between we can adjust the system to have several composite and some imprinted patterns stable. This band gets narrower as the number of modules m increases and vanishes for networks with many modules.

We can also study the reverse case with

$$s_j(t) = \begin{cases} \xi_j^1 & \text{for } j \in m(i) \\ \eta_j & \text{otherwise} \end{cases} \qquad (11.16)$$

as starting state. With this starting state we can ask how low the g-factor has to be in order for the substate ξ^1 in one module to be stable while all the other modules are pulling in other directions. The state of the network after one update with this starting state is

$$s_i(t+1) = \text{sign}(\frac{N'}{N}\xi_i^1 + \frac{g}{N}\xi_i^i \sum_{j \notin m(i)} \xi_j^1 \eta_j + \text{other noise-terms for } \mu \neq 1). \quad (11.17)$$

We can again neglect the additional noise term and use

$$\begin{aligned} \text{signal: } S &= \frac{N'}{N} = \frac{1}{m} \\ \text{noise: } \sigma &\approx \sqrt{\alpha z_1} \end{aligned} \quad (11.18)$$

in the signal-to-noise analysis. Thus, if we want to be able to keep a state of a module stable (free from interferences from the other modules) the g-factor has to obey

$$g^2 < \frac{1 - 2\alpha z_2 m}{2\alpha z_2 m(m-1)}. \quad (11.19)$$

The upper limit on g from these considerations is plotted in Fig. 11.6B as the line restricting the white area from above. With g-factors somewhere in the white area we could find points where patterns in different modules are relatively stable while still responding to some influence from the other patterns.

In the analysis we have assumed that intermodular weights have been trained with all patterns. In more realistic circumstances we would train intermodular connections for only specific subpattern combinations and could thus allow the white area to extend to higher values of m. Nevertheless, what is striking from our analysis is that there is a limit on the number of modules in the system when we want to have a system with a sensible balance of modular independence and modular interactions. This restricts the modularity that can be utilized in complex systems at each level. To be able to utilize many modules to encapsulate certain functionalities we have to combine these many modules within a hierarchical structure where on each level we have a restricted number of branches. We conjecture that the sensible balance of cooperation versus independence between modules is a driving force behind possible structures of complex systems such as the brain.

11.4 Working memory

We will now start to outline some specific examples of system level models of brain functions with a particular view to modular organizations in the brain. We start with a specific hypothesis on the implementation of working memory in the brain. Working memory is a construct that is hard to define precisely, and many different views of this concept in terms of definition and functional roles exist in the literature. The reason that it appears often in cognitive science literature is that it is a very appealing concept when speculating about complex cognitive processes that seem to rely on some form of workspace that provides the necessary information to solve complex tasks. Examples of such complex cognitive tasks include language comprehension, mental arithmetic, and all forms of reasoning that are necessary for problem-solving and decision-making. Working memory is different from the specific form of short-term memory (STM) discussed in Chapter 8 because many complex cognitive tasks also rely on intermediate-term memory (ITM) as, for example, provided by hippocampal structures and even on long-term memory (LTM), for example, some motor skills that enable us to perform specific tasks.

11.4.1 Distributed model of working memory

We outline here a conceptual model of working memory proposed by *Randall O'Reilly*, *Todd Braver*, and *Jonathan Cohen*, which is based on a modular structure as summarized in Fig. 11.7. Three modules are included in this model, each having some characteristic functionalities. It is often difficult to distribute specific functionalities in a distributed system as there are interactions within the system and the system properties frequently rely on the whole system. Nevertheless, what we have in mind here is to assign to each module a specific memory functionality in terms of short-, intermediate-, and long-term memory. The module labelled PFC has many (mainly) independent recurrent subsystems represented as single recurrent nodes. Each such node could hold a specific, previously presented, item of information over a short period of time through reverberating neural activity. This module mimics basic operations found in the *prefrontal cortex*, which we associated with short-term memory in Chapter 8. The structure labelled HCMP represents structures such as the *hippocampus* that we identified as suitable for rapid learning of associations as required for episodic memory (see Chapter 8). This form of memory is weight-based and can be thought of as intermediate-term memory when compared with the long-term memory component of the last module labelled PMC. This label stands for *perceptual and motor cortex*, and this module is included because such systems are necessary to solve many complex tasks. These include basic abilities such as performing necessary motor actions. Such skills are thought to be formed through slow learning and are imprinted deeply into the associated cortical areas (for example, riding a bike).

O'Reilly and collaborators used this conceptual model to highlight how the interactions of these modules enable the brain to utilize the kind of information that is necessary to perform complex mental tasks. From the distributed nature of this form of working memory it is clear that different mental tasks can use the different components of working memory with various strength, while it is the interaction between the modules that enables the use of memory in a way that would not be possible with isolated processors. The authors demonstrated the use of such a memory system with some simulations, which we recommend as further reading. Here we will only outline a specific issue surrounding working memory that is repeatedly discussed in the literature.

11.4.2 Limited capacity of working memory

It is easy to remember very quickly a small list of numbers such as '31, 27, 4, 18'. However, the ability to recall a list of numbers (without repeated learning) brakes down drastically when the list has a few more items. Try '62, 97, 12, 73, 27, 54, 8'. The memory required to perform such a task depends strongly on short-term memory mechanisms. However, in light of the discussion above we think of them more as limitations of working memory. The limited capacity of working memory was

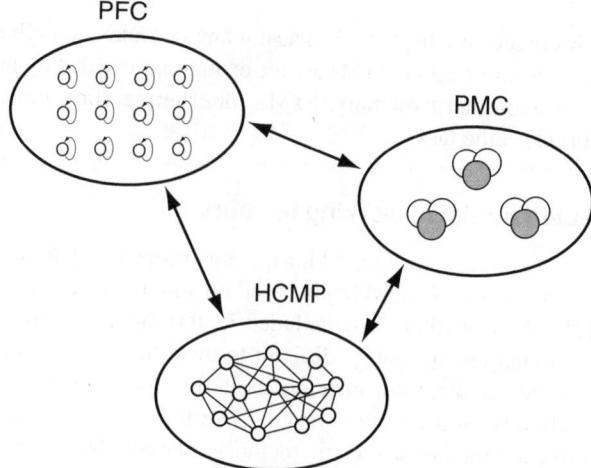

Fig. 11.7 A modular system of short-, intermediate-, and long-term memory, which are associated with functionalities of the prefrontal cortex (PFC), the hippocampus and related areas (HCMP), and the perceptual and motor cortex (PMC), respectively [adapted from O'Reilly, Braver, and Cohen, in *Models of working memory*, Miyake and Shah (eds.), Cambridge University Press, 1999].

described in 1928 by *H.S. Oberley. G. Miller* coined the phrase 'magical number 7 ± 2' for this limit. It is now clear that the precise limit depends strongly on the specific task that has to be performed, and the apparent limit can be modified by other processes. For example, a waiter is often highly skilled in remembering quickly long lists of orders. This is made possible by utilizing 'mental tricks' such as grouping items together, and the severe limit on the capacity of basic working memory is still apparent when controlling such possibilities in experiments.

The very limited capacity of working memory is puzzling if we remember that we can easily build systems that can rapidly store many thousand of items. It has been shown that the individual capacity of working memory is strongly correlated to such things as success in completing courses, indeed it is more strongly correlated than classical measurements of IQ factors. Hence, a larger storage capacity should make us fitter to survive and we can ask why evolution was not able to increase the working memory capacity considerably. The search for the reasons behind the limited capacity of working memory is therefore prominent in cognitive neuroscience.

11.4.3 Spurious synchronization

Many suggestions have been made in the literature for the reasons behind the limited capacity of working memory, and we will only discuss one suggestion as it is possible to quantify this suggestion using a short calculation and compare it to experimental data. Experimental results on the performance of a typical working memory task are shown in Fig. 11.8. In the corresponding experiments, conducted by *Steven Luck* and *Ed Vogel*, an image with N^{obj} simple objects such as coloured squares was followed by a second image with coloured squares after a 900 ms delay period, and the subjects had to judge if the two images were identical or not. The performance of the subjects, as measured by the average percentage of correct responses, is plotted in Fig. 11.8A with solid squares as function of the number of objects (squares) in the images. A sharp breakdown of the performance with an increasing number of objects in the image can be seen. With similar experiments the authors demonstrated in addition that: (1) the capacity limit does not depend on the number of features of objects relevant for making the decision; (2) the visual working memory is independent of the load in the verbal working memory; (3) the capacity limit is not due to an increase of the number of decisions that have to be made for an increasing sample set.

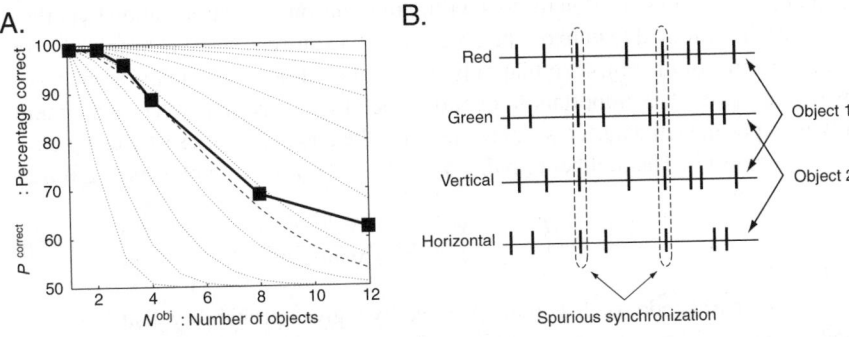

Fig. 11.8 (A) The percentage of correct responses (recall ability) in human subjects in a sequential comparison task of two images with different numbers of objects N^{obj} (solid squares). The dotted lines illustrate examples of the functional form as suggested by the synchronization hypothesis. The dashed line corresponds to the results with $P^{SS}(2) = 0.04$, where P^{SS} is the probability of spurious synchronization. (B) Illustration of the spurious synchronization hypothesis. The features of the object are represented by different spike trains, so that the number of synchronous spikes within a certain resolution increases with increasing an number of objects [adapted from Luck and Vogel, *Trends Cogni. Scie* 2: 78–80, 1998].

Steven Luck and Ed Vogel have suggested a possible reason for the limited capacity of working memory. Their idea is based on the assumption that a feature is represented by a particular spike train[31] as illustrated in Fig. 11.8B. We assume that a feature is coded by a spike train with an arbitrary code (that is, a random spike train), and that the decoding of the spike trains is limited by a finite resolution of the spike times (finite time windows). The different features of different objects could be represented by *synchronized spike trains*. This is a popular way of solving the *binding problem*, the problem of relating the different features of one object when those features are represented in different brain areas. Of course, we have to assume that different objects are represented by different spike trains. However, it is possible that some spikes in the different spike trains for the different objects fall within the same time window as illustrated in Fig. 11.8B. As this is based purely on random coincidences we call this type of synchronization *spurious synchronization*. The number of coincident spikes between some objects does increase if we increase the number of objects in a set. At some point the level of spurious synchronization would become very large, possibly

---

[31] This assumption is already questionable as we have seen in Chapter 5 that there is little evidence that supports the representation of features by specific spike patterns.

large enough to destroy the ability to distinguish between the objects. This would explain the breakdown of performance in the experiments described by Luck and Vogel.

11.4.4 Quantification of the spurious synchronization hypothesis

The idea of spurious synchronization as the reason behind the capacity limit of working memory is very appealing. It would, for example, predict that the capacity limit does not depend on the number of features that an object has, which was indeed also found experimentally. The explanation for this within the spurious synchronization hypothesis is that each additional feature of one object would in any case be synchronized with the other features of the object so that only the number of objects with different spike trains would matter. It is important to quantify such hypotheses and to compare them closely to experimental data. This can be done for the above hypothesis by using only basic combinatorics. The number of pairs N^{PAIRS} in a set of N^{obj} objects is given by

$$N^{\text{pairs}} = \binom{N^{\text{obj}}}{2} = \frac{N^{\text{obj}}!}{2! \ (N^{\text{obj}} - 2)!}. \tag{11.20}$$

With a given neural code, such as that specified by a given probability distribution of the spikes, there will be a given probability of spurious synchronization between two spike trains, which we symbolize as $P^{\text{ss}}(2)$. The probability of not having spurious synchronization between two spike trains is then $1 - P^{\text{ss}}(2)$. The probability of spurious synchronization between at least two spike trains in a set of N^{obj} spike trains (pattern) is then given by

$$P^{\text{SS}}(N^{\text{obj}}) = 1 - \left(1 - P^{\text{SS}}(2)\right)^{N^{\text{pairs}}}, \tag{11.21}$$

because the spike trains are independent so that the probabilities simply multiply. An increasing probability of spurious synchronization decreases the performance in the object comparison task. We can thus quantify the hypothesis of Luck and Vogel with a functional expectation of the percentage of correct recall given by

$$P^{\text{correct}} = \left(1 - P^{\text{SS}}(N^{\text{obj}})\right) 100 + \left(P^{\text{SS}}(N^{\text{obj}})\right) 50. \tag{11.22}$$

The second term on the right-hand side is added because a random answer can be expected if objects cannot be correctly separated. The recall performance from eqn 11.22 for various values of the probabilities of spurious synchronization between two spike trains is shown as dotted lines in Fig. 11.8A. The dashed line corresponds to the spurious synchronization hypothesis with $P^{\text{ss}}(2) = 0.04$. We see that we can use the quantification of the hypothesis to extract some parameters that can be further compared to experimental data. For example, the probability of spurious synchronization between two spike trains can be compared to models of spike trains that take realistic values for spike distributions, firing rates, and time precision into account.

The comparison of experimental and theoretical data also indicates some possible discrepancies that should be analysed further. For example, the experimentally observed transition seems a little bit sharper than the one derived from the spurious synchronization hypothesis. Also, there seems to be a strong deviation for large numbers of objects that could indicate a separate effect on working memory. We will not follow this line of research further as the discussion was only intended to illustrate that the quantification of hypotheses is often important and should enable us to develop better models in the future.

11.4.5 Other hypotheses on the capacity limit

Several other thoughts about and models of the reasons behind the limited capacity of working memory have been proposed in the literature. For example, *Lisman* and *Idiart* have suggested that the limit might be solely due to the limiting ability of reverberating neural activity for short-term memory when the representation of different objects is kept in different high-frequency subcycles of low-frequency oscillations found in the brain. (Such a hypothesis would explain that synchronization of spike trains with representations of features for the same object can be found because the related neurons are locked to a specific subcycle.) Another suggestion, advanced by *Nelson Cowan*, is based on the limits of an attentional system that is often thought to be a necessary ingredient in working memory models.

The discussion of the limits of modular networks in the previous section suggests another possibility. We have seen that the number of processes that we will be able to maintain is quite limited if we demand a sensible interaction between them. Solving tasks for which working memory is essential relies on some form of interaction between the entities in working memory, and for these reasons a capacity limit seems therefore quite possible.

11.5 Attentive vision

Most of the human brain is to some degree involved in visual processing, and vision research is a strong domain within neuroscience. It is beyond the scope of the book to introduce this fascinating area, but we will briefly outline a system level model that can shed some light on the way in which we think visual object recognition is achieved within the ventral visual pathway of the brain. This includes an outline of how visual scenes are represented in early visual areas, and how translation-invariant object recognition is supported by associative memories of the kind discussed in Chapter 8, which may be implemented in the inferior-temporal cortex. The model we outline here specifies how attentional mechanisms might interact with this recognition system. We will therefore distinguish between object-based attention and spatial attention.

11.5.1 Object recognition versus visual search

Visual input is an important source of information on which humans act in various ways. We can, for example, ask a subject to look for a particular object in a visual scene, such as the top of the Eiffel tower in the example shown in Fig. 11.9. Such a visual search demands the top–down influence of an object bias that specifies what to look for. In contrast, we can ask the subject to identify an object at a particular location (for example, what is in the lower right corner of the picture). This demands a spatial attention which is usually accompanied by directing the gaze towards this location. However, such tasks can also be performed with the gaze fixed at a central point of the picture, and more generally we thus have to consider spatial attention.

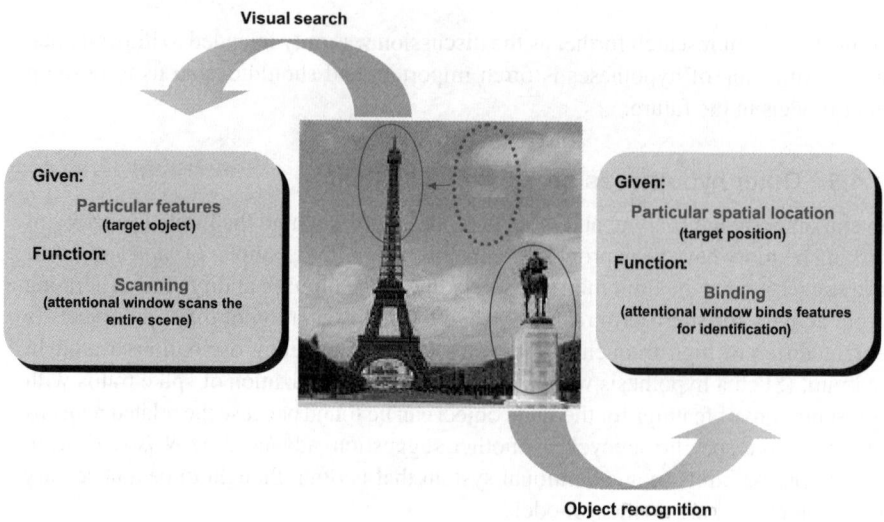

Fig. 11.9 Illustration of a visual search and an object recognition task. Each task demands a different strategy in exploring the visual scene [figure courtesy of Gustavo Deco].

11.5.2 The overall model

A lot of research is carried out with the objective of understanding how our visual system works. *Gustavo Deco* and collaborators have gathered some of the known specifics of the primate visual system and developed a model that can shed some light on the processes involved in visual search and object recognition. The overall scheme of their model is outlined in Fig. 11.10. The model basically has three important parts: one that is labelled 'V1–V4', one that is labelled 'IT', and one that is labelled 'PP'. We will outline the specific architectures and roles of these parts separately in the following. It is, however, good to keep in mind right from the beginning that the different parts do not work in isolation. Here we are particularly interested in how the different parts influence each other.

11.5.3 Object representation in early visual areas

The first part of the model labelled 'V1–V4' represents early visual areas such as the striate cortex (V1) and adjacent visual areas in the occipital lobe. The primary visual area V1 is actually the main cortical area that receives visual input from the *lateral geniculate nucleus* (LGN) of the thalamus, which itself is the major target of the optic nerves from the eyes. Neuronal responses to visual input from the eyes can be parametrized with *Gabor functions*, and a representation of the visual image decomposed into components with Gabor functions is thus taken as input into the first part of the model. The principal role of this part of the model is the decomposition of the visual field into features. This is known to occur at this early stage of visual processing in primates. For example, V1 neurons respond mainly to simple features such as the orientation of edges, and later areas are often specialized to represent other features such as colour, motion, or the combination of basic features.

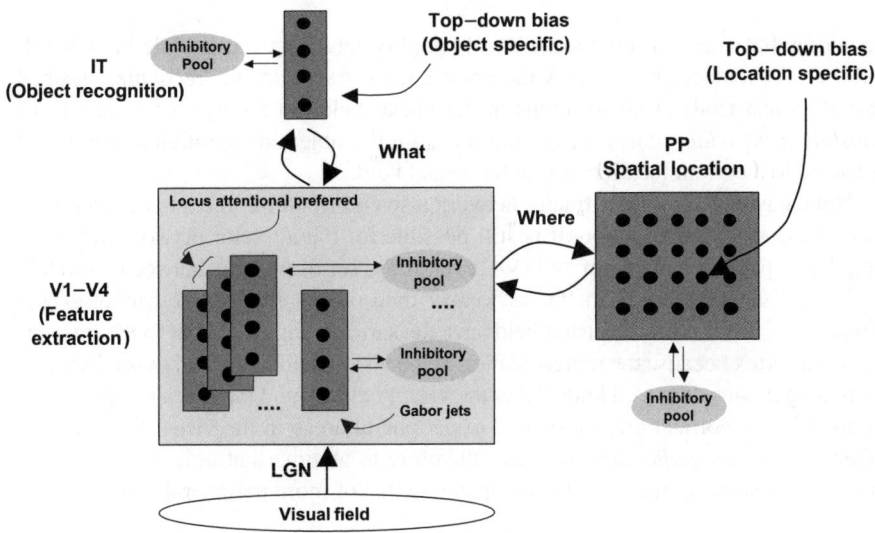

Fig. 11.10 Outline of a model of the visual processing in primates to simulate visual search and object recognition. The main parts of the model are inspired by structural features of the cortical areas thought to be central in these processes. These include early visual areas (labelled 'V1–V4') that represent the content of the visual field topographically with basic features, the inferior-temporal cortex (labelled 'IT') that is known to be central for object recognition, and the posterior parietal cortex (labelled 'PP') that is strongly associated with spatial attention [figure courtesy of Gustavo Deco].

From the modelling point of view it is not essential that we precisely rebuild all the details of the visual field representation in the brain as we are primarily interested in other aspects of the visual processing. It is only important here that the feature representation in this part of the model is topographic in that features are represented in modules that correspond to the location of the object in the visual field. This is analogous to the brain. Each module represents a feature of a part of the visual field as a vector of node activities within this module. Note that there is no global competition between the modules; only within each module is there an inhibitory pool that keeps the sparseness of the activity in each module roughly constant as discussed in Chapter 8.

11.5.4 Translation-invariant object recognition with ANN

The representation of the visual field in 'V1–V4' in this model feeds into the part labelled 'IT'. This part is aimed at modelling processes in the inferior-temporal cortex, which is known to be involved in object recognition. This recognition process can be modelled as associative memory as outlined in Chapter 8. Point attractors of specific objects are then formed through the collateral connections within this structure that can be trained with Hebbian learning. The connections between the 'V1–V4' and the 'IT' parts of the model can also be trained with Hebbian learning. We can thus place an object at all locations in the visual field that are covered by the modules in 'V1–V4'. By training each set of weights between one module in 'V1–V4' and the nodes in 'IT' we end up basically with the same set of weights between each module in 'V1–V4' and the 'IT' nodes. This enables the point attractor network to 'recognize' trained objects in test trials at all locations in the visual field, an ability that is said to be *translation-invariant object recognition* because the object recognition ability is not restricted to a certain location within the visual field.

Note, however, that the attractor network also contributes to the translation invariance of the object recognition. It is still possible for the attractor network in 'IT' to complete a pattern input from 'V1–V4' even if the set of weights between a module in 'V1–V4' and the nodes in 'IT' is weaker than others. Indeed, we cannot expect that all locations within the visual field have the same strength of input to the inferior-temporal cortex because the representation of the visual field in the early visual areas is not homogeneous. It is well known that the visual field around the fovea is represented within a larger cortical area compared to peripheral areas in the visual field. This is called *cortical magnification*. We have therefore to assume that objects close to the fovea have a stronger input to 'IT' compared to that of more peripheral objects.

11.5.5   Size of the receptive field

The contribution of the attractor network in 'IT' to translation-invariant object recognition already explains some recent experimental findings that showed that the size of the receptive field of inferior-temporal neurons depends on the content of the visual field and the specifics of the task. For example, if a single object is shown on a screen with blank background, then it was found that the receptive field of a neuron that responds to this object can be very large (as much as 30 deg or more). An example of the firing rate of one inferior-temporal neuron in a non-human primate engaged in a visual search task is shown by a solid line in Fig. 11.11A. Interestingly, if two objects are presented simultaneously, or if the target object is shown on top of a complex background (which can be viewed as a scene with many objects), then the size of the receptive field shrinks markedly (see dashed line in Fig. 11.11A for the case of an object on a complex background).

A possible explanation of the experimental findings is given by the model discussed in this section, which is based on the attractor dynamics of the auto-associator network in 'IT'. If only one object is shown, then this object would trigger the right point attractor and thus the recall of the object regardless of its location which corresponds to large receptive fields. If, however, two or more trained objects are shown, then it is likely that the final state of the attractor network is mainly dominated by the object closest to the fovea, which gets the most weight due to cortical magnification. Simulation results, shown in Fig. 11.11B, confirm this hypothesis.

11.5.6   Attentional bias in visual search and object recognition

Visual search can be simulated within this model by supplying an object bias input to the attractor network in 'IT' that tells the system what to look for. Such top–down information is thought to originate in the frontal areas of the brain. Simulations have shown that the additional input of an object bias to 'IT' can speed-up the recognition process in 'IT'. The object bias also supports the recognition ability of the input from 'V1-V4' that corresponds to the target object in visual search. Correspondingly, it was

A.   Neuron recording results   B.   Model simulation results

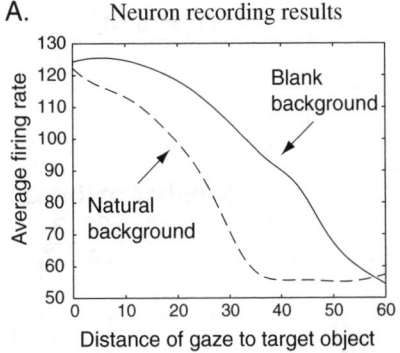 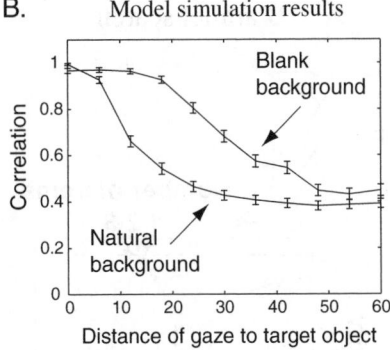

Fig. 11.11 (A) Example of the average firing rate from recordings of a neuron in the inferior-temporal cortex of a monkey in response to an effective stimulus that is located at various degrees away from the direction of gaze [experimental data courtesy of Edmund Rolls]. (B) Simulation results of a model with the essential components of the model shown in Fig. 11.10. The correlation is thereby a measure of overlap between the activity of 'IT' nodes with the representation of the target object that was used during training [adapted from Trappenberg, Rolls, and Stringer, *Advances in Neural Processing Systems 14 (NIPS 2001)*, 2002].

found, in simulations as well as in primate experiments, that the receptive fields of target objects are larger than receptive fields of non-target objects.

Parallel conclusions can be drawn in an object recognition task in which top–down input to a specific location in 'PP' is given. This activity can enhance the neural activity in 'V1–V4' for the features of the object that is located at the corresponding location in the visual field. Consequently, it will be easier for the 'IT' network to complete the input patterns (and hence to 'recognize' objects) for objects at the corresponding location. This can lead to an increase in the speed of recognition as well as an increase of the size of the corresponding receptive fields.

11.5.7   Parallel versus serial search

Note that the attentional biases have different origins in the model shown in Fig. 11.10. The attentional bias used in visual search originates from top–down input to 'IT' and is object-based. The attentional bias in the object recognition task acts on 'PP' and is location-based. However, it may be difficult to separate the different forms of attention in experiments because all parts of the model are bidirectional, in agreement with anatomical findings. For example, an object bias to 'IT' in the visual search task will ultimately trigger some activity in 'PP' that corresponds to the location where the object is, and this activity itself will help the recognition process of that object.

The time elapsed until the activity of a node in 'PP' reaches a certain threshold can be taken as an indication of the reaction time needed to find a specific object in visual search. *Gustavo Deco* has simulated a visual search task in which the visual scene had a target (the letter 'E') and a number of distractors. The distractors (the letter 'X') in one experiment were visually very different from the target pattern. The simulated reaction time was in this case independent of the number of objects in the visual scene as indicated in Fig. 11.12A. In contrast, the reaction time increased linearly with the

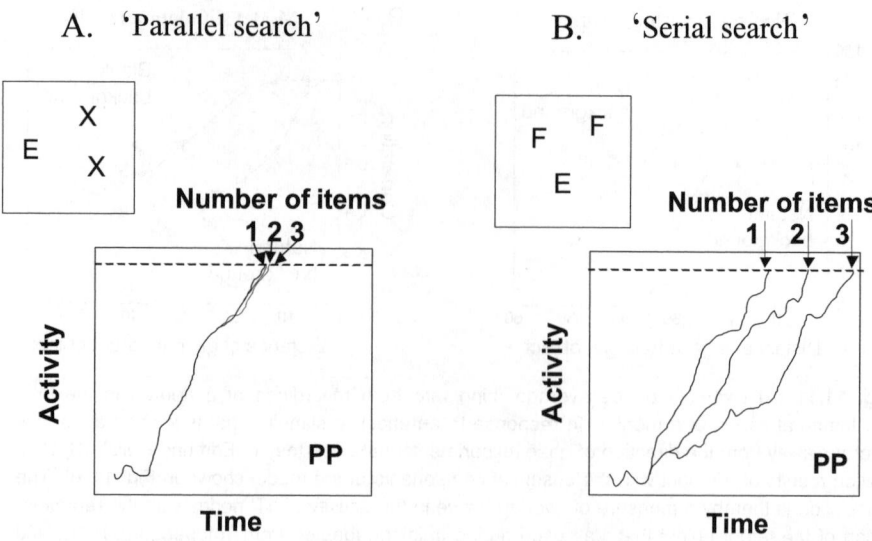

Fig. 11.12 Numerical experiments in which the model simulated a visual search task of a target object (the letter 'E') in a visual scene with visual distractors. (A) In one experiment the distractors consist of the letters 'X' that are visually very different from the target letter. The activity of a 'PP' node that corresponds to the target location increases in these experiments independently of the number of distractors, implying parallel search. (B) The second experiment was done with distractors (letter 'F') that were visually similar to the target letter. The reaction times, as measured from 'PP' nodes, depends linearly on the number of objects, a feature that is also characteristic of serial search. Both modes are, however, present in the same 'parallel architecture' [adapted from Gustavo Deco, personal communication].

number of distractors when the distractor objects (the letter 'F') were visually similar to the target (see Fig. 11.12B). A linear increase of the reaction time with the number of objects in the visual scene is commonly attributed to a serial search mechanism in contrast to the lack of dependence of the reaction time on the number of objects that indicates parallel search.

Similar findings in psychophysical experiments with humans have often been interpreted as evidence for different search strategies thought to be implemented by different neural mechanisms. However, the simulation results demonstrate that such psychophysical results are consistent with a single parallel neural machinery, and that the apparent serial search is only due to the more intense conflict-resolution demand in the recognition process. The model is able to make many more predictions that can be verified experimentally, on a behavioural level, with brain imaging techniques or with cell recordings. This model is therefore a good example of the type of model that we hope will emerge for many more brain processes.

11.6   An interconnecting workspace hypothesis

It is increasingly clear that complex cognitive tasks can only be solved with the flexible cooperation of many specialized modules. In contrast to most existing models of brain functions, it is highly fascinating to observe how flexible we humans are in coping with the complex world around us. In contrast to most models, which produce very stereotyped behaviour, humans display the ability to master highly skilled functions while still maintaining the ability to produce novel solutions that have not been trained extensively. For example, we can drive a car with such ease that little attention and effort seems necessary to do so. This is despite the fact that we have to react to the unknown environment, for example, in following a road that we have not driven on before. However, when very novel and critical circumstances occur we are able to engage ourselves with many resources on this problem, which typically goes along with very attentive and often very arduous mental activity.

11.6.1   The global workspace

From a system level perspective we expect that specialized processors can have a high degree of autonomy while still being able to engage, if necessary, in more global system activities. It is, however, still largely unknown how such flexibility can be achieved in a robust way. In the remainder of this chapter we will review one speculative yet far-reaching idea as to how this could be achieved in the brain, which was proposed by *Stanislas Dehaene, Michel Kerzberg*, and *Jean-Piere Changeux*. In Fig. 11.13A we have reproduced their illustration of the principal idea behind their proposal. In this the authors divided brain functions into five basic subsystems, ranging from a perceptual system, which is our window to the world, to the motor system responsible for executing desired goals. The three remaining subsystems are a memory system, an evaluative system, and an attentional system, all of which are basic ingredients for complex cognitive tasks as viewed by many cognitive scientists. The main point made by Dehaene and colleagues is that these subsystems must be able to work alone or in combinations in a flexible manner, and they speculate that this may be achieved through some form of network between the subsystems. Such an interconnecting network would form a *global workspace*, which would form the basis of our ability to solve complex tasks in a flexible manner.

In contrast to the more localized basic processing networks that make up the five basic subsystems as illustrated in Fig. 11.13A, the workspace has to be a more global computational space in the brain. Projections between cortical areas are indeed abundant, and associated fibres of cortico-cortical connections predominantly originate from pyramidal neurons in layers II and III as mentioned in Chapter 4. The authors suggest that a large portion of the global workspace could hence be localized within these layers. This suggestion has interesting consequences for the interpretation of anatomical findings concerning these layers. We have mentioned that the extent of layers within the cortex varies considerably. Layers II and III are, for example, elevated in the dorsolateral prefrontal cortex, an area that has been implicated in a variety of types of complex mental processing. For example, it was shown to become active in early stages of learning a sequence of numbers, and patients with lesions in this area are thought to have difficulties in switching between different rules in more challenging

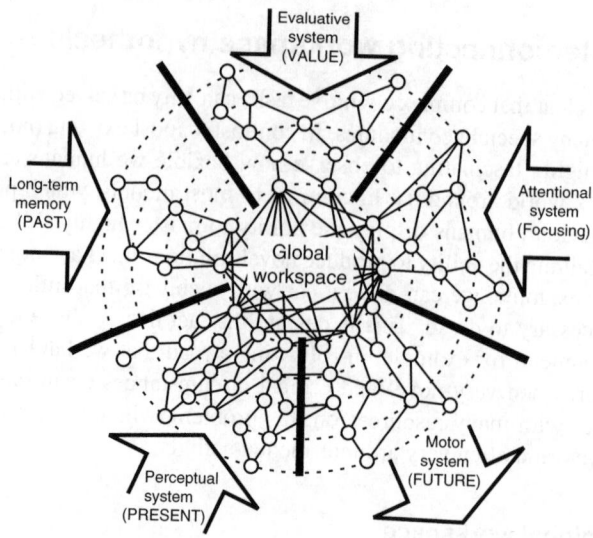

Fig. 11.13 Illustration of the workspace hypothesis. Two computational spaces can be distinguished, the subnetworks with localized and specific computational specialization, and an interconnecting network that is the platform of the global workspace [adapted from S. Dehaene, M. Kerszberg, and J. Changeux, *Proc. Natl. Acad. Sci.* 95: 14529–34, 1998].

categorization tasks. It is difficult to record brain activity from specific layers, and direct verifications of the model discussed below are therefore challenging. In such situations it becomes necessary to work out other predictions that can be verified experimentally.

11.6.2 Demonstration of the global workspace in the Stroop task.

Dehaene and colleagues demonstrated the principal idea in much more detail with a model that sheds light on the processing of the *Stroop task* when following the global workspace hypothesis. The Stroop task is illustrated in Fig. 11.14A. In a common form it consists of a set of words that name colours, where each word can be printed in a different colour. In the Stroop task a subject is asked to either read the word or name the colour in which the word is written (all relatively rapidly). We are highly trained in reading, so that the display of a word does not render any problems in pronouncing the word. However, the naming of the colour can initially cause some problems as the image of the word prompts us to read the word rather than to ignore the content and to report the colour of the letters. It initially takes some effort to suppress the reading of the word until we have automated our response after a few trials. The explanation offered by Dehaene and colleagues is that the global workspace has to become active to 're-wire' the commonly active word-naming configuration of the brain.

To demonstrate this idea further the authors developed a model that can be tested on a Stroop task. This is illustrated in Fig. 11.14B. The model includes three specialized processors, one that indicates the meaning of the word that is displayed, one that indicates the colour of the word that is displayed, and one that indicates the response

A.

B.

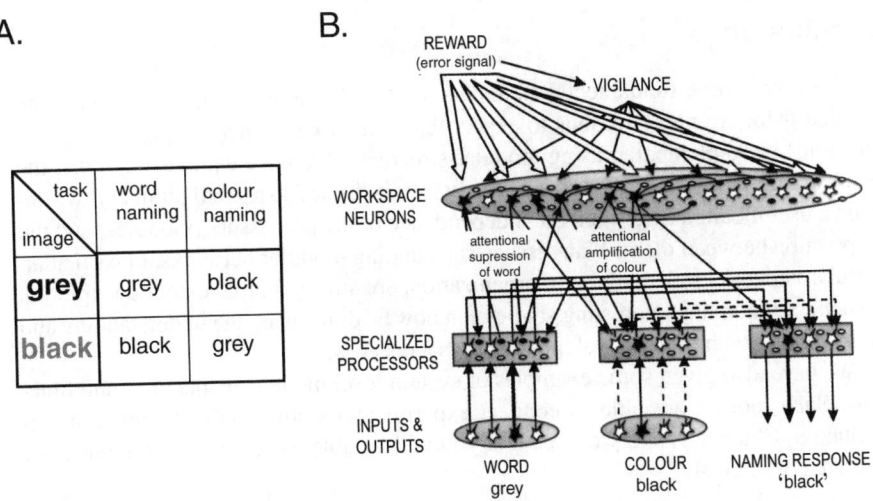

image \ task	word naming	colour naming
grey	grey	black
black	black	grey

Fig. 11.14 (A) In a Stroop task a word for a colour, written in a colour that can be different from the meaning of the word, is shown to a subject who is ask to perform either a word-naming or colour-naming task. (B) Global workspace model that is able to reproduce several experimental findings in the Stoop task [adapted from S. Dehaene, M. Kerszberg, and J. Changeux, *Proc. Natl. Acad. Sci.* 95:14529–14534, 1998].

of the system. In the more standard word-reading task we can assume that there is a tight coupling between the first input processor and the response processor so that the system could report initially easily the content of the word. The second task, that of naming the colour in which the word is written, then requires a suppression of these connections and the enhancement of the connections between the 'colour' input processor and the response processor. This is achieved by the top–down influence of workspace nodes on the nodes in the specialized processors.

What prompts the workspace neurons to become active or to change their behaviour in the first place? Basically, the system has to be told what to do. This is achieved in the form of a reward signal that only indicates a mismatch between the desired response and actual response of the system. The large error signal then triggers a vigilance parameter to increase, which in turn allows the workspace neurons to become active that effectively reconfigure the system to allow a correct response within the required task. The vigilance parameter decreases once the system responds correctly. The authors argue that an increased vigilance (and therefore an increased activity of workspace nodes) could parallel the increased mental effort that is felt by subjects during the initial period after switching the task.

The model can make several predictions that can be tested experimentally. The model also demonstrates an interesting flexibility of a system to solve different tasks, and we can imagine that the flexible, and possibly novel, coupling of many highly specialized subsystems in the brain could be responsible for our ability to cope with the often challenging environment we live in.

Conclusion

We discussed some fundamental examples of modular networks in this chapter and saw that there are many reasons to expect that complex information-processing systems must be implemented using modular structures. We have demonstrated that the number of modules in coupled attractor networks is limited to a small number if we demand a useful balance between the independence of the processors (modules) and the cooperation between them. Many issues surrounding modular networks, in particular, modular networks reflecting brain organization, are still to a large extent unexplored. We believe that many interesting studies can now be done using the understanding and the modelling techniques developed in the last decades.

We have also given some examples of system level models of specific brain functions. Such models are able to connect experimental studies on different levels as outlined in Chapter 1, and we hope that you will be able to develop such models for your area of interest.

Further reading

Claus-C. Hilgetag, Mark A. O'Neill, and Malcolm P. Young

> Indeterminate organization of the visual system, in *Science* 271: 776–7, 1996.
> The authors investigated possible hierarchical models of the visual cortex with numerical methods in this paper. Such maps were previously prepared by hand (see references in this paper). There is a longer write-up from which Fig. 11.1 is taken (see reference in figure caption), and some details can also be found on the web at
> http://www.psychology.ncl.ac.uk/hierarchy.html.

Robert A. Jacobs, Michael I. Jordan, and Andrew G. Barto

> Task decomposition through competition in a modular connectionist architecture: the what and where tasks, in *Cogn. Sci.* 15: 219–250, 1991.

Robert A. Jacobs and Michael I. Jordan

> Computational consequences of a bias toward short connections, in *J. Cogn. Neurosci.* 4: 323–36, 1992.

N. J. Nilsson

> *Learning machines: foundations of trainable pattern-classifying systems*, McGraw-Hill, 1965.

O. G. Selfridge

> Pandemonium: a paradigm of learning, in the mechanization of thought processes, in *Proceedings of a Symposium Held at the National Physical Laboratory*, November 1958, 511–27, London HMSO, 1958.

Marvin Minsky

> *The society of mind*, Simon & Schuster, 1986.

Geoffrey Hinton

Products of experts, in *Proceedings of the Ninth International Conference on Artificial Neural Networks* ICANN '99, 1:1–6, 1999.

You can find follow-ups of this idea on Hinton's web site.

Yaneer Bar-Yam

Dynamics of complex systems, Addison-Wesley, 1997.

This book is on complex systems with neural networks as a prominent example. The two chapters on neural networks (chapters 2 and 3) include a nice discussion of modular networks as well as some speculation about the role of sleep in training such structures. The principal idea of the restrictions on the number of modules in modular architectures is similar to the arguments outlined in this chapter, although some details in the derivation are different.

Edmund T. Rolls and Simon M. Stringer

A model of the interaction between mood and memory, in *Networks: Compt. Neural Syst.* 12: 89–109, 2001.

An example of the possible relation of interacting recurrent networks in some brain processes.

Akira Miyake and Priti Shah (eds.)

Models of working memory, Cambridge University Press, 1999.

This book describes and compares several models of working memory. This includes the model by O'Reilly, Braver, and Cohen outlined in this chapter.

Edmund T. Rolls and Gustavo Deco

Computational neuroscience of vision, Oxford University Press, 2002.

A new book on computational neuroscience that discusses many issues in vision, including the model outlined in this chapter.

Stanislas Dehaene, Michel Kerszberg, and Jean-Pierre Changeux

A neuronal model of a global workspace in effortful cognitive tasks, in *Proc. Natl. Acad. Sci., USA*, 95: 14529–34, 1998.

12   A MATLAB guide to computational neuroscience

MATLAB® is an interactive programming environment for scientific computing, including numerical calculations and visualization, that is very useful in simulating models in computational neuroscience. In this chapter we highlight some of the features that are essential to simulate most of the models discussed in the book, and mention some 'tricks' that might be useful in such simulations. This chapter includes most of the details of the models and simulations mentioned in the book, and outlines some numerical methods that are useful for computational studies.

12.1   Introduction to the MATLAB programming environment

MATLAB[32] is a collection of programs available for many computer systems (including Windows-based personal computers and UNIX workstations) that is very efficient in writing programs for numerical simulations, executing them, and visualizing the results. Complex numerical problems can often be solved in a fraction of the time needed to do so in other programming languages such as Fortran or C. The name MATLAB is derived from MATrix LABoratory, which highlights a major feature of this tool, that it is mostly based on matrix operations. Matrix notations are compact and are very convenient for network descriptions. The formulas in this book can often be translated directly into MATLAB code without writing extensive programs. Even if you don't have access to MATLAB it is useful to follow the examples in this chapter as MATLAB code can be regarded as a useful metalanguage to describe the algorithms, and translations to other programming languages should be fairly straightforward.

MATLAB includes many useful functions and algorithms to solve linear algebra and systems of differential equations on which we can base our simulations. Many more specialized collections of functions and algorithms, called a toolbox in MATLAB, can be purchased in addition to the basic MATLAB package or imported from third parties.[33] There is, for example, a Neural Network Toolbox available that incorporates functions for building and analysing standard neural networks. This toolbox includes many algorithms particularly suitable for connectionist modelling and neural network applications and in particular includes advanced back-propagation algorithms. However, all the models described in this book can easily be implemented with the basic MATLAB package. We thus program most of the network models from 'scratch'

---

[32] MATLAB and Simulink are registered trademarks, and MATLAB Compiler is a trademark of The MathWorks, Inc.

[33] Many researchers publish their algorithms in MATLAB on the web.

rather then rely on network simulators that have implemented specific architectures and methods. The programs of the basic models in this book often consist of only a few lines of code. Using the basic programming approach provides great transparency and flexibility of code.[34]

12.1.1 Starting a MATLAB session

Once MATLAB is installed it can be started by typing `matlab` in a UNIX command window or by selecting the program from a program list in Windows systems. This starts the MATLAB desktop as shown in Fig. 12.1 for MATLAB version 6. All of the basic features used here can be found in earlier versions of MATLAB, while later versions are mainly enhanced in terms of the user interfaces to functionalities that can always be executed by command lines in earlier versions. The MATLAB desktop is comprised of several windows of which we use only the essential *MATLAB Command Window* in the following. We can thus close the other windows if they are open (such as the *Launch Pad* or the *Current Directory Window*); you can always get them back from the *view* menu. Alternatively we can undock the *Command* window by clicking the arrow button on the *Command Window* bar. An undocked *Command Window* is illustrated on the left in Fig. 12.2. Older versions of MATLAB start directly with a *Command Window* or simply with a MATLAB command prompt >> in a standard system window. The *Command Window* is our central control centre for accessing the essential MATLAB functionalities.

12.1.2 Working with MATLAB

The MATLAB programming environment is interactive in that all commands can be typed into the command window behind the command prompt (see Fig. 12.2). The commands are interpreted directly, and the result returned and displayed in the *Command Window*. Type a=3 into the *Command Window*. The system should respond with

```
a =

    3
```

Ending a command with semicolon ';' suppresses the printing of the result on screen. It is therefore generally included in our programs unless we want to view some intermediate results.

If we want to repeat a series of commands it is convenient to write this list of commands into an ASCII file with extension '.m'. Any ASCII editor (for example; WordPad, Emacs, etc.) can be used. The MATLAB package contains an editor that has the advantage of colouring the content of the files for better readability. The list of commands in the ASCII file (e.g. *filename*.m) is called a *script* in MATLAB and

---

[34] This does not mean that I advocate always writing every algorithm from scratch. If implementations of algorithms (or whole simulation packages) exist that are sufficient for the problem I would use them. The power of MATLAB originates partly from the many implementations of standard methods that are provided, and we will rely on algorithms implemented as part of the standard MATLAB package (such as solvers for numerical integrations) in some simulations.

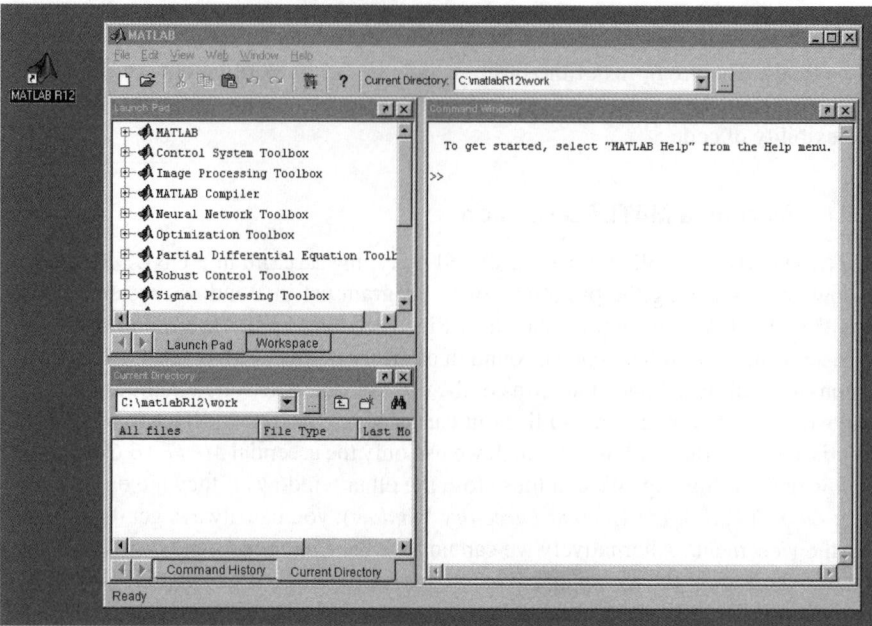

Fig. 12.1 The MATLAB *Desktop Window* of MATLAB Version 6.

makes up a MATLAB program. This program can be executed by calling the name of the file within the command window (for example, by typing *filename*). [35]

All the variables which are created by a program are kept in a buffer called *workspace* that can be viewed with the command who s (or displayed in the 'Workspace' window in version 6). The result of this commands after declaring the variable a that we have done above results should look like.

```
Name       Size          Bytes  Class

a          1x1               8  double array
```

It displays beside the name of the variable the size that is often useful to find out the size of matrices as discussed later. The variables in the workspace can be used as long as MATLAB is running and as long as it is not cleared with the command `clear`. The workspace can also be save with the command `save` *filename*, which creates a file *filename*.mat in internal MATLAB format. The saved workspace can be reloaded into MATLAB with the command `load` *filename*. It is also possible to export and import data in ASCII format. The workspace is very convenient as we can run a program within a MATLAB session and can then work interactively with the results for example to plot some of the generated data.

---

[35] We assumed here that the program file is in the current directory of the MATLAB session or in one of the directory paths that can be specified in MATLAB. In version 6 there is a 'Current Directory' window that can be opened under the view menu; some older versions have a 'Path Browser'. One can also change directories with UNIX-style commands like cd in the command window.

We can also write functions that are programs kept in a separate file (again with extension '.m'). The only difference between a script and a MATLAB function is that functions have a specific first line in the file that specifies the variables that are passed to the function and the variables that are passed back to the calling program. The variables within the function are local to the function and are not included in the workspace of the calling program (except the variables that are specified in the return list as long as they are assigned to variables in the calling program). Functions are useful for encapsulating specific parts of the program. They have the additional advantage that they can be pre-compiled with the MATLAB Compiler^{TM} that is available from The MathWorks, Inc. The compiler can translate the function into a compiled MATLAB function, or translate the function into C or C++ code. The first option can be used to optimize the code,[36] while the second option can be used to generate stand-alone applications within MATLAB.

12.1.3 Basic variables in MATLAB

Variables in MATLAB are basically matrices (or data arrays), which we will see is very convenient for most of our purposes. Note that this includes vectors ($1 \times N$ matrix) and scalars (1×1 matrix). The variable type of the elements (such as an integer, real number of text strings) is determined by the values assigned to the elements, and we do not have to declare the variable types separately. Values can be assigned to matrix elements in several ways.[37] The most basic one is using square brackets and separating rows by a semicolon within the square brackets, for example (see Fig. 12.2),

```
a=[1 2 3; 4 5 6; 7 8 9]
```

Also, MATLAB functions often return matrices that are frequently used to assign values to matrix elements. For example, a random 3×3 matrix can be generated with the command

```
b=rand(3)
```

The multiplication of two matrices (following the matrix multiplication rules) can be done in MATLAB by typing

```
c=a*b
```

This is equivalent to

```
c=zeros(3);
for i=1:3
   for j=1:3
      for k=1:3
         c(i,j)=c(i,j)+a(i,k)*b(k,j);
      end
   end
end
```

---

[36] The advantages of optimized code are often only minor when the program is written efficiently as specified further below.

[37] Version 6 includes an array editor, and data in ASCII files can be assigned to matrices when loading the data.

which is the common way of writing matrix multiplications in other programming languages. Formulating operations on whole matrices rather than on the individual components separately is not only more convenient and clear, but the performance advantages of the programs are usually much higher. [38] Whenever possible, operations on whole matrices should be used. This is likely to be the major change in your programming style when converting from other programming languages to MATLAB. The performance disadvantage of an interpreted language is often negligible when using operations on whole matrices, and a compilation of the program is then not necessary.

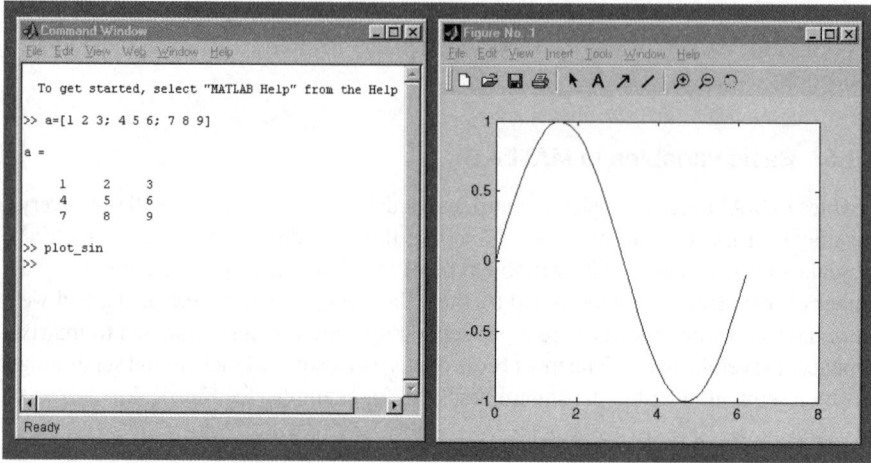

Fig. 12.2 A MATLAB *Command Window* (left) and a MATLAB *Figure Window* (right) displaying the results of the function plot_sin developed in the text.

12.1.4 Basic MATLAB commands

Basic programming constructs such as variable assignments and control flow commands are similar to those in other programming languages. Some examples are listed in Table 12.1. On top of these common programming constructs, MATLAB is rich in special functions implementing many algorithms, mathematical constructs, and advanced graphic handling, as well as general information and help functions. You can always search for some keywords using the useful command lookfor followed by the keyword. This command lists all the names of the functions that include the keywords in a short description in the function file within the first *comment lines* after the function declaration in the function file. A comment line, which is a line with possible text that is not interpreted by MATLAB, is a line that starts with the symbol '%', for example,

    % This is a comment not interpreted by MATLAB

The command help followed by the function name displays the first block of comment lines in a function file. This information in the functions provided by MATLAB is usually sufficient to get enough information to use these functions in your application.

---

[38] The reason for this is that MATLAB routines are implemented very efficiently.

Table 12.1 Some examples of commands in MATLAB. The help command for such commands links you to other related commands in MATLAB.

Programming construct	Command	Syntax
Assignment	=	a=b
Arithmetic operations	add	a+b
	multiplication	
	matrix	a*b
	element-wise	a.*b
	power	a∧b
Relational operators	equal	a==b
	not equal	a∼=b
	less than	a<b
Logical operators	AND	a & b
	OR	a‖b

Conditional command	if statement	if *logical expressions*
		    statement
		 elseif *logical expressions*
		    statement
		 else
		    statement
		 end

Loop	for	for *index=start:increment:end* statement end
	while	while *expression* statement end
Function		function [x,y,...]=*name*(a,b,...)

12.1.5 Graphics

MATLAB is also advanced and flexible in producing scientific graphics. We want to illustrate this by writing our first program in MATLAB: calculating and plotting the sine function. The program is

```
x=0:0.1:2*pi;
y=sin(x);
plot(x,y)
```

The first line assigns elements to a vector **x** starting with $x(1) = 0$ and incrementing the value of each further component by 0.1 until the value 2π is reached (the variable pi has the appropriate value in MATLAB). The last element is $x(63) = 6.2$. The second line calls the MATLAB function sin with the vector x and assigns the results to a vector y. The third line calls a MATLAB plotting routine. You can type these lines into an ASCII file that you can name plot_sin.m. The code can be executed by typing plot_sin as illustrated in the *Command Window* in Fig. 12.2.[39] The execution of this program starts a *Figure Window* with the plot of the sinus function as illustrated on the right in Fig. 12.2.

The appearance of the plot can easily be changed by changing the attributes of the plot. There are several functions that help in performing this task, for example, the function axis that can be used to set the limits of the axis. New versions of MATLAB provide window-based interfaces to the attributes of the plot. However, there are also two basic commands, get and set, that we find useful. The command get(gca), where gca is the axis handle and stands for 'get current axis' returns a list with the axis properties currently in effect. This command is useful for finding out what properties exist. The attributes of the axis can then be changed with the set command. For example, if we want to change the size of the labels we can type set(gca,'fontsize',18). There is also a handle for the current figure gcf that can be used to get and set other attributes of the figure. MATLAB provides many routines to produce various special forms of plots including plots with error-bars, histograms, three-dimensional graphics, and multi-plot figures. We will see some of them in action in the simulation programs of this book discussed in the remainder of this chapter.

12.2 Spiking neurons and numerical integration in MATLAB

The programs discussed here are listed in tables and can be downloaded from the Oxford University Press web page at

http://www.oup.co.uk/isbn/0-19-851583-9

The subfolder spikes contains the code of several simulations with spiking neurons that we will discuss in this section.

12.2.1 Integrating Hodgkin–Huxley equations with the Euler method

The file hh.m (see Table 12.2) contains the integration of the Hodgkin–Huxley equations as discussed in Chapter 2. Have a look at this code. We first defined some parameters, the maximal conductances and battery voltages as used by Hodgkin and Huxley. We placed them into vectors, which makes it possible to write the later equations in compressed form using matrix notation. After the initialization of some more variables we enter a big loop over the time steps over which we want to integrate the Hodgkin–Huxley equations. For each time step the voltage of the membrane potential may have changed and we have first to calculate the parameters τ_x and x_0 of equation 2.6 for each of the parameters m, n, and h. These are functions that were chosen by Hodgkin and Huxley. They actually defined them in terms of functions α_x and β_x that are directly related to τ_x and x_0. The particular values of some parameters in these functions are the ones originally used by Hodgkin and Huxley.

After we have calculated τ_x and x_0 for the given voltage of the membrane we are ready to integrate eqn 2.6 using a simple Euler method (see Appendix C) by replacing this continuous equation with the discrete version

---

[39] Provided that the MATLAB session points to the folder in which you placed the code

Table 12.2 Program hh.m

```
%%%%%%%%%%%%%%%%%%%%%%%%%%%%%%%%%%%%%%%%%%%%%%%%%%%%%%%%%%%%%%
% Integration of Hodgkin--Huxley equations with Euler method
%%%%%%%%%%%%%%%%%%%%%%%%%%%%%%%%%%%%%%%%%%%%%%%%%%%%%%%%%%%%%%
clear; clf;
% maximal conductances (in units of mS/cm^2); 1=K, 2=Na, 3=R
   g(1)=36; g(2)=120; g(3)=0.3;
% battery voltage ( in mV); 1=n, 2=m, 3=h
   E(1)=-12; E(2)=115; E(3)=10.613;
% Initialization of some variables
   I_ext=0; V=-10; x=zeros(1,3); x(3)=1; t_rec=0;
% Time step for integration
   dt=0.01;
% Integration with Euler method
   for t=-30:dt:50
     if t==10; I_ext=10; end  % turns external current on at t=10
     if t==40; I_ext=0; end    % turns external current off at t=40
% alpha functions used by Hodgkin-and Huxley
     alpha(1)=(10-V)/(100*(exp((10-V)/10)-1));
     alpha(2)=(25-V)/(10*(exp((25-V)/10)-1));
     alpha(3)=0.07*exp(-V/20);
% beta functions used by Hodgkin-and Huxley
     beta(1)=0.125*exp(-V/80);
     beta(2)=4*exp(-V/18);
     beta(3)=1/(exp((30-V)/10)+1);
% Tau_x and x_0 are defined with alpha and beta
     tau=1./(alpha+beta);
     x_0=alpha.*tau;
% leaky integration with Euler method
     x=(1-dt./tau).*x+dt./tau.*x_0;
% calculate actual conductances g with given n, m, h
     gnmh(1)=g(1)*x(1)^4;
     gnmh(2)=g(2)*x(2)^3*x(3);
     gnmh(3)=g(3);
% Ohm's law
     I=gnmh.*(V-E);
% update voltage of membrane
     V=V+dt*(I_ext-sum(I));
% record some variables for plotting after equilibration
     if t>=0;
        t_rec=t_rec+1;
        x_plot(t_rec)=t;
        y_plot(t_rec)=V;
     end
   end  % time loop
   plot(x_plot,y_plot); xlabel('Time'); ylabel('Voltage');
```

$$x(t + \Delta t) = (1 - \frac{\Delta t}{\tau_x})x(t) + \frac{\Delta t}{\tau_x}x_0, \tag{12.1}$$

which is directly translated into MATLAB as

```
x=(1-dt./tau).*x+dt./tau.*x_0.
```

Notice that we have used dots before the multiplication and division symbols. This indicates that the following operation should act on each individual component instead of the usual matrix multiplications and divisions. The final steps are to use Ohm's law to calculate the corresponding synaptic currents and to update the membrane potential from the synaptic currents and the external current that flow within the small time step Δt for which all of these calculations are done. We have included several time steps before recording the membrane potential over time in order to equilibrate the system first so that our results do not depend on the initial conditions. The results of these simulations are shown in Fig. 12.3. We started the external current sufficiently early to elicit a spike at $t = 10$, and the model neuron started to depolarize and to spike shortly thereafter. This is followed by a hyperpolarization phase, and the constant external current can only elicit the next spike after some time. The external input was removed at $t = 40$ before the model neuron was able to generate a third spike.

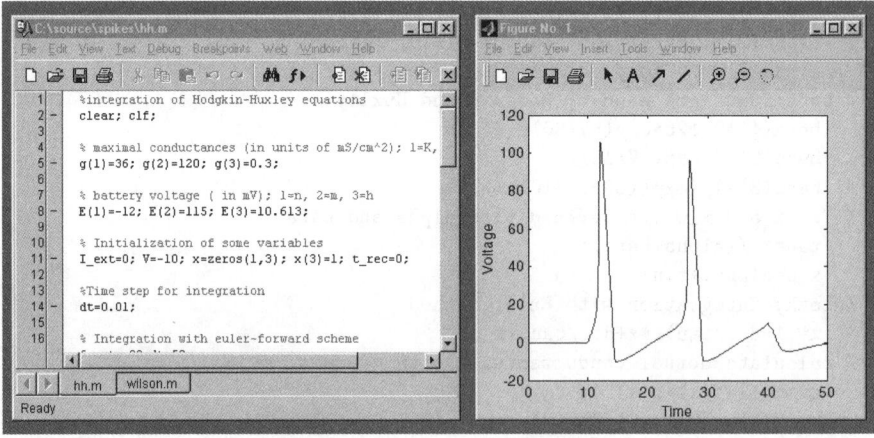

Fig. 12.3 MATLAB *Editor Window* (left) and a *Figure Window* (right) for simulating the model by Hodgkin and Huxley that can be found in the program file `hh.m` .

12.2.2 The Wilson model and advanced integration

Every set of ordinary differential equations (ODE) can easily be integrated with the Euler method. In program `wilson_euler.m` (see Table 12.3) we have done this for the Wilson model discussed in Chapter 2. The low-order Euler method can, however, be numerically unstable as discussed in Appendix C. In short, the low-order Euler method only represents the continuous differential closely if we use a very small time step in the numerical integration. Try, for example, to use a larger time step of `dt=0.1`, in the program, and don't be surprised to see different results from before (for example, no firing). The question of how small the time step should be becomes obvious. In practice we can make some numerical experiments in which we alter the size of the time step considerably. If the numerical results are stable with regard to the change of the step width we can gain some confidence that the results are generic. However, there are many advanced integration methods. For example, we can use some information on the curvature of the function instead of using a linear interpolation between integration points. Such higher-order methods generally produce more accurate results when integrated with the same time step. The time step should also be chosen properly in each domain of the functions we want to integrate. If the integral is not changing very much over time then it is reasonable to use a large time step. In contrast, if the integral is changing rapidly we have to use a smaller time step. There are several methods to choose the time step as a function of the performance of the algorithm. Such methods are generally known as *adaptive time step methods*.

All we have said is common knowledge in numerical mathematics and details can be found in many books. An advantage of MATLAB is that many of these algorithms are implemented as functions that can be used directly within MATLAB. A demonstration using a MATLAB-native ODE solver to integrate the Wilson equations is included in the file `wilson.m` (see Table 12.4). The beginning of the program is the same as in the previous program. The only difference is that we have replaced the time loop with the function call

```
[t,y]=ode15s('wilson_odefile',tspan,y0,[],I_ext,g,E,tau);
```

This function returns the value of the function y, defined through the differential equation

$$\frac{dy}{dt} = f(y). \tag{12.2}$$

The function f, which incorporates all the details of our model, is specified in a function file whose name is passed as the first argument in the call to the ODE solver. In our case we encapsulate the right-hand side of eqn 12.2 in the file `wilson_odefile.m` (see Table 12.5). Translating the continuous formulas in the text to programs is very easy. We no longer have to calculate the corresponding discrete version of the equation and can write a function that very much resembles our equations in the book. The ODE solvers are generic in MATLAB, so it is possible to change the ODE solver by simply changing the name in the function call, for example, `ode15a`→`ode45` (type `help ode45` in the MATLAB command window for more information). The graph produced by the program `wilson.m` is shown on the left side of Fig. 12.4. It also runs much faster than the program `wilson_euler` with the necessary small time step specified in the listing.

12.2.3 MATLAB function files

A *function* is a general construct in MATLAB. You can turn every MATLAB program into a function by including a line at the top that has the keyword `function` and specifies the input arguments and the return values. The first block of comment lines that follows the function-definition line is displayed[40] when calling help with the function name. It is therefore very convenient to build libraries and special functions

---

[40] In earlier versions of MATLAB only five lines are displayed.

in MATLAB, and many collections of special functions are available from research groups in computational neuroscience. Functions can be compiled with a MATLAB Compiler, which has to be licensed separately. With the MATLAB Compiler it is possible to compile a function into a mex file that can be called like any other MATLAB function, or to translate it into a C or C++ source code that can be compiled with native C compilers to build stand-alone applications.

Table 12.3 Program wilson_euler.m

```
%%%%%%%%%%%%%%%%%%%%%%%%%%%%%%%%%%%%%%%%%%%%%%%%%%%%%%%%%%%%%%
%  Integration of Wilson model with the Euler method
%%%%%%%%%%%%%%%%%%%%%%%%%%%%%%%%%%%%%%%%%%%%%%%%%%%%%%%%%%%%%%
clear; clf;

% parameters of the model; 1=K,R  2=Ca,T   3=KCa,H   4=Na
g(1)=26; g(2)=2.25; g(3)=9.5; g(4)=1; E(1)=-.95; E(2)=1.20;
E(3)=E(1); E(4)=.50;

dt=0.01; I_ext=0; V=-1; x=zeros(1,4); tau(1)=dt./4.2;
tau(2)=dt./14; tau(3)=dt./45; tau(4)=1;

%Integration
t_rec=0;

for t=-100:dt:200
  switch t;
      case 0; I_ext=1;
  end

  x0(1)=1.24  +  3.7*V + 3.2*V^2;
  x0(2)=4.205 + 11.6*V + 8  *V^2;
  x0(3)=3*x(2);
  x0(4)=17.8  + 47.6*V +33.8*V^2;

  x=x-tau.*(x-x0); %rem x(4)=x0(4) because tau(4)=1
  I=g.*x.*(V-E);
  V=V+dt*(I_ext-sum(I));

  if t>=0;
      t_rec=t_rec+1;
      x_plot(t_rec)=t;
      y_plot(t_rec)=V;
  end
end % time loop

plot(x_plot,100*y_plot)
```

Table 12.4 Program wilson.m

```
%%%%%%%%%%%%%%%%%%%%%%%%%%%%%%%%%%%%%%%%%%%%%%%%%%%%%%%%%%%%%%%%%
% Integration of Wilson model with multistep ode solver
%%%%%%%%%%%%%%%%%%%%%%%%%%%%%%%%%%%%%%%%%%%%%%%%%%%%%%%%%%%%%%%%%
clear; clf;

% parameters of the model; 1=K,R;  2=Ca,T;  3=KCa,H; 4=Na;
g(1)=26; g(2)=2.25; g(3)=9.5; g(4)=1; g=g'; E(1)=-.95; E(2)=1.20;
E(3)=E(1); E(4)=.50; E=E'; tau(1)=1/4.2; tau(2)=1/14; tau(3)=1/45;
tau=tau';

%Integration:
 %1: Equilibration: no external input;
 y0=zeros(4,1); y0(4)=-1; param=0; I_ext=0; tspan=[0 100];
 [t,y]=ode15s('wilson_odefile',tspan,y0,[],I_ext,g,E,tau);

 %2: Integration with external input;
 y0=y(size(t,1),:); param=0; I_ext=1; tspan=[0 200];
 [t,y]=ode15s('wilson_odefile',tspan,y0,[],I_ext,g,E,tau);

plot(t,100*y(:,4));
```

Table 12.5 Function wilson_odefile.m

```
function ydot=wilson(t,y,flag,I_ext,g,E,tau)
% odefile for wilson equations
% parameters of the model; 1=K,R;  2=Ca,T;  3=KCa,H; 4=Na;

  V=y(4); y3=y(1:3);

  x0(1)=1.24  +  3.7*V + 3.2*V^2;
  x0(2)=4.205 + 11.6*V + 8  *V^2;
  x0(3)=3*y(2);

  x0=x0';

  ydot=tau.*(x0-y3);

  y3(4)=17.8  + 47.6*V +33.8*V^2;

  I=g.*y3.*(y(4)-E);
  ydot(4)=I_ext-sum(I);
return
```

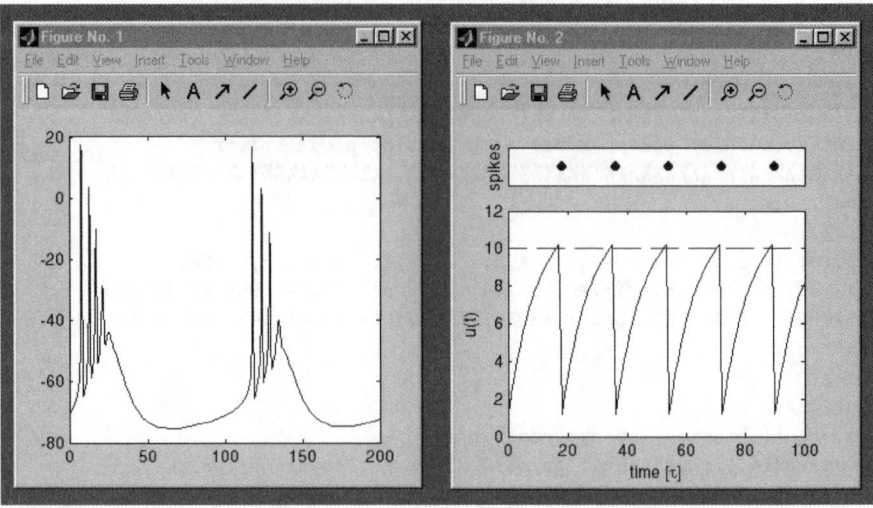

Fig. 12.4 (Left) MATLAB *Figure Window* produced with program `wilson.m`. The program `wilson_euler.m` produces a similar figure. (Right) MATLAB *Figure Window* produced by program `if_sim.m`.

12.2.4 Leaky integrate-and-fire neuron

A short example of a program for a leaky IF-neuron is included in file `if_sim.m` (see Table 12.6). It is straightforward to implement noise in various ways in these simulations. An example of a noisy integration is included in the line

```
uu=x*(1-tau_inv)*uu+tau_inv*I_e xt%+ ra ndn
```

as a comment. The function `randn` returns a normally distributed random variable with mean zero and variance one. This random variable can be included in the simulations by removing the two symbols ';%'. Such 'noisy' simulations can be compared to other noise models discussed in Chapter 3, such as the noisy reset.

The sample program also demonstrates how conditions can be used efficiently. For example, we included the firing threshold of the IF-neuron with the line

```
x=uu<theta;
```

This statement assigns values to each component of matrix x, which are one for all the elements of the matrix uu that are smaller than the scalar value `theta`, and which are zero for all the others. This matrix (or vector) is used to indicate a spike with the variable s by using `s(t_step)=1-x`. Finally, the command `subplot` is useful to place several plots into one figure window. The `help subplot` command gives more details. The Figure window produced by the program `if_sim.m` without noise is shown on the right side of Fig. 12.4.

12.2.5 Poisson spike trains

Before leaving this section we want to demonstrate with the help of program `poisson_spiketrain.m` (see Table 12.7) how to generate Poisson spike trains. A poisson spike train is simply defined by exponentially distributed interspike intervals.

Table 12.6 Program if_sim.m

```
%%%%%%%%%%%%%%%%%%%%%%%%%%%%%%%%%%%%%%%%%%%%%%%%%%%%%%%%%%%%%%%
% Simulation of (leaky) integrate-and-fire neuron
%%%%%%%%%%%%%%%%%%%%%%%%%%%%%%%%%%%%%%%%%%%%%%%%%%%%%%%%%%%%%%%
clear; clf;

% parameters of the model
tau_inv=0.1;     % inverse time constant
uu=0;            % initial membrane voltage
I_ext=12;        % constant external input
tspan=[0 100];   % integration interval
theta=10;        % firing threshold

%Integration with Euler method
t_step=0; for it=0:100;
    t_step=t_step+1;
    x=uu<theta;
    uu=x*(1-tau_inv)*uu+tau_inv*I_ext;%+randn;
    u(t_step)=uu;
    t(t_step)=it;
    s(t_step)=1-x;
end

subplot('position',[0.13 0.13 1-0.26 0.6])
  plot(t,u);
  hold on; plot([0 100],[10 10],'--');
  axis([0 100 0. 12])
  xlabel('time [\tau]');
  ylabel('u(t)')

subplot('position',[0.13 0.8 1-0.26 0.1])
  plot(t,s,'.','markersize',20);
  axis([0 100 0.5 1.5])
  set(gca,'xtick',[],'ytick',[])
  ylabel('spikes')
```

Exponentially distributed random numbers can be generated from uniformly distributed random numbers by taking the negative logarithm of the uniform random number and multiplying it by the parameter of the distribution (see Appendix B on how to transform random numbers). This can be implemented with the MATLAB command line

```
r=-lambda*log(rand(n1,n2))
```

To simulate the refractoriness of the simulated neuron producing such a spike train we deleted in the program the firing times with a certain probability that depends on the interspike interval. This does not affect most of the spikes in a low-frequency spike train, but removes some of the unrealistically large number of small interspike intervals. This program also features another specialized plotting routine, that of plotting histograms

Table 12.7 Program poisson_spiketrain.m

```
%%%%%%%%%%%%%%%%%%%%%%%%%%%%%%%%%%%%%%%%%%%%%%%%%%%%%%%%%%%%%%%%%%
% Generation of Poisson spike train with refractoriness
%%%%%%%%%%%%%%%%%%%%%%%%%%%%%%%%%%%%%%%%%%%%%%%%%%%%%%%%%%%%%%%%%%
clear; clf; hold on;  % parameters of the model;
fr_mean=15/1000;      % mean firing rate per ms

% generating presynaptic poisson spike trains
lambda=1/fr_mean;           % time interval/firing rate in interval
ns=1000;                    % number of spikes to be generated
isi1=-lambda.*log(rand(ns,1)); % generation of exponential distr. ISIs
% Delete spikes that are within refractory period
is=0; for i=1:ns;
    if rand>exp(-isi1(i)^2/32);
        is=is+1;
        isi(is)=isi1(i);
    end
end

hist(isi,50);          % Plot histogram of 50 bins
cv=std(isi)/mean(isi)  % coefficient of variation
```

with the command hist. Using this function without assigning it to a variable bins data and produces a bar plot. The functions hist can, however, be used more generally to bin values, and a separate function the MATLAB function bar can be used to produce bar plots. The program produces figures similar to the one used in Fig. 3.4B.

12.2.6 Netlet formulas by Anninos *et al.*

The formulas derived by P. A. Anninos, B. Beek, T. J. Csermely, E. M. Harth, and G. Pertile, Dynamics of neural structures, *J. Theor. Biol.* 26: 121–8 (1970), are coded into MATLAB in program anninos.m shown in Table 12.8. The constant h is the fraction of inhibitory nodes, and the variables c_in and c_ex are the average number of effective synapses (inputs) to an inhibitory or excitatory node, respectively. This program was used to produce Fig. 4.7.

12.3 Associators and Hebbian learning

The programs of this section can be found in the associator folder on the web.

12.3.1 Hebbian weight matrix in rate models

The basic Hebbian rule for rate models states that a term

$$\Delta \mathbf{w} = \mathbf{ba}'$$

(12.3)

Table 12.8 Program anninos.m

```
%%%%%%%%%%%%%%%%%%%%%%%%%%%%%%%%%%%%%%%%%%%%%%%%%%%%%%%%%%%%%%%%
% plot netlet formula of Anninos et al.
%%%%%%%%%%%%%%%%%%%%%%%%%%%%%%%%%%%%%%%%%%%%%%%%%%%%%%%%%%%%%%%%
clear; clf; hold on;

h=0.; c_in=10; c_ex=10;

for theta=1:7;
    rec=0;
    for x=0.:0.01:1.;
        rec=rec+1;
        tmp1=(1-h)*x*c_ex; tmp2=0;
        for i=0:theta-1; tmp2=tmp2+tmp1^i/prod(1:i); end;
        tmp3=h*x*c_in; tmp4=0;
        for i=0:20; tmp4=tmp4+tmp3^i/prod(1:i); end;
        f0(rec)=x;
        f1(rec)=(1-x)*exp(-tmp3)*tmp4*(1-exp(-tmp1)*tmp2);
    end
    plot(f0,f1);
end
plot([0 1],[0 1],'k')
plot([0 1],[1 0],'k')
box on; axis square
```

is added to the previous weights **w**, where **a** is a column vector with components a_i of firing rates of the presynaptic nodes of one example used for training, and **b** is the firing rate of the postsynaptic node. If we consider only one postsynaptic node the $\Delta\mathbf{w}$ (as well as **w**) is a row vector. We can collect all the nodes in a network into a matrix by adding corresponding rows in the weight matrix. Equation 12.3 is then the corresponding formula where **b** is now a column vector with the rates of the postsynaptic nodes. We can also collect all the column vectors of different training examples into an array, so that we get matrices **a** and **b** with components $a_{i\mu}$ and $b_{i\mu}$ for training example μ. The Hebbian weight matrix after training all the examples is then given by adding to the initial matrix (which can be the zero matrix) the sum over all the training examples, that is,

$$w_{ij} = w_{ij}^0 + \sum_{\mu} b_{i\mu} a_{i\mu}. \tag{12.4}$$

The second term on the right-hand side corresponds to a matrix multiplication of the form **ba'**. The weight change can thus still be written in the form of eqn 12.3. Hebbian training can thus be written very compactly in MATLAB and is given by the line

```
w=(r1-afr)*(r2-afr)';
```

in program hebb_dis.m printed in Table 12.9. The only difference is that we used there a Hebbian co-variance rule in which the mean firing rates are subtracted from the

Table 12.9 Program hebb_dis.m

```
%%%%%%%%%%%%%%%%%%%%%%%%%%%%%%%%%%%%%%%%%%%%%%%%%%%%%%%%%%%%%%%%%
% Weight distribution of Hebbian synapses in rate model
%%%%%%%%%%%%%%%%%%%%%%%%%%%%%%%%%%%%%%%%%%%%%%%%%%%%%%%%%%%%%%%%%
clear; clf; hold on;

nn=500; %try also nn=5000
npat=1000;

%random pattern; firing rates are exponential distributed
afr=40;
r1=-afr.*log(rand(nn,npat)); % exponential distr. presyn. firing rates
r2=-afr.*log(rand(1,npat));  % exponential distr. postsyn. firing rate

w=(r2-afr)*(r1-afr)';
w=w/npat;

x=-260:20:260;
[n,x]=hist(w,x);
n=n/sum(n)/20;
h=bar(x,n); set(h,'facecolor','none');

% fit normal to data
a0=[0 40];
a=lsqcurvefit('normal',a0,x,n);
n2=normal(a,x);
plot(x,n2,'r')
```

firing rates of the pre- and postsynaptic nodes. This program was utilized to produce Fig. 7.8 with 5000 presynaptic nodes (nn=5000;) to get a reasonable sampling. We have used exponentially distributed firing rates. It is straightforward to replace the corresponding lines with other rate vectors (for example r1=rand(nn,npat); etc.), and to test that the resulting weight distributions do not depend on the choice.

The program also demonstrates how a function, specified in a MATLAB function file as illustrated in Table 12.10, can be fitted to data points using the Statistics Toolbox routine lsqcurvefit (nlinfit in newer versions of the toolbox).

We mentioned in Section 7.6 that the weight distributions of basic Hebbian weight matrices are the same in networks in which we separate excitatory and inhibitory neurons in agreement with Dale's principle. You could test this yourself with the program printed in Table 12.11. We have there defined a random sign vector

    inex=2*rand(nn,1)-1;

that specifies the attribute (sign) of each synapse. After updating the weight matrix for each pattern we check if the sign in the weight matrix is still consistent with the attribute of the synapse, and set the weight value to zero if this is not the case. This is achieved with the line

Table 12.10 Function normal.m

```
function  y=fitfun(a,x);
% function for fitting normal distribution

y=1/(sqrt(2*pi)*a(2))*exp(-(x-a(1)).^2./(2*a(2).^2));

return
```

Table 12.11 Program hebb_dis_exin.m

```
%%%%%%%%%%%%%%%%%%%%%%%%%%%%%%%%%%%%%%%%%%%%%%%%%%%%%%%%%%%%%
% Weight distribution of Hebbian synapses
% with separate excitatory and inhibitory synapses
%%%%%%%%%%%%%%%%%%%%%%%%%%%%%%%%%%%%%%%%%%%%%%%%%%%%%%%%%%%%%
clear; clf; hold on

nn=500; npat=1000;

%random pattern and neuron types; firing rates are exp. distributed
afr=40;
inex=2*rand(nn,1)-1;
r1=-afr.*log(rand(nn,npat)); % exponential distr. presyn. firing rates
r2=-afr.*log(rand(1,npat));  % exponential distr. postsyn. firing rate


w=zeros(nn,1);
for ipat=1:npat
  w=w+(r1(:,ipat)-afr).*(r2(ipat)-afr);
  w=w.*((w.*inex)>0);
end

w=w/npat;

x=-260:20:260;
[n,x]=hist(w,x);
n=n/sum(n)/20;
h=bar(x,n); set(h,'facecolor','none');

% fit normal to data
a0=[0 40];
a=lsqcurvefit('normal',a0,x,n);
n2=normal(a,x);
plot(x,n2,'r')
```

```
w=w.*((w.*inex)>0);
```

The matrix w.*inex has components with positive signs for consistent synapses and negative signs for inconsistent ones. The matrix (w.*inex)>0 is a matrix that has zeros for negative components of w.*inex and ones otherwise. The component by component multiplication of this matrix with the weight matrix after learning therefore sets all the inconsistent values in the weight matrix to zero. The rest is the same as in program hebb_dis.m.

12.3.2 Hebbian learning with weight decay

An example of using an associative node with Hebbian learning and weight decay to perform a principal component analysis is discussed in Section 7.7.3. The corresponding program oja.m is given in Table 12.12. The network for this particular two-dimensional problem has only one linear node with two input channels. This architecture is specified by the initial weight matrix. In the program we go on and define a rotation matrix that is used to generate training examples. For each training example we update the network and change the weights using the term proposed by Oja (see eqn 7.14). The learning rate is therein set to the value 0.1. Within the training loop we record the weight matrix after each training step to plot the corresponding trajectory later in the figure. The program produces figures similar to the one shown in Fig. 7.11B (we have only later added an arrowhead, and a large x indicating the initial weight values, and some labels).

Table 12.12 Program oja.m

```
%%%%%%%%%%%%%%%%%%%%%%%%%%%%%%%%%%%%%%%%%%%%%%%%%%%%%%%%%%%%%%%
% Linear associator with Hebb and weight decay: PCA a la Oja
%%%%%%%%%%%%%%%%%%%%%%%%%%%%%%%%%%%%%%%%%%%%%%%%%%%%%%%%%%%%%%%
clear; clf; hold on

y=0; w=[0.1;0.4]; a=-pi/6;
rot=[cos(a) sin(a);-sin(a) cos(a)] % rotation matrix

for i=1:1000
    x=0.05*randn(2,1); x(1)=4*x(1); %training examples
    x=rot*x; % rotation of training examples
    y=w'*x;  % network update
    plot(x(1),x(2),'.')
    w=w+0.1*y*(x-y*w); % training
    w_traj(:,i)=w; % recording of weight history
end

plot(w_traj(1,:),w_traj(2,:),'r')
plot([0 w(1)],[0 w(2)],'k','linewidth',2)
axis([-1 1 -1 1]);
plot([-1 1],[0 0],'k'); plot([0 0],[-1 1],'k')
```

12.4 Recurrent networks and network dynamics

12.4.1 Example of a complete network simulation

We discussed auto-associator networks extensively in Chapter 8, and we will show here how these models can be simulated with MATLAB. We will run in this section through a whole simulation task: writing a program and running the numerical experiments. The minimal simulation program for an auto-associative neural network has to include the following five steps.

1. Generating the pattern to be stored;
2. Training the network with Hebbian learning;
3. Initializing the network;
4. Updating the network;
5. Plotting the results.

Each step can be programmed with simple instructions, and we will write a single line for each of the steps.

12.4.1.1 Generating the pattern to be stored

The first task is to generate a pattern set that consists of a set of random variables for each node in the network. We want to use binary random numbers with values of either -1 or 1. We can generate such equally distributed random numbers with the `rand` function in MATLAB. The statement `rand(500,10)` returns a matrix with 500 rows, one for each node in the network of 500 nodes used in this example, and 10 columns, each column representing a pattern in a set of 10 different patterns. The `rand` function returns a uniformly distributed random variable between 0 and 1. To transform this into binary random numbers we use the `floor` function that rounds its argument toward the next lower integer value. So, if we multiply variables in the range 0 to 1 by the factor 2, we get variables in the range 0 to 2. The floor function then returns either a 0 or a 1. By multiplying this number by 2 and subtracting 1 we get our anticipated random variable of either -1 or 1. Thus, the binary pattern set can be generated with the command line

```
pat=2*floor(2*rand(500,10))-1;            % Random binary pattern
```

12.4.1.2 Training the network with Hebbian learning

The next step is to produce a 500×500 weight matrix from the training pattern using Hebbian learning. Each pattern k contributes to each weight element with `w(i,j)=pat(i,k)pat(j,k)'`. We could then write three loops over i, j, and k, adding up all the contributions of all the patterns, to produce the Hebbian weight matrix. However, explicit program loops are not only lengthy to write and difficult to read, but can also slow down the processing time of the program considerably. However, the loops represent just a matrix multiplication (as discussed in Section 12.3) that we do not have to code explicitly in MATLAB. We recommend that matrix operations should be used whenever possible. Thus, we can code Hebbian learning in MATLAB with the simple line.

```
w=pat*pat';                               % Hebbian learning
```

The difference between this code and the one used in program hebb_dis.m (see eqn 12.4) is that matrix **a** and **b** are now the same as the connections are between all nodes in the network.

12.4.1.3 Initializing the network

We have now defined the weight matrix, and by doing so have defined the network architecture. The remaining tasks are to initialize the network and to test its performance. We make tests with random initial states, with a firing rate r for each node in the range of $-0.5 \leq r < 0.5$. The rate of the nodes at the first time step r(:,1), which is equivalent to r in this case, is set to be

```
r=rand(500,1)-0.5;                          % Initialize network
```

12.4.1.4 Updating the network

We are now ready to update the network, which consists of a loop over the time steps that we want to simulate surrounding the discrete updating rule. We can write this in MATLAB as

```
for t=2:10; r(:,t)=tanh(w*r(:,t-1)); end   % Update network
```

The update rule specifies the firing rates at the new time step as a function of the firing rates of the previous time step that are left-multiplied by the weight matrix and transformed with the activation function tanh. The matrix **r** represents the firing rates of all 500 nodes for time steps $t = 1, 2, ..., 10$ after the loop is completed.

12.4.1.5 Plotting the results

Finally, we want to display the results of the simulations graphically. At this time we have calculated the firing rates for all the neurons at each time step during the simulation interval of the network. Plotting all these values is too much information to comprehend. Instead we want to test how close the state of the network at each time step is to each of the training patterns. We can quantify this with the normalized dot product (see Appendix A) between the state of the network at each time step and each of the training vectors. This is achieved with the final command line

```
plot(r'*pat/500)                            % Plot overlap
```

that produces a plot with 10 lines (see Fig. 12.5), each representing the normalized dot product (overlap) of the network over time with each of the 10 memory states.

12.4.1.6 The complete program

The complete auto-associator network simulation program has only five lines, corresponding to the steps outlined at the beginning of this section, and is listed in Table 12.13 and included in the ann folder on the web.

Run this program several times. An example of the Figure window produced with this program is shown on the left side of Fig. 12.5. Of course, each run of the program produces different results due to the random nature of the stored pattern and the random initial conditions of the network. The network state came very close to one of the imprinted patterns (one scaled dot product is very close to 1) in the example shown in the figure. Sometimes you might also find a normalized dot product of -1 corresponding to one of the learned patterns. This corresponds to the exact inverse of a trained pattern and demonstrates that the inverse pattern is also a point attractor of the system as mentioned in Chapter 8.

Table 12.13 Program rnn_short.m

```
pat=2*floor(2*rand(500,10))-1;          % Random binary pattern
w=pat*pat';                             % Hebbian learning
r=rand(500,1)-0.5;                      % Initialize network
for t=2:10; r(:,t)=tanh(w*r(:,t-1)); end  % Update network
plot(r'*pat/500)                        % Plot overlap
```

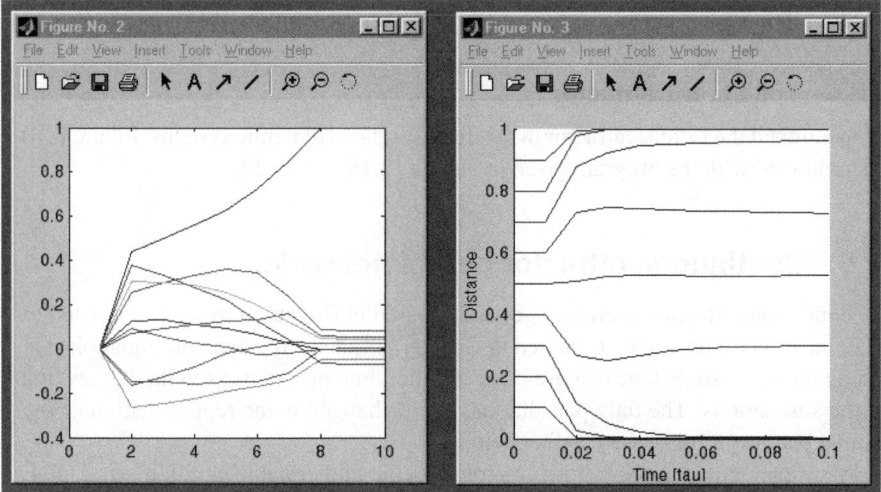

Fig. 12.5 Examples of results from simulations of auto-associative memory models. (Left) Example produced with the ann_short.m program. The normalized dot product of the network states during update with all of the ten patterns that were imprinted with Hebbian learning into the network is plotted. The network came close to one of the trained patterns for the line that comes close to 1. (Right) Example produced with the ann.m program. The distance of the network state to the first of 40 trained patterns for different initial conditions of the network is plotted. The distance is zero if the network state is equivalent to this trained pattern, in contrast to the simulation on the left where we displayed the normalized dot product.

12.4.2 Quasicontinuous attractor network

The short demonstration program that we just discussed is intended to demonstrate the principal idea behind such network simulations, and different kinds of experiments can be performed as discussed in the corresponding chapters. A similar simulation program, which includes smaller time steps in the network dynamics as well as a more controlled initialization of the network, is given in file ann.m in the ann folder and is listed in Table 12.14. In this program we can choose initial states of the network that have a predefined distance from the first pattern in the pattern set that was used for training the network. Some results of this program are shown in the right window of Fig. 12.5. Note that we plotted there the distance of the network state to the first training pattern, which is zero when these two vectors agree. We used a training set of 40 patterns in this simulation. The network was able to complete a noisy version of the pattern when the noise was small enough.

12.4.3 Networks with a random asymmetric weight matrix

In Table 12.15 we have listed the program that was used to generate Fig. 8.12. The program again contains the simple Euler method to update the network. The only difference to the previous programs is the choice of the weight matrix which is here only a simple random matrix not trained with Hebbian learning. Note that this program runs for a relatively long time compared to the other programs in this tutorial.

12.4.4 The Lorenz attractor

We mentioned the Lorenz attractor in the discussions on dynamic systems. Figure 8.10 was produced with the program given in Tables 12.16 and 12.17.

12.5 Continuous attractor neural networks

The continuous attractor neural networks discussed in Chapter 9 are closely related to the point attractor networks discussed in Chapter 8. Indeed, the dynamic equations are exactly the same so that we can use basically the same programs as in the last section for the simulations. The only part that has to be changed is the representation of the training patterns as discussed further below.

We implemented in the following programs the numerical integration using ODE solvers provided by MATLAB, which could in turn also be used in the simulations of point attractor networks. As explained before (see section 12.2), to use the ODE solvers provided in MATLAB we have to write a MATLAB function that contains the dynamic equations. The function that implements the dynamic equations of recurrent networks is shown in Table 12.18, which is a direct translation of eqns 8.1 or 9.1.

The main program is listed in Table 12.19. It mainly specifies some parameters, calls a training procedure that is discussed below, and then runs an experiment by updating the network first with an external input, and then updates the network without external input. This program was used to make Fig. 9.5. We have there utilized a three-dimensional plotting routine called surf, and set the view of the plot so that it is seen from above. MATLAB displays the z-values in a colour scheme that can easily be changed with the command colormap. We used the command colormap(1-gray) to produce a grey scale version of the plot to print it in the book. We also suppressed the grid lines. Note, however, that the linestyle parameter can only be specified within the surf command in MATLAB version 5.3 or higher. In earlier versions one has to return the result of the plot to a pointer, for example, h=surf(t',1:nn,r','linestyle' and then to change the linestyle with the set command like set(h,'linestyle','none');.

Table 12.14 Program ann.m

```
%%%%%%%%%%%%%%%%%%%%%%%%%%%%%%%%%%%%%%%%%%%%%%%%%%%%%%%%%%%%%%%
% fully connected continues recurrent ANN, with s in [-1,1]
%%%%%%%%%%%%%%%%%%%%%%%%%%%%%%%%%%%%%%%%%%%%%%%%%%%%%%%%%%%%%%%
clear; clf; hold on;

%%%%%% Network parameters %%%%%%
  nn=500;      % number of nodes
  npat=50;     % number of pattern
  dt=.01 ;     % time step
  tau=1;       % time constant
  tend=0.1;    % end of simulation
%%%%%% random binary pattern with elements [-1;1] %%%%%%%%%%
  pat=2*floor(2*rand(nn,npat))-1;

%%%%%%%%% weight matrix: Hebb learning %%%%%%%%%
  w=pat*pat';
  %w=w./npat-diag(1); %if we want to remove self couplings

for dist0=0.1:0.1:.9; %loop over different initial distances
  %%%% initial pattern: pattern(1) with nflip flips %%%%%%
  u=pat(:,1);
  nflips=floor(dist0*nn);
  flag=zeros(nn,1);
  iflip=1;
  while iflip<=nflips;
     ii=floor(nn*rand)+1;
     if flag(ii)==0; u(ii)=-u(ii); flag(ii)=1; iflip=iflip+1; end
  end
  s=tanh(u);
  x(1)=0;
  y(1)=(1-pat(:,1)'*s/(sqrt(pat(:,1)'*pat(:,1))*sqrt(s'*s)))/2;
  %%%%%%%%% Update %%%%%%%%%
  irec=1;
  for t=dt:dt:tend;
      s=tanh(u);
      u=(1-dt/tau)*u+dt/tau*(w*s);
      irec=irec+1;
      x(irec)=t;
      y(irec)=(1-pat(:,1)'*s/(sqrt(pat(:,1)'*pat(:,1))*sqrt(s'*s)))/2;
  end   %time
  plot(x,y);
end % dist0
axis([0 tend 0 1]); xlabel('Time [tau]'); ylabel('Distance')
```

Table 12.15 Program ann_chaos.m

```
%%%%%%%%%%%%%%%%%%%%%%%%%%%%%%%%%%%%%%%%%%%%%%%%%%%%%%%%%%%%%%%%%%%%
% recurrent network model with a mixture of symmetric and
% antisymmetric weights of unit absolute or random components
%%%%%%%%%%%%%%%%%%%%%%%%%%%%%%%%%%%%%%%%%%%%%%%%%%%%%%%%%%%%%%%%%%%%
clear; nodes=251; dt=.1 ;  tend=100; y=zeros(1,tend/dt+1);

wa=randn(nodes); ws=randn(nodes);  %random
%wa=ones(nodes); ws=ones(nodes);      %unit
for i=2:nodes; for j=1:i; wa(i,j)=-wa(j,i); end; end
for i=2:nodes; for j=1:i; ws(i,j)=ws(j,i); end; end
for i=1:nodes; wa(i,i)=0; ws(i,i)=0; end

for a1=1:41;
  a=(a1-21)*.5;
  for b1=1:41;
    b=(b1-21)*.5; disp([a,b]);
    w=a*ws+b*wa;

    %update
    u=2*rand(nodes,1)-1;
    irec=1; x(irec)=0; y(irec)=norm(u);
    for t=dt:dt:tend;
        s=tanh(u);
        u=(1-dt)*u+dt*w*s;
        irec=irec+1;
        x(irec)=t;
        y(irec)=norm(u);
    end %loop over time steps

    xa(a1)=a; xb(b1)=b;
    u_dif(a1,b1)=abs(y(irec)-y(irec-1));
  end %xa
end %xb
```

Table 12.16 Program lorenz_attr.m

```
%Plot trajectory of lorenz system
clear; clf;
a=10; b=28; c=8/3;
u0 = zeros(3,1)+0.5; param=0; tspan=[0,100];
[t,u]=ode45('lorenz_odefile',tspan,u0,[],a,b,c);
plot3(u(:,1),u(:,2),u(:,3))
```

Table 12.17 Function lorenz_odefile.m

```
function udot=lorenz(t,u,flag,a,b,c)
% odefile for lorenz system
  udot(1)=a*(u(2)-u(1));
  udot(2)=u(1)*(b-u(3))-u(2);
  udot(3)=u(1)*u(2)-c*u(3);
  udot=udot';
return
```

Table 12.18 Function rnn_ode.m

```
function udot=rnn(t,u,flag,nn,tau_inv,dx,beta,alpha,w,I_ext)
% odefile for recurrent network
  r=1./(1+exp(-beta.*(u-alpha)));
  sum=w*r*dx;
  udot=tau_inv*(-u+sum+I_ext);
return
```

The training of the network that specifies the weight matrix w is now done in the function hebb.m. This function implements the standard Hebbian learning rule used before, although we have used a loop over different training patterns in this implementation. The important difference to the point attractor networks discussed before is that we use training patterns with Gaussian rate profiles, centred around all the different nodes in the network. These training profiles are returned by the function in_signal_pbc.m listed in Table 12.21.

12.5.1 Path integration

A very interesting feature of the representation of states within continuous attractor models is that the states can be updated with differential information as discussed in Section 9.4 and demonstrated in Fig. 9.10. The program that was used for this figure is listed in Table 12.22. The 'path integration' mechanism is based on the use of sigma–pi neurons that have the effect of producing an effective weight matrix that consists of a mixture of a symmetric and an asymmetric component. This is coded in the MATLAB program, for example, in the experiment for counter-clockwise rotation, with the line w=(ws-w_inh).*(1+2*wa(:,:,1)); towards the end of the program. The first part is the regular symmetric weight matrix resulting from the standard Hebbian learning and from which the inhibition constant is subtracted. The second term is a modulation factor (because it is multiplied) that alters the strength of the regular weight matrix with the form specified by the asymmetric weight matrix wa. The factor in front of this weight matrix specifies the strength of the modulation, which can, for example, be proportional to the firing rate of rotation cells feeding into the CANN.

The symmetric and asymmetric weight matrices are set up in a training session as

Table 12.19 Program cann.m

```
%%%%%%%%%%%%%%%%%%%%%%%%%%%%%%%%%%%%%%%%%%%%%%%%%%%%%%%%%%%%%%%
% 1-d Continuous Attractor Neural Network with Hebbian learning
%%%%%%%%%%%%%%%%%%%%%%%%%%%%%%%%%%%%%%%%%%%%%%%%%%%%%%%%%%%%%%%
clear; clf; hold on;

nn = 100; dx=100/nn; % number of nodes and resolution in deg
tau_inv = 1./1;      % inverse of membrane time constant
beta =.07; alpha=.0; % transfer function is 1/(1+exp(-beta*(u-theta)))

%weight matrices
    sig = 5/dx;
    w_sym=hebb(nn,sig);
    w_inh=3*(sqrt(2*pi)*sig)^2/nn;
    w=w_sym-w_inh;


%%%%%%%%%%%%%%%%%%%%%%%%%%%%%%%%%%%%%%%%%%%%%%%%%%%%%%%%%%%%%%%
%%%%%%%%%%%%%%%%%%%%  Experiment   %%%%%%%%%%%%%%%%%%%%%%%%%%%%
%%%%%%%%%%%%%%%%%%%%%%%%%%%%%%%%%%%%%%%%%%%%%%%%%%%%%%%%%%%%%%%

%%%% external input to initiate bubble
    u0 = zeros(nn,1)-10;
    I_ext=zeros(nn,1); for i=-20:20; I_ext(nn/2+i)=50; end
    param=0;
    tspan=[0,10];
    [t,u]=ode45('rnn_ode',tspan,u0,[],nn,tau_inv,dx,beta,alpha,w,I_ext);
    r=1./(1+exp(-beta.*(u-alpha)));
    surf(t',1:nn,r','linestyle','none'); view(0,90);

%%%% no external input to equilibrate
    u0 = u(size(t,1),:);
    I_ext=zeros(nn,1);
    param=0;
    tspan=[10,20];
    [t,u]=ode45('rnn_ode',tspan,u0,[],nn,tau_inv,dx,beta,alpha,w,I_ext);
    r=1./(1+exp(-beta.*(u-alpha)));
    surf(t',1:nn,r','linestyle','none');
    [max_val,max_ind1]=max(r(size(t,1),:))
```

specified in the function listed in Table 12.23. The symmetric part is analogous to the training in function 12.20 used in the basic CANN model. The important addition to enable path integration is the use of a trace rule that enables the temporal association between the co-firing of rotation cells and the direction of the activity packet in the CANN model as explained in Section 9.4. The Hebbian learning thus acts on the trace of the firing rates stored in the variable r_trace.

Table 12.20 Function hebb.m

```
function w = hebb(nn,sig)
% self organization of symmetric cann interactions
    lrate=1;    % learning rate
    w=zeros(nn);
    %%%%%% learning session
    for loc=1:nn;
        r=in_signal_pbc(loc,1,sig,nn);
        dw=lrate.*r*r';
        w=w+dw;
    end
return
```

Table 12.21 Function in_signal_pbc.m

```
function y = in_signal_1d(loc,ampl,sig,nn)
% !!! REM sig should be smaller than nn/6
% assigns an input signal with gaussian shape
    if sig>nn/6;
        disp('Warning: The width of input should be less than nn/6 !!!');
    end
    y=zeros(nn,1);
    for i=floor(loc-3*sig):ceil(loc+3*sig);
        imod=mod(i-1,nn)+1;
        dis=(i-loc)^2/(2.*sig^2);
        y(imod)=ampl*exp(-dis);
    end
return
```

12.6 Error-back-propagation network

The final program that we want to outline in this tutorial is the standard error-back-propagation algorithm mentioned in Chapter 10. We demonstrate the utilization of this algorithm by training a sigmoidal network with two hidden nodes and one output node on the XOR function discussed in Chapter 6. The program is listed in Table 12.24 and included in the folder others on the web. This program defines the weight matrices with small random initial values and then defines the training pattern that represents the XOR function. In a loop over 10,000 training examples we select randomly one of the input–output patterns, propagate it through the network, calculate the delta values, update the weight matrices, and then test all the patterns on the current version of the network.

A typical training curve produced with this program is shown in Fig. 12.6 (each time the program is called a different learning curve is produced because different random initial weight values are used and a different order of the training examples is presented

Table 12.22 Program cann_pathinte.m

```
%%%%%%%%%%%%%%%%%%%%%%%%%%%%%%%%%%%%%%%%%%%%%%%%%%%%%%%%%%%%%%%%%%%%
% 1-d Continuous Attractor Neural Network for path integration
%%%%%%%%%%%%%%%%%%%%%%%%%%%%%%%%%%%%%%%%%%%%%%%%%%%%%%%%%%%%%%%%%%%%
clear; clf; hold on; view(0,90);
  nn = 100; dx=100/nn; % number of node an resolution in deg
  tau_inv = 1./1;      % inverse of membrane time constant
  beta =.07; alpha=.0; % transfer function is 1/(1+exp(-beta*(u-theta))
%weight matrices
  sig = 5/dx;
  [ws,wa]=hebb_trace(nn,sig);
  w_inh=7*(sqrt(2*pi)*sig)^2/nn;
%%%%%%%%%%%%%%%%%%%%%%%%%%%%%%%%%%%%%%%%%%%%%%%%%%%%%%%%%%%%%%%%%%%%
%%%%%%%%%%%%%%%%%%%%  Experiment   %%%%%%%%%%%%%%%%%%%%%%%%%%%%%%%%
%%%%%%%%%%%%%%%%%%%%%%%%%%%%%%%%%%%%%%%%%%%%%%%%%%%%%%%%%%%%%%%%%%%%
%%%% external input to initiate bubble, no idiothetic cue, w sym.
  u0 = zeros(nn,1)-10; tspan=[0,10]; param=0;
  w=ws-w_inh;
  I_ext=zeros(nn,1); for i=-20:20; I_ext(nn/2+i)=50; end
  [t1,u1]=ode45('rnn_ode',tspan,u0,[],nn,tau_inv,dx,beta,alpha,w,I_ext);
%%%% no external input to equilibrate, no idiothetic cue.
  u0 = u1(size(t1,1),:);  tspan=[10,20];
  I_ext=zeros(nn,1);
  [t2,u2]=ode45('rnn_ode',tspan,u0,[],nn,tau_inv,dx,beta,alpha,w,I_ext);
%%%% clockwise rotation node on --> weight matrix asymetric
  u0 = u2(size(t2,1),:); tspan=[20,40];
  w=(ws-w_inh).*(1+wa(:,:,2));
  [t3,u3]=ode45('rnn_ode',tspan,u0,[],nn,tau_inv,dx,beta,alpha,w,I_ext);
%%%% no idiothetic cue --> weight matrix symetric
  u0 = u3(size(t3,1),:); tspan=[40,50];
  w=ws-w_inh; I_ext=zeros(nn,1);
  [t4,u4]=ode45('rnn_ode',tspan,u0,[],nn,tau_inv,dx,beta,alpha,w,I_ext);
%%%% counter-clockwise rotation node on --> weight matrix asymetric
  u0 = u4(size(t4,1),:); tspan=[50,70];
  w=(ws-w_inh).*(1+2*wa(:,:,1)); I_ext=zeros(nn,1);
  [t5,u5]=ode45('rnn_ode',tspan,u0,[],nn,tau_inv,dx,beta,alpha,w,I_ext);
%%%% no idiothetic cue --> weight matrix symetric
  u0 = u5(size(t5,1),:); tspan=[70,80];
  w=ws-w_inh; I_ext=zeros(nn,1);
  [t6,u6]=ode45('rnn_ode',tspan,u0,[],nn,tau_inv,dx,beta,alpha,w,I_ext);
%%%% plot results
  u=[u1; u2; u3; u4; u5; u6]; t=[t1; t2; t3; t4; t5; t6];
  r=1./(1+exp(-beta.*(u-alpha)));
  %surf(t',1:nn,r','linestyle','none')
  h=surf(t',1:nn,r');
  set(h,'linestyle','none');
```

Table 12.23 Function hebb_trace.m

```
function [ws,wa] = hebb_trace(nn,sig)
% self organization of idiothetic connections
  lrate=1;      % learning rate
  ws=zeros(nn); wa=zeros(nn,nn,2);
  r_trace=0; eta=0.9;

  %%%%%% learning session
  for loc=1:nn; % counter-clockwise rotation
      r=in_signal_pbc(loc,1,sig,nn);
      r_trace=eta*r_trace+(1-eta)*r;
      dws=lrate.*r*r';
      dwa(:,:,1)=lrate.*r*r_trace';
      dwa(:,:,2)=0;
      ws=ws+dws;
      wa=wa+dwa;
  end
  r_trace=0; eta=0.9;
  for loc=nn:-1:1; % clockwise rotation
      r=in_signal_pbc(loc,1,sig,nn);
      r_trace=eta*r_trace+(1-eta)*r;
      dws=lrate.*r*r';
      dwa(:,:,1)=0;
      dwa(:,:,2)=lrate.*r*r_trace';
      ws=ws+dws;
      wa=wa+dwa;
  end
 return
```

during learning). Due to the small initial weights the network always responds initially with values around $r^{out} \approx 0.5$ which makes the mean square error relatively small from the start. It then takes many iterations, even with the relatively large learning rate of 0.7, to finally reduce the error significantly and to reach a level of performance that is sufficient to represent the function correctly. The network responded after the 10,000 learning steps shown in the example with the values

   0.8872   0.2443   0.2587   0.4252

to the four different training vectors, respectively. Such a response pattern is indeed sufficient to represent the XOR function. Just imagine that we use a final output node with a threshold activation function with a firing threshold at 0.3.

The basic error-back-propagation is known to be very inefficient in training neural networks, and many advances have been made to improve the performance considerably as mentioned in the text. These include higher-order gradient methods, hybrid methods, the smart choice of initial conditions, and the proper use of adaptive learning rates. MATLAB provides a neural network toolbox that implements many such advanced algorithms, and shareware versions of different algorithms can also be found

Table 12.24 Program mlp.m

```
%%%%%%%%%%%%%%%%%%%%%%%%%%%%%%%%%%%%%%%%%%%%%%%%%%%%%%%%%%%%%%%%%
% Back-propagation network: XOR Example
%%%%%%%%%%%%%%%%%%%%%%%%%%%%%%%%%%%%%%%%%%%%%%%%%%%%%%%%%%%%%%%%%
clear; clf;

N_i=2; N_h=2; N_o=1;

w_h=1*(rand(N_h,N_i)-0.5); w_o=1*(rand(N_o,N_h)-0.5);

% training vectors (XOR)
r_i=[0 1 0 1;
     0 0 1 1];
r_d=[1 0 0 1];

%Updating and training network with sigmoid activation function
for sweep=1:10000;
    i=ceil(4*rand);
    r_h=1./(1+exp(-w_h*r_i(:,i)));
    r_o=1./(1+exp(-w_o*r_h));
    d_o=(r_o.*(1-r_o))'.*(r_d(:,i)-r_o);
    d_h=(r_h.*(1-r_h))'.*(w_o*d_o);
    w_o=w_o+0.7*(r_h*d_o)';
    w_h=w_h+0.7*(r_i(:,i)*d_h)';
    % test all pattern
      r_o_test=1./(1+exp(-w_o*(1./(1+exp(-w_h*r_i)))));
      d(sweep)=0.5*sum((r_o_test-r_d).^2);
end

plot(d)
```

on the web. We have not utilized this toolbox in this book as we have mainly focused on biologically plausible models that are not the main focus of the MATLAB Neural Network toolbox, and the utilization of the basic programming primitives enables us to specify models with great flexibility. It should, however, be mentioned that there are additional MATLAB packages, such as a graphical platform to run simulations called Simulink®, which might be useful in your specific applications.

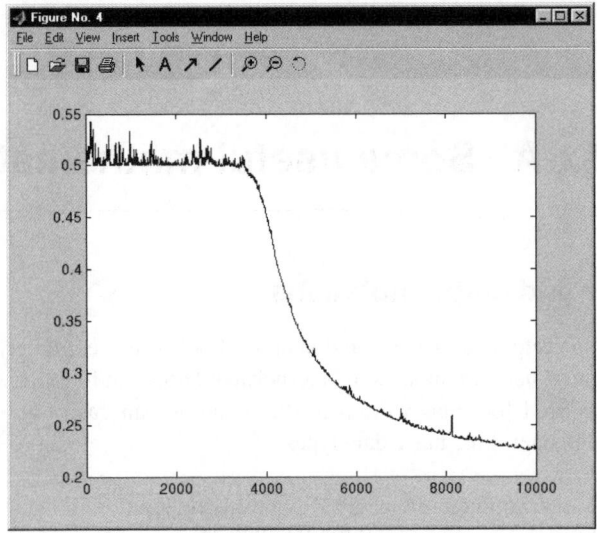

Fig. 12.6 Training curve of a sigmoidal feed-forward network with two hidden nodes with the basic error-back-propagation algorithm.

Further reading

For MATLAB product information, please contact:

> The MathWorks, Inc.
> 3 Apple Hill Drive
> Natick, MA, 01760-2098 USA
> Tel: 508-647-7000
> Fax: 508-647-7101
> E-mail: info@mathworks.com

The original MATLAB literature is very useful. The basic manuals are:

1. *Getting Started with MATLAB*, The Math Works Inc., second printing, 2000
2. *Using MATLAB*, The Math Works Inc., fifth printing, 2000
3. *Using MATLAB Graphics*, The Math Works Inc., fourth printing, 2000

There are many introductions available on the internet. Here are some examples:

1. Graeme Chandler http://www.maths.uq.edu.au/ gac/mlb/mlb.html
2. Mark S. Gockenbach, http://www.math.mtu.edu/ msgocken/

There are also many book on MATLAB including

Duane Hanselman and Bruce Littlefield
> *Mastering MATLAB®5: a comprehensive tutorial and reference*
> Prentice Hall, 1998.

The standard reference for numerical methods is

William H. Press, Saul A. Teukolsky, William T. Vetterling, and Brian P. Flannery
> *Numerical recipes in C: the art of scientific computing*, 2nd edition, Cambridge University Press, 1992.

Appendix A   Some useful mathematics

A.1   Vector and matrix notations

We frequently use vector and matrix notation in this book as it is extremely convenient for specifying neural network models. It is a shorthand notation for otherwise lengthy looking formulas, and formulas written in this notation can easily be entered into MATLAB. We consider three basic data types:

1. Scalar:

$$a \text{ for example } 41 \tag{A.1}$$

2. Vector:

$$\mathbf{a} \text{ or component-wise } \begin{pmatrix} a_1 \\ a_2 \\ a_3 \end{pmatrix} \text{ for example } \begin{pmatrix} 41 \\ 7 \\ 13 \end{pmatrix} \tag{A.2}$$

3. Matrix:

$$\mathbf{a} \text{ or component-wise } \begin{pmatrix} a_{11} & a_{12} \\ a_{21} & a_{22} \\ a_{31} & a_{32} \end{pmatrix} \text{ for example } \begin{pmatrix} 41 & 12 \\ 7 & 45 \\ 13 & 9 \end{pmatrix} \tag{A.3}$$

Note that we have used bold face to indicate both a vector and a matrix because the difference is usually apparent from the circumstances. A matrix is just a collection of scalars or vectors. We talk about an $n \times m$ matrix where n is the number of rows and m is the number of columns. A scalar is thus a 1×1 matrix, and a vector of length n can be considered as an $n \times 1$ matrix. We will thus summarize the following rules for matrices, which generalize directly to scalars and vectors.

We have to define how to add and multiply two matrices so that we can use them in algebraic equations. The *sum of two matrices* is defined to be the sum of the individual components

$$\mathbf{a} + \mathbf{b} = \begin{pmatrix} a_{11} + b_{11} & a_{12} + b_{12} \\ a_{21} + b_{21} & a_{22} + b_{22} \\ a_{31} + b_{31} & a_{32} + b_{32} \end{pmatrix} \tag{A.4}$$

Matrix multiplication is only defined for matrices $\mathbf{a}$ and $\mathbf{b}$ where the number of rows of the matrix $\mathbf{a}$ is equal to the number of columns of matrix $\mathbf{b}$ and vice versa. When specifying the rules for matrix multiplication with components we will see that we get lengthy expressions. Such lengthy expressions often occur when dealing with systems of coupled equations, and the matrix notation was indeed invented to simplify such expressions. The general rule for matrix multiplications is illustrated in Fig. A.1. For two square matrices with two rows and two columns this is given by

$$\mathbf{a} * \mathbf{b} = \begin{pmatrix} a_{11}b_{11} + a_{12}b_{21} & a_{11}b_{12} + a_{12}b_{22} \\ a_{21}b_{11} + a_{22}b_{21} & a_{21}b_{12} + a_{22}b_{22} \end{pmatrix} \qquad (A.5)$$

Each component in the resulting matrix is therefore calculated from the sum of two multiplicative terms. The rule for multiplying two matrices is tedious but straightforward and can easily be implemented in a computer and is the default when multiplying variables of the matrix type in MATLAB. If we want to multiply each component of a matrix by the corresponding component in a second matrix, we just have to include the operator '.*' between the matrices in MATLAB where the dot in front of the multiplication sign indicates 'component by component'.

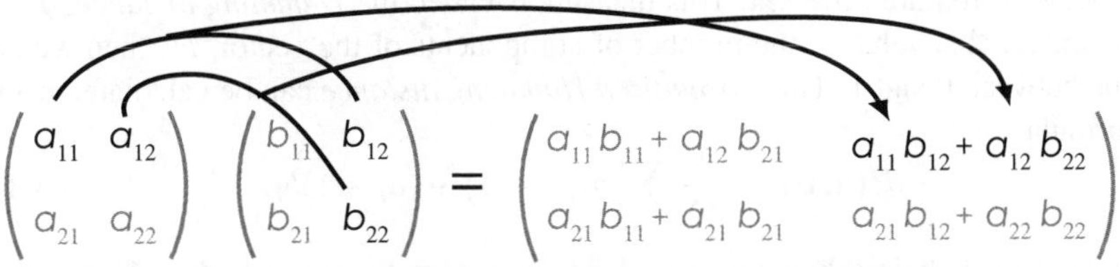

$$\begin{pmatrix} a_{11} & a_{12} \\ a_{21} & a_{22} \end{pmatrix} \begin{pmatrix} b_{11} & b_{12} \\ b_{21} & b_{22} \end{pmatrix} = \begin{pmatrix} a_{11}b_{11} + a_{12}b_{21} & a_{11}b_{12} + a_{12}b_{22} \\ a_{21}b_{11} + a_{21}b_{21} & a_{21}b_{12} + a_{22}b_{22} \end{pmatrix}$$

Fig. A.1 Illustration of a matrix multiplication. Each element in the resulting matrix consists of terms that are taken from the corresponding row of the first matrix and column of the second matrix. Thus in the example we calculate the highlighted element from the components of the first row of the first matrix and the second column from the second matrix. From these rows and columns we add all the terms that consist of the element-wise multiplication of the terms.

A final useful definition is the *transpose* of a matrix. This operation, indicated usually by a superscript t or a prime ($'$), means that the matrix should be rotated 90 degrees, that is, the first row becomes the first column, the second row becomes the second column, etc. For example, the transpose of the example in A.3 is

$$\mathbf{a}' = \begin{pmatrix} 41 & 7 & 13 \\ 12 & 45 & 9 \end{pmatrix} \qquad (A.6)$$

The transpose of a vector therefore transforms a column vector into a row vector and vice versa.

Finally, we want to show that a system of coupled algebraic equation can be easily written in matrix notation. Consider, for example, the system of three equations

$$41x_1 + 12x_2 = 17 \qquad (A.7)$$
$$7x_1 + 45x_2 = -83 \qquad (A.8)$$
$$13x_1 + 9x_2 = -5. \qquad (A.9)$$

This can be written as

$$\mathbf{ax} = \mathbf{b} \qquad (A.10)$$

with the matrix $\mathbf{a}$ as in the example of A.3, the vector $\mathbf{x} = (x_1 \ x_2)'$, and the vector $\mathbf{b} = (17 \ -83 \ -5)'$. Solutions of such equation systems can be formulated by using matrix notation, and many corresponding routines are implemented in MATLAB.

A.2 Distance measures

We are often confronted with comparing two vectors. By comparing them component by component it is easy to decide if they are equivalent or different. However, if they are different we would like to put a value, d, on the difference that gives us some measure of how different they are. There are many different possible definitions of such a difference measure. The only restriction that is obvious to impose is that the value should be $d = 0$ if the vectors are the same, and that it should be positive otherwise, $d > 0$. We do not allow negative values because we wish to interpret this value as a distance.

We sometimes have to evaluate the distance between two binary vectors **a** and **b**, in which the components have only two possible values for example $a_i, b_i \in \{0, 1\}$. A possible value for the distance of such vectors is obtained by counting how many components (bits) are different. This measure is called the *Hamming distance*, d^h. If we normalize this value to the number of components of the vector, N, then we get a value between 0 and 1. This *normalized Hamming distance* can be calculated using the formula

$$d^h(\mathbf{a}, \mathbf{b}) = \frac{1}{2N} \sum_i a_i(b_i - 1) + (a_i - 1)b_i. \tag{A.11}$$

It is not however obvious how to generalize this measure to vectors with real numbered components as we would have to judge the amount of the difference among the components using their possible values, which we might not know *a priori*. One possible definition is to use the *normalized dot product* between vectors. The *dot product* or *inner product* of two column vectors **a** and **b** is only a special case of more general matrix multiplication and is given by

$$\mathbf{a}\mathbf{b}' = \sum_i a_i b_i. \tag{A.12}$$

Note that we used the transpose of the second vector. Within a geometrical interpretation of vectors (see Fig. A.2) this number is proportional to the cosine of the angle between the two vectors,

$$\mathbf{a}\mathbf{b}' = ||\mathbf{a}||\,||\mathbf{b}'||\cos(\alpha), \tag{A.13}$$

where $||\mathbf{a}||$ is the *length* or *norm* of vector **a**. The length of the vector is defined as

$$||\mathbf{a}|| = \sqrt{\sum_i a_i^2}, \tag{A.14}$$

which can also be written with the help of a dot product as

$$||\mathbf{a}|| = \sqrt{\mathbf{a}\mathbf{a}'}. \tag{A.15}$$

The cosine of the angle α between two vectors is a number between -1 and 1 that is only zero when the vectors are pointing in the same direction. The absolute value of this number is therefore a possible definition of the similarity of two vectors,

$$d^{\text{dot}} = \left| \frac{\mathbf{a}\mathbf{b}'}{||\mathbf{a}||\,||\mathbf{b}'||} \right|. \tag{A.16}$$

Other definitions, such as taking the square of the normalized dot product, are also valid. Another definition that is sometimes used as measure of the distance between two vectors with positive real numbered components $a_i, b_i \in \mathbf{R}^+$ has the form of the *Pearson correlation coefficient* from statistics, namely

$$d^{\text{Pearson}} = \frac{\|\mathbf{ab}\| - \|\mathbf{a}\| \|\mathbf{b}\|}{\sqrt{(\|\mathbf{a}^2\| - \|\mathbf{a}\|^2)(\|\mathbf{b}^2\| - \|\mathbf{b}\|^2)}}. \tag{A.17}$$

We call this distance measure the *Pearson distance*.

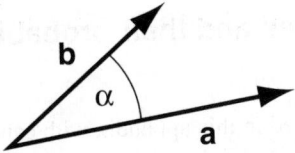

Fig. A.2 Graphical representation of two vectors in a two-dimensional space with an angle α between them.

A.3 The δ-function

The δ-function is a very convenient notation, which, however, drives mathematicians mad (they call it formally a functional). You can think of it as a density function that is zero except for its arguments for which it is infinite. However, the integral over the δ-function is one, that is,

$$\int_{-\infty}^{\infty} \delta(x = x_1)\mathrm{d}x = 1. \tag{A.18}$$

It is usually used as an integration kernel with other functions as in

$$\int_{-\infty}^{\infty} \delta(x_1)f(x)\mathrm{d}x = f(x_1). \tag{A.19}$$

The delta function is useful for writing discrete events in a continuous form.

Appendix B Basic probability theory

B.1 Random variables and their probability (density) function

A *random variable* X (indicated in this appendix with capital letters to distinguish it from a regular variable written with a lower-case letter) is a variable that can have a different value each time we assign a value to this variable by a process or experiment. In the case of discrete numbers for the possible values we speak of a *discrete random variable*, whereas a *continuous random variable* is a random variable that has possible values in a continuous set of numbers. There is, in principle, not much difference between these two kinds of random variables except that the mathematical formulation has to be slightly different to be mathematically correct. For example, the *probability function* $P_X(x) = P(X = x)$ describes the frequency with which each possible value x of a discrete variable X occurs. Note that x is a regular variable, not a random variable. The value of $P_X(x)$ gives the fraction (sometimes also written as a percentage) of the times we get a value x for the random variable X if we draw many examples of the random variable. From this definition it follows that the frequency of having any of the possible values is equal to one, which is often stated as the normalization condition

$$\sum_x P_X(x) = 1. \tag{B.1}$$

In the case of continuous random variables we have an infinite number of possible values x so that the fraction for each number becomes infinitesimally small. It is then more appropriate to write the probability distribution function as $P_X(x) = p_X(x)\mathrm{d}x$, where $p_X(x)$ is the *probability density function* (pdf). The sum in eqn B.1 then becomes an integral and we can write the equivalent of eqn B.1 for a continuous random variable as

$$\int_x p_X(x)\mathrm{d}x = 1. \tag{B.2}$$

We will formulate the rest of this section in terms of continuous random variables. The corresponding formulas for discrete random variables can easily be deduced by replacing the integrals over the probability density functions with sums over the probability function. It is also possible to use the δ-function, outlined in Appendix A, to write discrete random processes in a continuous form.

B.2 Examples of probability (density) functions

In this section we briefly outline some specific probability distributions.

B.2.1 Bernoulli distribution

A Bernoulli random variable is a variable from an experiment that has two possible outcomes, success with probability p or failure with probability $(1 - p)$.

> Probability function:
> $$P(\text{success}) = p; P(\text{failure}) = 1 - p$$
> mean: p
> variance: $p(1 - p)$

B.2.2 Binomial distribution

The number of successes in n Bernoulli trials with probability of success p is binomially distributed. Note that the binomial coefficient is defined as

$$\binom{n}{x} = \frac{n!}{x!(n - x)!} \tag{B.3}$$

and is given by the MATLAB function nchoosek.

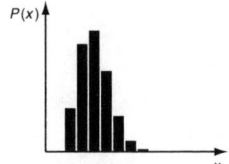

> Probability function:
> $$P(x) = \binom{n}{x} p^x (1 - p)^{n-x}$$
> mean: np
> variance: $np(1 - p)$

B.2.3 Chi-square distribution

The sum of the squares of normally distributed random numbers is chi-square distributed and depends on a parameter ν that is equal to the mean. Γ is the gamma function included in MATLAB as gamma.

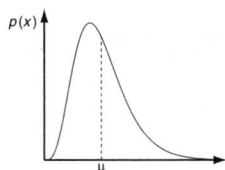

> Probability density function:
> $$p(x) = \frac{x^{(\nu-2)/2} e^{-x/2}}{2^{\nu/2} \Gamma(\nu/2)}$$
> mean: ν
> variance: 2ν

B.2.4 Exponential distribution

Distributions of time in between events that are Poisson distributed. Depends on one parameter λ, which is called the hazard function.

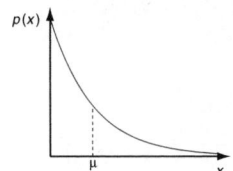

> Probability density function:
> $$p(x) = \lambda e^{-\lambda x}$$
> mean: $1/\lambda$
> variance: $1/\lambda^2$

B.2.5 Lognormal distribution

Distribution that is constrained to be zero at $x = 0$ and has a few large numbers. It depends on two parameters, the scale parameter m (median) and the shape parameter σ.

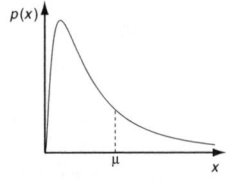

Probability density function:
$$p(x) = \frac{1}{x\sigma\sqrt{2\pi}}e^{\frac{-[log(x/m)]^2}{2\sigma^2}}$$
mean: $me^{\frac{1}{2}\sigma^2}$
variance: $m^2 e^{\sigma^2}\left(1 - e^{\sigma^2}\right)$

B.2.6 Multinomial distribution

Generalization of the binomial distribution where n trials can have k outcomes, each with different probabilities p_i.

Probability function:
$$P(x_i) = n! \prod_{i=1}^{k}(p_i^{x_i}/x_i!)$$
mean: np_i
variance: $np_i(1 - p_i)$

B.2.7 Normal (Gaussian) distribution

Limit of the binomial distribution for a large number of trials. Depends on two parameters, the mean μ and the standard deviation σ. The importance of the normal distribution stems from the central limit theorem outlined below.

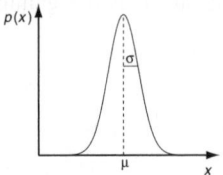

Probability density function:
$$p(x) = \frac{1}{\sigma\sqrt{2\pi}}e^{\frac{-(x-\mu)^2}{2\sigma^2}}$$
mean: μ
variance: σ^2

B.2.8 Poisson distribution

This discrete distribution is frequently used to describe the number of rare but open-ended events and to model spike trains of cortical neurons. It is closely related to the exponential distribution given above. The parameter λ is equal to the mean and the variance.

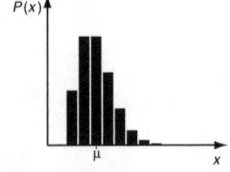

Probability function:
$$P(x) = \frac{\lambda^x}{x!}e^{-\lambda}$$
mean: λ
variance: λ

B.2.9 Uniform distribution

Equally distributed random numbers in the interval $a \leq x \leq b$. Pseudo-random variables with this distribution are often generated by routines in many programming languages.

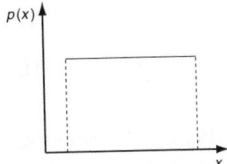

Probability density function:
$$p(x) = \frac{1}{b-a}$$
mean: $(a+b)/2$
variance: $(b-a)^2/12$

B.3 Cumulative probability (density) function and the Gaussian error function

The probability of having a value x for the random variable X in the range of $x_1 \leq x \leq x_2$ is given by

$$P(x_1 \leq X \leq x_2) = \int_{x_1}^{x_2} p(x)\mathrm{d}x. \tag{B.4}$$

Note that we have shortened the notation by replacing the notation $P_X(x_1 \leq X \leq x_2)$ by $P(x_1 \leq X \leq x_2)$ to simplify the following expressions. In the main text we often need to calculate the probability that a normally (or Gaussian) distributed variable of has values between $x_1 = \mu$ and $x_2 = y$. The probability of eqn B.4 then becomes a function of y. This defines the *Gaussian error function*

$$\frac{1}{\sqrt{2\pi}\sigma} \int_{\mu}^{y} e^{-\frac{(x-\mu)^2}{2\sigma^2}} \mathrm{d}x = \frac{1}{2}\mathrm{erf}(\frac{y-\mu}{\sqrt{2}\sigma}). \tag{B.5}$$

This Gaussian error function (erf) for normally distributed variables (Gaussian distribution with mean $\mu = 0$ and variance $\sigma = 1$) is commonly tabulated in books on statistics. Programming libraries also frequently include routines that return the values for specific arguments. In MATLAB this is implemented by the routine erf, and values for the inverse of the error function are returned by the routine erfinv.

Another special case of eqn B.4 is when x_1 in the equation is equal to the lowest possible value of the random variable (usually $-\infty$). The integral in eqn B.4 then corresponds to the probability that a random variable has a value smaller than a certain value, say y. This function of y is called the *cumulative density function* (cdf),[41]

$$P^{\mathrm{cum}}(x < y) = \int_{-\infty}^{y} p(x)\mathrm{d}x, \tag{B.6}$$

which we will utilize further below.

---

[41] Note that this is a probability function, not a density function.

B.4 Moments: mean and variance

Random variables are fully specified by the probability (density) function. However, we frequently do not know the precise form of these functions and we may only be able to measure certain characteristics of a random variable. Such a characteristic quantity is the *mean* of a random variable, also called the *expected value* or *expectation value*, defined as

$$\mu = \int_{-\infty}^{\infty} x f(x) \mathrm{d}x. \tag{B.7}$$

This quantity is, of course, not enough to characterize the probability function uniquely. This is only possible if we know all moments of a distribution, where the nth *moment about the mean* is defined as

$$m^n = \int_{-\infty}^{\infty} (x - \mu)^n f(x) \mathrm{d}x. \tag{B.8}$$

The second moment about the mean is called the *variance*,

$$\sigma^2 = \int_{-\infty}^{\infty} (x - \mu)^2 f(x) \mathrm{d}x, \tag{B.9}$$

which is a measure of the variation of the random variable around the mean. Higher moments specify further characteristics such as the kurtosis and skewness. The square root of the variance is the *standard deviation*, and the ratio of the standard deviation and the mean is called the *coefficient of variation*,

$$C_V = \frac{\sigma}{\nu}. \tag{B.10}$$

The mean can be estimated from a sample of measurements by the *sample mean*

$$\bar{x} = \frac{1}{n} \sum_{i=1}^{n} x_i, \tag{B.11}$$

and the variance either from the *biased sample variance*

$$s_1^2 = \frac{1}{n} \sum_{i=1}^{n} (\bar{x} - x_i)^2 \tag{B.12}$$

or the *unbiased sample variance*

$$s_2^2 = \frac{1}{n-1} \sum_{i=1}^{n} (\bar{x} - x_i)^2. \tag{B.13}$$

A statistic is said to be biased if the mean of the sampling distribution is not equal to the parameter that is intended to be estimated.

B.5   Functions of random variables

A function of a random variable X,

$$Y = f(X), \tag{B.14}$$

is also a random variable, Y, and we are often interested in what the probability (density) function of this new random variable is. However, we have to be a bit careful in calculating this probability (density) function. Let us illustrate how to do this with an example. Say we have an equally distributed random variable X as commonly approximated with pseudo-random number generators on a computer. The probability density function of this variable is given by

$$p_X(x) = \begin{cases} 1 \text{ for } 0 \le x \le 1, \\ 0 \text{ otherwise.} \end{cases} \tag{B.15}$$

We are seeking the probability density function $p_Y(y)$ of the random variable

$$Y = e^{-X^2}. \tag{B.16}$$

The random number Y is **not** Gaussian distributed as we might think naively. To calculate the probability density function we can employ the cumulative density function egn B.6 by noting that

$$P(Y \le y) = P(e^{-X^2} \le y) = P(X \ge \sqrt{-\ln y}) \tag{B.17}$$

This probability can be calculated from the probability density function of $p_X(x)$ and is given by $p_X(x)$

$$P(X \ge \sqrt{-\ln y}) = \begin{cases} \int_{\sqrt{-\ln y}}^{1} 1 \, dy = 1 - \sqrt{-\ln y} \text{ for } e^{-1} \le y \le 1, \\ 0 \qquad\qquad\qquad\qquad \text{otherwise.} \end{cases} \tag{B.18}$$

This probability density function is indeed very different from the one for a Gaussian distribution.

A special function of random variables that is of particular interest is the sum of many random variables. For example, such a sum occurs if we calculate averages from measured quantities, that is,

$$\bar{X} = \frac{1}{n} \sum_{i=1}^{n} X_i, \tag{B.19}$$

and we are interested in the probability density function of such random variables. This function depends, of course, on the specific density function of the random variables X_i. However, there is an important observation summarized in the *central limit theorem* that we can often utilize. The central limit theorem states that the average (normalized sum) of n random variables that are drawn from any distribution with mean μ and variance σ is approximately normally distributed with mean μ and variance σ/n for a sufficiently large sample size n. The approximation is in practice often very good also for small sample sizes. For example, the normalized sum of only seven uniformly distributed pseudo-random numbers is often used as a pseudo-random number for a normal distribution.

Further reading

Dennis D. Wackerly, William Mendenhall III, and Richard L. Scheaffer

Mathematical statistics with applications, Duxbury Press, 6th edition, 2002.
There are many books of statistics on an introductory and advanced level, and it should be easy to find a book that suits you. Many books are specifically on model evaluation and hypothesis testing, which is of course essential in all scientific work. The applications of statistics in this book are mainly based on the probability distributions that are commonly found in books on mathematical statistics. This is a good example of such a book.

Merran Evans, Nicholas Hastings, and Brian Peacock

Statistical distributions, John Wiley & Sons, 3rd edition, 2000.
A useful collection of probability functions. Not an introduction to statistics.

Appendix C   Numerical integration

C.1   Initial value problem

The time dynamics of several models in this book are defined with differential equations that specify the change of a quantity $\mathrm{d}x$ for an infinitesimal time step $\mathrm{d}t$. This change is then specified by a function $f(x,t)$, which may depend on the quantity x itself and may depend explicitly on the time. Mathematically this is expressed by

$$\frac{\mathrm{d}x}{\mathrm{d}t} = f(x,t), \tag{C.1}$$

which is a first-order differential equation with one independent variable. We will outline the principles of some numerical integration techniques for this simple case. Generalizations of these methods can also be applied to systems of coupled differential equations and differential equations that contain higher-order differentials. We are concerned with calculating the values of the quantity $x(t)$ for specific times $t > t_0$ when the value of the quantity at time $t = t_0$ is known,

$$x(t_0) = x_0. \tag{C.2}$$

This is called an *initial value problem.*[42]

C.2   Euler method

Digital computers can only represent discrete numbers and not infinitesimally small changes. We must therefore express the continuous dynamic given by eqn C.1 in discrete time steps. We call this procedure *discretization*. The simplest form is to use small but finite time steps between two consecutive times t_1 and t_2. We write this time step as

$$\Delta t = t_2 - t_1. \tag{C.3}$$

The continuous process is recovered in the limit $\lim_{\Delta t \to 0}$, which we will call the *continuum limit*. The change of the quantity between the two time steps can then be expressed as

$$\Delta x(t + \Delta t) = x(t + \Delta t) - x(t). \tag{C.4}$$

Note that we have assigned the change of the quantity to the time $t + \Delta t$ which is a little bit arbitrary. We could also have assigned this change to the time t, and such

---

[42] We often refer to the initial value problem in the text as numerical integration. Numerical integration can also refer to other numerical tasks such as numerically approximating integrals such as $\int_{x1}^{x2} f(x)\mathrm{d}x$, or to finding solutions of differential equations with other constraints such as fixed values of the variable for different time steps.

discrete models would indeed be different. However, we have to make all the following choices in the light of the continuum in which we should recover the same answers for the different methods.

The differential eqn C.1 can thus be discretized by writing it as a difference equation

$$\frac{\Delta x(t + \Delta t)}{\Delta t} = f(x(t), t), \tag{C.5}$$

and after including the expression eqn C.4 we get

$$x(t + \Delta t) = x(t) + \Delta t f(x(t), t). \tag{C.6}$$

This equation can be used to calculate the value of the quantity x at the next time step following t if we know the value of the quantity at time t. This method is known as the *Euler method*.

C.3 Example

To illustrate this procedure we follow throughout this appendix a specific example, given by a particular choice of the function $f(x, t)$ that we choose to be

$$f(x, t) = t - x + 1. \tag{C.7}$$

The differential equation C.1 can be solved analytically. With the initial condition $x(0) = 1$ this is given by

$$x(t) = t + e^{-t}, \tag{C.8}$$

which can easily be verified by inserting this function into eqn C.1. Values for this function at different time steps are shown in Table C.1 in the column labelled 'exact solution'. The approximation values calculated with the Euler method with a time step of $\Delta t = 0.2$ are included in the column labelled 'Euler method'. The absolute difference between these two columns is plotted in Fig. C.1A. In this particular example the error did not accumulate with increasing time steps but reached a maximum for a time around $t = 0$. This depends, however, on the specific form of the function f.

C.4 Higher-order methods

In the Euler method just outlined we have only taken the first derivative of the function f (the slope of the function) into account and have used this to extrapolate linearly to the value of x at the next time step $t + \Delta t$. We can improve the accuracy of the approximation if we also take into account higher-order derivatives (curvature terms). We can formalize this by considering the *Taylor expansion* of the function around the time for which we want to estimate the value. The Taylor expansion is given by

$$x(t + \Delta t) = x(t) + \Delta t \frac{dx}{dt} + \frac{1}{2}(\Delta t)^2 \frac{d^2 x}{dt^2} + O\left((\delta t)^3\right), \tag{C.9}$$

where the last term stands for all possible higher-order terms. This equation with the first two terms is exactly the Euler method that we used before (eqn C.6) as we can

Table C.1 Values of different numerical solutions of the differential equation $dx/dt = t - x + 1$ with initial condition $x(0) = 1$.

time t	Exact solution	Euler method	Runge–Kutta (adaptive)
0	1	1	1
0.2	1.01873075	1	1.01873077
0.4	1.07032005	1.04	1.07032008
0.6	1.14881164	1.112	1.14881168
0.8	1.24932896	1.2096	1.24932901
1.0	1.36787944	1.32768	1.36787949
1.2	1.50119421	1.462144	1.50119426
1.4	1.64659696	1.6097152	1.64659701
1.6	1.80189652	1.76777216	1.80189656
1.8	1.96529889	1.93421773	1.96529893
2.0	2.13533528	2.10737418	2.13533532

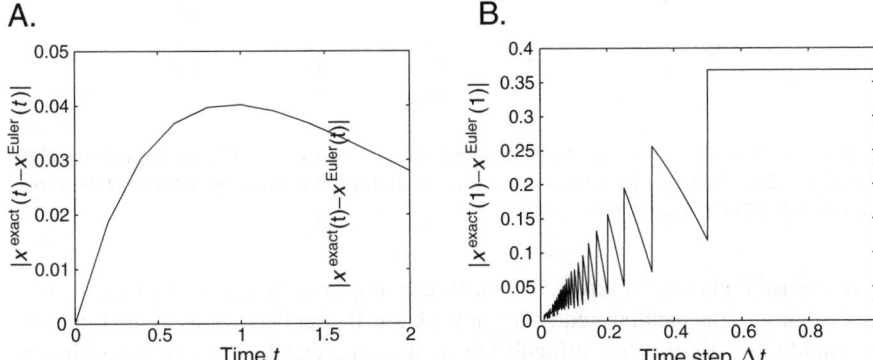

Fig. C.1 (A) The absolute difference between two numerical solutions, $x^{\text{exact}}(t)$ and $x^{\text{Euler}}(t)$, of the differential equation $dx/dt = t - x + 1$ with the initial condition $x(0) = 1$. $x^{\text{exact}}(t)$ is calculated from the analytical solution, while $x^{\text{Euler}}(t)$ is calculated using the first-order Euler method with step width $\Delta t = 0.2$. (B) Dependence of the absolute difference as used in (A) for $t = 1$ and different values of the integration time step Δt.

identify the derivative with $f(x, t)$ as stated by eqn C.1. For our sample function this equation reads

$$x(t + \Delta t) = x(t) + (x - t + 1)\Delta t, \qquad (\text{C.10})$$

which is linear in Δt. This solution is drawn in Fig. C.2 as a dashed line, where we have used the extrapolation from the exact value at $t = 0.1$. This deviation does deviate from the exact solution shown as a solid line for increasing time steps Δt, and we have stressed before that the time step is therefore to be chosen relatively small. If we consider the next order term in the Taylor expansion, we get the approximation

$$x(t + \Delta t) = x(t) + \Delta t(x - t + 1) + \frac{1}{2}(x - t + 1)(\Delta t)^2, \qquad (\text{C.11})$$

which adds a term proportional to $(\Delta t)^2$ to the previous expression. The extrapolation curve with this expression is therefore a linear plus quadratic curve and is plotted as a dotted line in Fig. C.2. As can be seen, this is a much better approximates to the exact solution, even if we use a larger time step Δt.

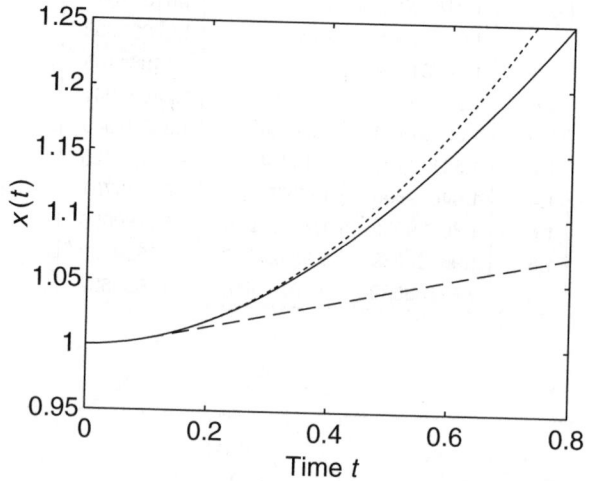

Fig. C.2 Solid line: Exact solution of differential eqn C.1 for the example C.7 as given by eqn C.8. Dashed line: Extrapolation of solution from $t = 0.1$ with the first-order Euler method. Dotted line: Extrapolation with the second-order method.

If we use higher-order terms we can further improve the approximations of the exact solution. This methods requires that we know higher derivatives of the function f explicitly. In case they are difficult to calculate analytically we can also estimate them numerically, although this introduces further numerical errors. This method is therefore not used directly but rather indirectly in the following way. If we use only the first derivative (slope) of the function at time t we assume that this is the same for subsequent times. If this is changing (for example, if the function has a curvature) then we use the slope at a time between t and $t + \Delta t$ as a better guess for the average slope in this interval. Thus, instead of using $f(x, t)$ in eqn C.6 we should use

$$f(x, t) \rightarrow f(\tilde{x}, t + \alpha \Delta t), \tag{C.12}$$

where α is a parameter. The value $\tilde{x}$ should be the value of x at $t + \alpha \Delta t$, that is, $\tilde{x} = x(t + \alpha \Delta t)$, which we don't know *a priori*. We could, however, guess this using the first-order Euler method, that is,

$$\tilde{x} = x(t) + \alpha \Delta t f(x(t), t). \tag{C.13}$$

A common choice for the parameter α is $\alpha = 1/2$, in which case this method is called the *midpoint method*. This method cancels error terms of first order and corresponds therefore to a second-order method. The midpoint method requires two function calls, but no higher derivatives are needed. Note that we made several assumptions that might often be appropriate but not always. Therefore, higher-order methods can usually improve the approximation of the solution $x(t)$, but this does not have to be the case in all circumstances.

C.5 Adaptive Runge–Kutta

It is not necessary to stop at the second-order and midpoint approximation and it is possible (though tedious) to work out solutions for higher-order approximations. The method most commonly used for numerical integrations is thereby a fourth-order methods known as the *Runge–Kutta method*. This method is given by the following set of equations

$$k_1 = \Delta t f(x(t), t) \tag{C.14}$$

$$k_2 = \Delta t f(x(t) + \frac{1}{2}k_1 + \frac{1}{2}\Delta t) \tag{C.15}$$

$$k_3 = \Delta t f(x(t) + \frac{1}{2}k_2, t + \frac{1}{2}\Delta t) \tag{C.16}$$

$$k_4 = \Delta t f(x(t) + k_3, t + \Delta t) \tag{C.17}$$

$$x(t + \Delta t) = x(t) + \frac{1}{6}k_1 + \frac{1}{3}k_2 + \frac{1}{3}k_3 + \frac{1}{6}k_4, \tag{C.18}$$

which is a direct generalization of the midpoint method. The conclusions we drew for the midpoint method still hold. This fourth-order method is often better than lower-order methods, although this cannot always be guaranteed.

Applying any of those numerical methods requires that we check that the results we get do not critically depend on the choice of the parameters such as the time step Δt. At the very least we should check that the numerical results do not change more than within a certain range that we demand of the solutions when modifying the time step parameter considerably (for example, double it and halve it). However, many numerical solvers have already implemented such strategies with algorithms that change the time step as long as the variations in the results are lower than the error bound we give to the system. Such algorithms are called *adaptive time step methods*. The numerical solution of our example solved with the MATLAB routine ode45, which is basically a Runge–Kutta of at least fourth order with adaptive time steps, is included in Table C.1. The maximum of the absolute difference is less than $5 * 10^{-8}$ with the standard parameters used by MATLAB. This is six orders of magnitude better than the Euler approximation.

There are several other methods for numerical integration, each having different strengths and weaknesses in certain application domains. For example, the Runge–Kutta method breaks down if there are very sudden changes in the exact solution for which the linear interpolations used in the midpoint approach break down. MATLAB provides implementations of several other solvers that should be considered for various types of problems. A summary of such solvers is provided in Table C.2. The different solvers are provided within MATLAB with the same generic function calls, so that a switch between solvers can easily be achieved by changing only the name of the function that is called.

Table C.2 Some MATLAB functions for numerical integrations of first-order differential equations [Adapted from D. Hanselman and B. Littlefield, Mastering MATLAB® 5, Prentice Hall, 1998].

MATLAB function	Method	Comments
ode23	Runge–Kutta (order 2–3)	for problems that are non-stiff or that are discontinuous, or where lower accuracy is acceptable
ode45	Runge–Kutta (order 4–5)	non-stiff problems, often tried first for new problem
ode15s	multistep (order 1–5)	stiff problems, try if ode45 fails or is too inefficient
ode113	Adams–Bashford–Moutlon (order 1–13)	non-stiff-problems, multistep method, where $f(x,t)$ is expensive to compute, not suitable for discontinuities.
ode23s	modified Rosenbrock (order 2)	stiff problems with discontinuities, lower accuracy is acceptable

Further reading

William H. Press, Saul A. Teukolsky, William T. Vetterling, and Brian P. Flannery
Numerical recipes in C: the art of scientific computing, Cambridge University Press, 2nd edition, 1992.

A standard reference you will find on the shelves of most people applying computer simulations.

B. H. Flowers
An introduction to numerical methods in C++, Oxford University Press, 1995.

A more teaching-oriented book with clear descriptions and explicit program examples in a more up-to-date programming approach.

Index

place field, 209
plasticity–stability dilemma, 230
point attractor, 182, 184, 198
Poisson
  distribution, 48, 322
  spike train, 51, 102–3, 296–8
population
  code, 107–18
  decoding, 110–2
  dynamic, 72–8
  rapid response, 75–6, 106
  vector, 110–2, 114
posterior distribution, 97
postsynaptic, 14
postsynaptic potential, 20
  excitatory, EPSP, 20
  inhibitory, IPSP, 20
presynaptic, 14
principal component, 169, 302
prior distribution, 97
probabilistic mapping network, 142–4
probability (density) function, 320–3
probability distribution
  Bernoulli, 321
  binomial, 321
  chi-square, 321
  exponential, 321
  Gaussian, 47, 322
  lognormal, 50, 322
  multinomial, 322
  normal, 47, 322
  Poisson, 48, 322
  uniform, 323
product of experts, 262–3
proprioreceptive feedback, 222, 234
prototype, 124, 171
pruning algorithm, 138
pseudo-inverse, 192
Purkinje cells, 236
pyramidal neuron, 14, 15, 59

R
radial basis function, 81
random
  network, 66–72, 204
  variable, 320
Ranvier node, 28
rate
  code, 90
  code entropy, 102–3
  model, 72–87
re-entry, 176, 179
recency effect, 174
receptive field, 91, 121, 214, 276–7
  size of, 276–7
recurrence, 175
recurrent network, 141, 174–232, 263–8, 303–6
refractory time, 27

absolute, 27
relative, 27
regularization, 137
reinforcement learning, 246–52
replica method, 187
representation
  distributed, 112, 113
  local, 112, 113
  sparse, 113
Rescorla–Wagner theory, 248
resting potential, 16
retrieval phase, 178
retrograde messenger, 154
reward
  chains, 249
  learning, 246–50
rotation nodes, 223
Runge–Kutta method, 331

S
sample
  mean, 324
  variance, 324
self-organizing map (SOM), 226–9
sequence learning, 226
serial search, 277
short-term
  depression (STD), 169
  memory, 142, 174–5, 268
shunting inhibition, 22, 86
sigma node, 79–84
sigma–pi node, 84–7
sigmoid function, 81
signal-to-noise analysis, 181–4
simple recurrent network, 141
simulated annealing, 245
soft competition, 143
softmax function, 143
SOM, 226–9
sparse
  pattern, 192–7
  representation, 113
sparseness
  control of, 193–7
  definition, 113–4
  of attractor networks, 192–7
  of object representation in IT, 117–8
spatial
  cross-talk, 261
  representation, 209
spike time accuracy, 93
spike train entropy, 100–2, 105
spike-response model, 42–4
spiking neurons, 24–8, 33–44, 290–8
spin glass, 187
spin model analogy, 186
spiny neuron, 59, 251, 252
spontaneous activity, 66